OPTICAL METHODS OF INVESTIGATING SOLID BODIES

OPTICHESKIE METODY ISSLEDOVANIYA STRUKTURY TVERDOGO TELA

ОПТИЧЕСКИЕ МЕТОДЫ ИССЛЕДОВАНИЯ СТРУКТУРЫ ТВЕРДОГО ТЕЛА

*Proceedings (Trudy) of the P.N. Lebedev Physics Institute*

Volume 25

# OPTICAL METHODS
## OF INVESTIGATING
## SOLID BODIES

Edited by
Academician D. V. Skobel'tsyn

*Authorized translation from the Russian*

CONSULTANTS BUREAU
NEW YORK
1965

The Russian text was published by Nauka in Moscow in 1964 for the USSR
Academy of Sciences as Volume XXV of the Proceedings (Trudy) of the
P. N. Lebedev Physics Institute

Оптические методы исследования структуры твердого тела

*Труды Физического института им. П. Н. Лебедева*
Том XXV

ISBN 978-1-4684-7208-0     ISBN 978-1-4684-7206-6 (eBook)
DOI 10.1007/978-1-4684-7206-6

# CONTENTS

# POLARIZED LUMINESCENCE OF MOLECULAR CRYSTALS

## N. D. Zhevandrov

# INTRODUCTION*

In molecular crystals, the forces of interaction between molecules are by definition considerably weaker than the forces between atoms within the molecule. For this reason, molecules in a lattice to a certain extent retain their individuality. This gives molecular crystals such properties as low melting points (and low boiling points of their melts) and the sublimation of complete molecules. The sublimation energy is a few kcal/mole, i.e., an order of magnitude less than the dissociation energy of the individual molecules. At the same time, interaction between molecules is the basis of the very existence of a crystal, and must thus play a major role in the various physical properties of crystals.

This leading characteristic of the structure of molecular crystals has also determined the main approach to their study. The essence of this lies in considering a crystal, to a first approximation, as an "oriented gas," formed by periodically distributed and regularly oriented noninteractiong molecules. The interaction of molecules is introduced later. Therefore, the theory uses the method of perturbations, and experimental data left unexplained by the oriented gas model may receive interpretation in this way.

The main feature of the zero approximation is the oriented state of the molecules, and since the separate molecules are anisotropic, so is the crystal as a whole. It is thus natural that various methods of studying anisotropy should be widely used in investigating molecular crystals, and these include optical methods. Such methods are indeed used, but unfortunately not so much as they deserve.

The value of using these methods increases as the problems posed go deeper and accuracy improves. In general, optical methods of investigating anisotropy have played by no means the least part in studying crystals. It is enough to cite double refraction, natural optical activity, pleochroism, and finally polarized luminescence.

In recent years, polarized luminescence methods have found much wider fields of application. Following their successful application to the study of resonance luminescence of vapor there appeared a series of papers, mainly from the Soviet school of luminescence established by Academician S. I. Vavilov, on the many-sided study of complex molecules in solutions by means of various methods of polarized luminescence, and later on cubic alkaline-earth halide crystals [1,2,3].

The present treatment is devoted to a study of molecular crystals by means of polarized luminescence. It will be seen that the polarization of luminescence can give valuable information, not only directly relating to crystal anisotropy (for example, the orientation of the molecules in a crystal lattice), but also regarding important processes connected with the interaction of molecules in the lattice, among these energy migration.

The line of investigation is as follows. As a first step, we study the anisotropy of separate molecules in solutions of substances which we later examine in the crystalline state. Then, on the basis of the oriented gas model, we consider the following problem: to determine the orientation of molecules in the lattice from the spatial distribution of the polarization of luminescence, and thus to discover how reliable this model is. Devia-

---

* Dissertation for the degree of Doctor of Physico-Mathematical Science. Defended at the Physics Faculty of Moscow State University, December 12, 1962. (Printed in somewhat shortened and altered form.)

tions from the model are subsequently used for the study of processes connected with molecular interactions, such as the migration of energy and exciton processes.

We found experimentally that the polarization of luminescence was independent of that of the exciting illumination, and this enabled us to observe and study the migration of excitation energy in one-component and impure molecular crystals. The spectral polarization investigations made it possible to follow the part played by free excitons in the luminescence of molecular crystals. Comparison of the polarization characteristics of photo- and radioluminescence provided information of special interest in studying the mechanism of hard excitation.

CHAPTER I

# POLYATOMIC MOLECULES AND MOLECULAR CRYSTALS

AIMS, METHODS, AND OBJECTS OF THE INVESTIGATION. STATEMENT OF PROBLEMS.

REVIEW OF LITERATURE

## §1. Molecular Crystals

Molecular crystals differ from the ionic and metallic variety in that the valence interaction forces between the atoms in the molecules are considerably greater than the intermolecular forces. By nature the intermolecular forces in molecular crystals are those of Van der Waals attraction [4,5], linked with the mutual polarization of molecules. These forces also arise between neutral molecules and between molecules without inherent dipole moments. But even for molecules with dipole moments the Van der Waals interaction is dominant. At first glance this seems paradoxical, but it is explained by the strong dependence of multipolar static interactions on direction. In contrast to the Van der Waals interaction, the static effect diminishes on averaging over different orientations of the molecules.

In view of the relatively small intermolecular forces, the molecules in a crystal to some extent retain their individuality, and to a first approximation we may suppose that the molecules in such crystals have just the same physical properties as do isolated molecules in a gas or solution.

As examples of molecular crystals we may cite crystals of inert gases (Ne, Ar, Kr, Xe) and crystals formed from molecules with saturated bonds ($H_2$, $O_2$, $CH_4$, and others). A vast number of molecular crystals are formed by organic compounds, for example, luminiscent aromatics: naphthalene, anthracene, phenanthrene, and others.

The molecules of aromatic and many other organic compounds as a rule have considerable optical anisotropy. The only exceptions are molecules of high symmetry, such as benzene. We may correlate linear absorption and radiation oscillators with specific electronic absorption bands and the luminescence band of isolated complex molecules (for example, in solution). Their directions coincide with those of the dipole moments of the corresponding electron transitions in the molecule.

This oscillator model of the molecule is in good agreement with extensive experimental material from the study of polarized luminescence of solutions, of which more will be said in a later chapter. These experimental data indicate the presence of a linear radiation oscillator in the molecule (and thus its complete optical anisotropy) and enable us to determine the orientation of the oscillator relative to the chemical structural elements of the molecules of specific compounds.

If all the molecules in a crystal were oriented in the same fashion, then clearly the crystal would have complete optical anisotropy. In particular, its luminescence would be linearly polarized. But in fact, as we shall see later, the luminescence of molecular crystals is always partly polarized. This and many other experimental facts indicate that the molecules in a crystal are oriented not by one mechanism, but by several.

In modern organic crystal chemistry, the principle of the close-packing of particles in a crystal is fundamental [6]. A feature of its application to molecular crystals is that, in this case, it is the close packing of arbitrarily shaped figures which has to be considered and analyzed, rather than that of spheres. These molecule-modeling figures are obtained by marking out spheres with radii equal to the intermolecular distances from the centers of the atoms. The outer surface of the spheres so constructed determines the shape and size of the molecule. The close packing principle asserts that the molecules in a crystal are disposed in such a way that the empty space between them should have the least possible volume. Furthermore, the "projection" of some molecules must fit into the "depressions" of others, i.e., a molecule will touch its neighbors via a number of atoms and in a number of places. This close packing principle is confirmed by a wealth of experimental material from modern organic crystallography and crystal chemistry. Thus it appears that the closest packing of molecules in a crystal lattice corresponds to the minimum internal energy of the system. Direct calculations, however, are impossible at the present time, even for much simpler systems than organic molecules and organic crystals. Hence the empirical principle of close packing is of great significance.

It is evident that, in general, molecules cannot be packed as closely as possible and still be oriented in this way. Usually there are several types of molecular orientation in the elementary crystal lattice cell. Thus, in the monoclinic crystal lattice of aromatic compounds of the naphthalene and anthracene type, the face-centered elementary cell has two kinds of molecular orientation. The existence of different molecular orientations is of great importance on the physical properties of crystals, as will be seen later.

Since the forces between atoms inside the molecule are considerably greater than intermolecular forces, a molecular crystal may to a first approximation be regarded as an oriented gas, consisting of molecules ordered and oriented in a definite manner, but not interacting. Clearly this simplified representation cannot explain all the accumulated experimental facts arising from the study of molecular crystals; in order to explain these, molecular interaction will subsequently have to be taken into consideration. In particular, such interaction will appear in the most important process of the interchange of electronic excitation energy between the molecules, i.e., the migration of energy about the crystal.

Energy migration also takes place in solutions on account of the inductive resonance interaction of molecules at fairly large concentration. Very efficient ways of studying this process are those based on the study of luminescence in solutions, as developed by Academician S. I. Vavilov and his school [1]. The most important of these are the concentration methods, or the study of luminescence characteristics (output, polarization) an functions of the concentration of the solution, and also fluorescence methods, based on the direct experimental measurement of the lifetime $\tau$ of the excited state of a molecule.

The migration of electronic excitation energy in molecular crystals has specific features in comparison with solutions. These features are connected with exciton processes, which are characteristic of a crystal, as an ordered and oriented system possessing translational symmetry.

Molecular organic crystals, we feel, are of special interest since they may be regarded as objects modeling elementary biological units. The point is that modern investigations of various biological objects (chloroplast facets, nerve fibers, rods and cones of the retina, various cell elements, and many others), carried out by physical methods (x-ray structural analysis, electron diffraction, optical methods, electron paramagnetic resonance, electron microscopy using ultramicrotomes, etc.) have shown that animate nature contains a wide distribution of molecular orderliness in its structure.

This extensive material from modern biology and biophysics forms an even firmer basis for our views on the important part played by order and orientation of molecules in elementary biological formations [7]. The quasi-crystalline structure of these formations has first order significance in biological processes, in particular, perhaps, because it is linked with the possibility of efficient migration of energy and exciton processes. It is very probable that the latter play a part in photosynthesis, sight mechanism, the propagation of nerve excitation, and probably many more phenomena.

Direct physical study of biological objects, however, is always made very difficult by the extreme complexity of their general construction, weakness, and instability so that, amid the general melée of the huge number of different processes and properties, it is hard to separate specific aspects out "in pure form" and study them in detail.

In view of this, there is considerable significance for biophysics in the study of physical models, which contain certain characteristic traits of complex animate structures, but still constitute sufficiently transparent and pure objects for careful and reliable physical experiments (especially optical) to be carried out on them. These circumstances excite great interest in the physical study of molecular crystals. Also very important for solid state physics is the fact that the retention of a certain molecular autonomy by different molecules permits the separation of "molecular" and "crystalline" effects, and the study of the latter in more or less pure form.

Besides this, molecular crystals have important practical applications (for example, as scintillators).

## §2. Excitons and Molecular Crystals; Davydov Splitting

The concept of the exciton was first formulated by Frenkel [8] as a quasi particle corresponding to a wave of excitation being propagated in a crystal.

Frenkel theoretically investigated only the cubic crystals of the noble monatomic gases (Ne, Kr, etc.). A. S. Davydov [9] applied the exciton representation to a molecular crystals formed from complex polyatomic molecules and having arbitrary lattice symmetry. Davydov theoretically investigated various forms of the migration of electron excitation energy in crystals and their characteristics, and also a series of properties of molecular crystal spectra which distinguish them from the spectra of isolated molecules. Davydov also studied theoretically the behavior of free and localized excitons in molecular crystals, the concept of these having been introduced by Frenkel.

The absorption of a light quantum of definite frequency by an isolated molecule transforms it into an excited electronic state. When molecules are situated in a crystal, separated from one another by the lattice spacing, the excitation will not remain for all time in one molecule, but will be transmitted from one to the other. This transfer of electronic excitation energy from molecule to molecule about a lattice with translational symmetry may be described as a propagation of excitation waves or excitons about the crystal.

When a molecule undergoes transition into an excited electronic state, its forces of interaction with surrounding molecules alter, so that the molecule shifts into a new equilibrium position. We must here distingush two cases.

1) The passage of excitation energy from molecule to molecule takes place so quickly that no shift of the molecule into a new equilibrium position can occur.

2) The excitation energy moves from one molecule to another slowly, so that the molecules are able to occupy their new equilibrium positions. Local lattice deformation occurs, and this moves about the crystal together with the excitation energy.

In the first case the exciton is called free, and in the second localized. Of course there is no absolute localization; this term is relative, and emphasizes that the velocity of motion is small compared with that of the free excitons.

Davydov calculated the splitting of the nondegenerate levels of the molecules in a crystal caused by their interaction, linking this phenomenon with exciton processes. He studied the spectra of molecular crystals by the methods of quantum mechanics, starting from the simplest case, in which molecules are firmly fixed in their equilibrium positions (this corresponds to very low temperature and very large molecular mass).

Suppose that the elementary cell of a crystal contains $\sigma$ molecules, each molecule containing S optical electrons. Let us assume certain wave functions $\varphi_{mol}^f$ and energy levels $E_{mol}^f$ for the stationary state of an isolated molecule. The crystal contains N cells. We suppose that the molecules have no degenerate states. The splitting of degenerate levels of a molecule in a crystal is analogous to that of atomic terms in an electric field, and will not be considered here. Davydov showed that, even in the absence of degeneracy, a particular splitting of levels took place as a result of the existence of differently oriented molecules in the cell. This subsequently became known in the literature as "Davydov splitting."

Schrodinger's equation determining the wave functions and energy levels of the stationary states of the whole crystal as a single system has the following form:

$$\left( \sum_{n\alpha} H_{n\alpha} + \frac{1}{2} \sum_{n\alpha, m\beta}' V_{n\alpha, m\beta} - E \right) \Phi = 0, \tag{L1}$$

7

where n is the index of the elementary cells, $\alpha$ the index of a molecule in a given elementary cell, $H_{n\alpha}$ the energy operator of the molecule, and $V_{n\alpha,m\beta}$ the operator for the interaction energy between molecules $n\alpha$ and $m\beta$.

The summation is taken over all $\sigma N$ molecules of the crystal. The prime indicates the absence from the summation of the term for which both molecules coincide.

If the interaction between the molecules is small, then in the zero approximation it may be neglected. Then the wave function of the ground state of the crystal may be presented in the form of antisymmetrized products of wave functions $\overline{\varphi}_{n\alpha}^0$ for the normal states of isolated molecules:

$$\Phi_0 = [(S\sigma N)!]^{-\frac{1}{2}} \sum_\nu (-1)^\nu P_\nu \psi_0, \tag{I.2}$$

where $\psi_0 = \prod_{n\alpha} \varphi_{n\alpha}^0$.

Since the characteristic wave functions $\varphi_{n\alpha}$ of the molecules are usually unknown, we may suppose that they are functions of the external as well as the internal coordinates of the molecules, i.e., allowance is made in them for the mutual polarization which causes the Van der Waals attraction between molecules in a crystal. Using (L1) and (L2) and the usual methods, we may find the energy of the ground state of the crystal in the first approximation ($E^0$). If not all the molecules are in the ground state, but one of them (for example, $n\alpha$) has moved over into the $f$-th excited state, then this excited state of the crystal will be described by the wave function

$$\chi_{n\alpha}^f = [(S\sigma N)!]^{-\frac{1}{2}} \sum_\nu (-1)^\nu P_\nu \varphi_{n\alpha}^f, \tag{I.3}$$

where $\psi_{n\alpha}^f = \varphi_{n\alpha}^f \prod_{m\beta \neq n\alpha} \varphi_{m\beta}^0$.

When the molecules are close together, as in a real crystal, the excitation will not be localized in molecule $n\alpha$, but will be transferred from one molecule to another by means of waves of excitation or excitons. States corresponding to the attachment of excitation energy to different molecules in the crystal have identical energies. Hence the wave functions (I.3) from a degenerate system of functions and the wave function of the excited state of the crystal may be written as a superposition of states (I.3):

$$\Phi^f = (\sigma N)^{-\frac{1}{2}} \sum_{n\alpha} a_{n\alpha} \chi_{n\alpha}^f, \tag{I.4}$$

where $|a_{n\alpha}|^2/\sigma N$ gives the probability that the $(n\alpha)$-th molecule is excited. Since the excitation of all molecules is equally probable, $|a_{n\alpha}|^2 = $ const.

A system of algebraic equations for determining $a_{n\alpha}$ and energy $E^f$ of the crystal in the excited state may be obtained from (I.4) and (I.1) in the usual way:

$$\sum_{n\alpha}' M_{n\alpha,\,m\beta}^f a_{n\alpha} - \varepsilon^f a_{m\beta} = 0. \tag{I.5}$$

Matrix $M_{n\alpha,m\beta}^f$ determines the exchange of excitation energy between molecules $n\alpha$ and $m\beta$. The system of equations for $E^f$ is very cumbersome and will not be given here. From these equations we may find $E^f$ and by subtracting the energy $E^0$ of the ground state we obtain the following expression for the excitation energy of the crystal:

$$\Delta E^f = E^f - E^0 = \Delta E_{m\beta}^f + D_{m\beta}^f + \varepsilon^f, \tag{I.6}$$

where $\Delta E^f = E_{m\beta}^f - E_{m\beta}^0$ is the excitation energy of one molecule, $D_{m\beta}^f$ gives the difference in interaction energy of the excited and unexcited molecules with surrounding molecules, and $\varepsilon^f$ is the solution of equation system (I.5).

Owing to the translational symmetry of the crystal, system (I.5) may be simplified by means of the substitution

$$a_{n\alpha} = B_\alpha e^{i\mathbf{k}\mathbf{n}}, \qquad (I.7)$$

where $\mathbf{k}$ is the wave vector of the excitation wave in the crystal, running through N discrete values. Thus

$$\mathbf{k} = \sum_{i=1}^{3} \frac{2\pi}{N_i} \mathbf{a}_i^{-1} \nu_i,$$

where the whole numbers $\nu_i$ satisfy the inequality

$$-\frac{N_i}{2} < \nu_i \leqslant \frac{N_i}{2}, \qquad i = 1, 2, 3,$$

and $\mathbf{a}_i^{-1}$ represents the reciprocal lattice vectors.

Coefficients $B_\alpha$ in (I.7) satisfy the following system of equations:

$$\sum_{\alpha=1}^{\sigma} L_{\alpha\beta}^f (\mathbf{k}) B_\alpha - \varepsilon^f B_\beta = 0, \qquad (I.8)$$

where

$$L_{\alpha\beta}^f (\mathbf{k}) = {\sum_{n}}' M_{n\alpha, m\beta}^f e^{i\mathbf{k}(\mathbf{m}-\mathbf{n})}. \qquad (I.9)$$

Equation system (I.8) has a nontrivial solution if

$$|L_{\alpha\beta}^f (\mathbf{k}) - \varepsilon^f \delta_{\alpha\beta}| = 0. \qquad (I.10)$$

This expression constitutes an equation of the $\sigma$th degree in $\varepsilon$. Solution of this gives $\sigma$ values of $\varepsilon$ as a function of wave number $\mathbf{k}$:

$$\varepsilon_\mu^f = \varepsilon_\mu^f (\mathbf{k}), \qquad \mu = 1, 2 \ldots \sigma, \qquad (I.11)$$

i.e., we have as many different values of $\varepsilon^f$ as there are forms of molecular orientation in the crystal. Thus there will be the same number of energy levels in the crystal corresponding to one and the same molecular level. And this is the "Davydov splitting."

The energy gaps between the $\varepsilon_\mu^f$ with different $\mu$ are determined by the matrix elements $L_{\alpha\beta}^f(\mathbf{k})$, which are connected by formula (I.9) with the matrix elements $M^f$ of the excitation energy transfer between two molecules. In molecular crystals these are always small. With the aid of the known values of $\varepsilon_\mu^f$ and equation (I.8) we may obtain $\sigma$ systems of coefficients $B_\alpha^\mu$, which will determine $\sigma$ wave functions of the excited states of the crystal, corresponding to the $f$-th excited state of a separate molecule:

$$\Phi_\mu^f (\mathbf{k}) = (\sigma N)^{-\frac{1}{2}} \sum_{n\alpha} B_\alpha^\mu \chi_{n\alpha}^f e^{i\mathbf{k}\mathbf{n}}. \qquad (I.12)$$

This plane wave constitutes an excitation wave being propagated about the crystal. The wave may be regarded as a quasi particle ("exciton") with a certain effective mass and quasi momentum $\mathbf{p} = \hbar\mathbf{k}$. In order to find the selection rules for quantum transitions under the action of light from the normal state $\Phi^0$ of the crystal into the excited state $\Phi_\mu^f(\mathbf{k})$, we must calculate the matrix element $\int \Phi_\mu^{*f} (\mathbf{k}) \, \mathbf{r} \, \Phi^0 d\tau$. Calculation shows that the matrix element will only differ from zero when $\mathbf{k} = 0$. Hence transitions may take place only to those points of the energy bands $\varepsilon_\mu(\mathbf{k})$, which correspond to values $\mathbf{k} = 0$. In classical language this means that only threshold frequencies are excited by light.

## §3. Polarization Component of the Davydov Splitting

Determination of the polarization component of Davydov splitting may be made by means of group theory. In many molecular crystals possessing luminescence, especially media of the monoclinic class (naphthalene, anthracene, stilbene, diphenylacetylene, and many more), the elementary cell contains two differently oriented molecules, i.e., $\delta = 2$. In these cases there will be two Davydov splitting components of the corresponding molecular term.

By means of the methods of group theory, we may determine the symmetry properties of wave functions (I.12), and, knowing these, find the polarization of the corresponding transitions.

In order to calculate coefficients $B_\alpha^\mu$ occurring in equations (I.8) and (I.10), we must write wave functions (I.12) for the case $\sigma = 2$ and $k = 0$ in the following form:

$$\Phi_1^f = (2N)^{-\frac{1}{2}} (\chi_{n_1}^f B_1^1 + \chi_{n_2}^f B_2^1),$$

$$\Phi_2^f = (2N)^{-\frac{1}{2}} (\chi_{n_1}^f B_1^2 + \chi_{n_2}^f B_2^2).$$

Taking account of the orthogonality and normalization of these functions, we may easily bring them to the following form:

$$\Phi_1^f = (2N)^{-\frac{1}{2}} (\chi_{n_1}^f + \chi_{n_2}^f),$$
$$\Phi_2^f = (2N)^{-\frac{1}{2}} (\chi_{n_1}^f - \chi_{n_2}^f). \tag{I.13}$$

The ground state wave functions $\varphi_{m\beta}^0$ of the molecules belong to the holosymmetrical representation. Hence the symmetry properties of wave functions $\chi_{n\alpha}^f$ must be determined by the symmetry of the wave functions of the excited state of the molecules $\varphi_{n\alpha}^f$. Thus, in order to find the irreducible representations of the transformations of functions (I.13), we must know the irreducible representations of the wave functions of the excited states of the molecules from which the crystal is formed.

We shall give an account of a specific calculation, using the methods of group theory, carried out by Davydov for a monoclinic crystal of the naphthalene type with space group $C_{2h}^5$, containing two molecules in the elementary cell ($\sigma = 2$). Besides naphthalene, many luminescent crystals used as subjects for our investigations (in particular anthracene) belong to this type. X-ray structural analysis of naphthalene appeared in [10] and anthracene in [11].

Group $C_{2h}^5$, the fifth Fedorov group of the monoclinic prismatic class, has four elements of symmetry: second-order screw axes $c_2$ parallel to the crystal $b$ axis; centers of symmetry $i$ set at the corners, in the middles of the sides, and at the center of the $ab$ face, and also in the center of a plane parallel to the $ab$ face at a dis-

TABLE 1

| Irreducible representation of group $C_{2h}^5$ | Elements of symmetry | | | |
|---|---|---|---|---|
| | $E$ | $c_2$ | $i$ | $\sigma$ |
| $A_g$ | 1 | 1 | 1 | 1 |
| $r_b$, $A_u$ | 1 | 1 | $-1$ | $-1$ |
| $B_g$ | 1 | $-1$ | 1 | $-1$ |
| $r_a$, $r_c'$, $B_u$ | 1 | $-1$ | $-1$ | 1 |

TABLE 2

| Irreducible representation of group $D_{2h}$ | Elements of symmetry | | | | | | | |
|---|---|---|---|---|---|---|---|---|
| | $E$ | $c_2^x$ | $c_2^y$ | $c_2^z$ | $i$ | $\sigma^x$ | $\sigma^y$ | $\sigma^z$ |
| $A_{1g}$ | 1 | 1 | 1 | 1 | 1 | 1 | 1 | 1 |
| $B_{1g}$ | 1 | −1 | −1 | 1 | 1 | −1 | −1 | 1 |
| $A_{2g}$ | 1 | 1 | −1 | −1 | 1 | 1 | −1 | −1 |
| $B_{2g}$ | 1 | −1 | 1 | −1 | 1 | −1 | 1 | −1 |
| $A_{1u}$ | 1 | 1 | 1 | 1 | −1 | −1 | −1 | −1 |
| $z$, $B_{1u}$ | 1 | −1 | −1 | 1 | −1 | 1 | 1 | −1 |
| $x$, $A_{2u}$ | 1 | 1 | −1 | −1 | −1 | −1 | 1 | 1 |
| $y$, $B_{2u}$ | 1 | −1 | 1 | −1 | −1 | 1 | −1 | 1 |

tance of c/2 from it; glide planes σ, perpendicular to the c-axis; and identical element E. The character of the irreducible representations of this group is shown in Table 1 with an indication of the properties of the transformations of radius vector components $r_a$, $r_b$, $r_c'$.

As an example, we consider the naphthalene crystal. The qualitative results obtained are also valid for other crystals of this type. The plane naphthalene molecule belongs to symmetry group $D_{2h}$ (just as the anthracene molecule). The irreducible representations of this group are shown in Table 2, together with an indication of the transformation properties of radius vector components (x, y, z).

Davydov showed in [12] that in the naphthalene molecule the excited states of the π-electrons belong to four irreducible representations $A_{1g}$, $A_{2g}$, $B_{1u}$, $B_{2u}$ of the symmetry group $D_{2h}$. The unexcited state of the molecules, however, corresponds to the irreducible representation $A_{1g}$. Optical transitions from the ground state to the excited state $B_{1u}$ (with polarization along the long axis of the molecule) and to the excited state $B_{2u}$ (with polarization along the short axis of the molecule) are permitted. Transitions to states $A_{1g}$ and $A_{2g}$ are forbidden.

Knowing the symmetry of the molecular wave functions, we may use group theory methods to find the symmetry of the corresponding wave functions of the molecular crystal, and hence the selection rules and polarization of transitions in the crystal.

The ground-state wave function $\Phi^0$ of the crystal transforms as the holosymmetrical representation $A_g$ of the symmetry group $C_{2h}^5$:

$$\Phi^0 \sim A_g.$$

The character of the excited state of the crystal is determined by the corresponding excited state of the molecule, so that for index $f$ of the excited crystal state we may use the symmetry symbol of the corresponding state in the molecule. By means of the character table we obtain the following laws for the transformations of crystal wave functions $\Phi_1^{B_{1u}}$:

$$E\Phi_1^{B_{1u}} = \Phi_1^{B_{1u}}, \qquad c_2\Phi_1^{B_{1u}} = -\Phi_1^{B_{1u}},$$
$$i\Phi_1^{B_{1u}} = -\Phi_1^{B_{1u}}, \qquad \sigma\Phi_1^{B_{1u}} = \Phi_1^{B_{1u}}.$$

If we compare these results with the character tables, we arrive at the conclusion that the crystal wave function $\Phi_1^{B_{1u}}$ transforms in the same way as irreducible representation $B_u$ of the crystal symmetry group:

$$\Phi_1^{B_{1u}} \sim B_u.$$

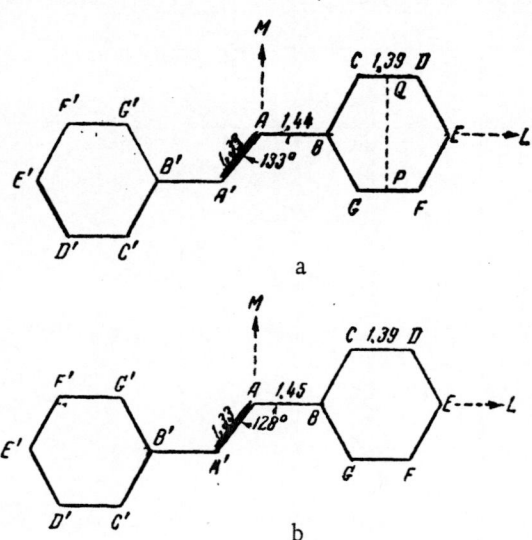

a

b

Fig. 1A. Stilbene. a) Molecule configuration
I; b) molecule configuration II.

It may be shown analogously that

$$\Phi_1^{A_{1g}}, \quad \Phi_2^{A_{2g}} \sim A_g,$$

$$\Phi_1^{A_{2g}}, \quad \Phi_2^{A_{1g}} \sim B_g,$$

$$\Phi_1^{B_{2u}}, \quad \Phi_2^{B_{1u}} \sim A_u,$$

$$\Phi_1^{B_{1u}}, \quad \Phi_2^{B_{2u}} \sim B_u.$$

The selection rules and polarization of the transitions in the crystal follow at once from group theory. Forbidden transitions are those from the ground state to states with crystal wave functions:

$$\Phi_1^{A_{1g}}, \quad \Phi_2^{A_{2g}}, \quad \Phi_1^{A_{2g}}, \quad \Phi_2^{A_{1g}}.$$

The remaining transitions are permitted. Thus transitions in the crystal are forbidden if they correspond to a forbidden transition in the molecule, and permitted when they correspond to a permitted molecular transition. In addition, on account of the Davydov splitting, there are two permitted transitions corresponding to one permitted transition in a molecule of the crystal. This splitting, as indicated above, is caused by the different orientation of molecules in the elementary cell and the resonance interaction between them. The two terms formed in the crystal differ one from another in their polarization. One term, belonging to irreducible crystal representation $A_u$, is so polarized that the electric vector vibrates parallel to the b axis; the other term is polarized in a plane perpendicular to the b axis. This conclusion is extremely important for the interpretation of a number of our experimental results to be recounted later.

Lubchenko [13] applied Davydov's theory to the calculation of stilbene crystals. The stilbene lattice, like that of naphthalene and anthracene, belongs to space group $C_{2h}^5$. X-ray structural analysis of stilbene was carried out be Robertson and Woodward [14]. The elementary cell contains four molecules. The cell structure comprises two symmetrically independent layers. The symmetrically independent molecules all have trans-configuration, but differ slightly in structure (slight difference in the valence angles and the inclination of the double bond to the plane of the benzene ring, Fig. 1A).

The characters of the irreducible representations of the stilbene molecule symmetry group in the trans-configuration (symmetry $S_2$) are shown in Table 3 [15].

In determining the wave-function symmetry of the excited state of the crystal, it is essential that the identical operation E and inversion i coincide in molecule and crystal. The symmetry operation $c_2$, i.e., rotation of the crystal around a second-order axis (coinciding with crystallographic axis b), corresponds to the molecular transposition $1 \rightleftarrows 2$, $3 \rightleftarrows 4$ and the rotation of each of these molecules in space by 180° around molecular axis M (Fig. 1B). On operation $\sigma_h$, molecules I and II change places, and each is reflected in a plane passing through the middle of the double bond, paralleled to the planes of the benzene rings. These two operations (rota-

TABLE 3

| Irreducible representation of group $S_2$ | | | Elements of symmetry | | Transition |
|---|---|---|---|---|---|
| | | | $E$ | $i$ | |
| $x^2,\ y^2,\ z^2,\ xy$ | $R_x,\ R_y,\ R_z$ | $A_g$ | 1 | 1 | Forbidden |
| $xz,\ yz$ | $x,\ y,\ z$ | $A_u$ | 1 | −1 | Permitted |

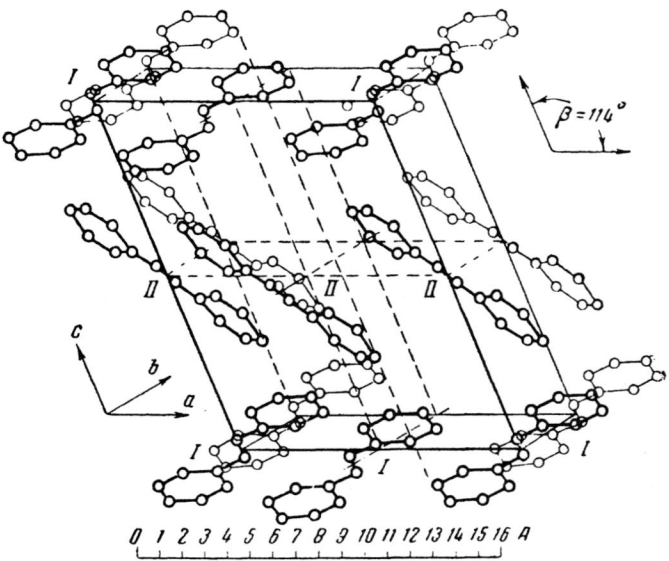

Fig. 1B. Stilbene. Diposition of molecules in the cell.

tion and reflection) do not consitute molecular symmetry operations. Hence the action of operators $c_2$ and $\sigma_h$ on the wave function of the excited state of the excited state of the crystal ($\psi_\mu^{ff'}$) may be written as follows:

$$c_2\psi_\mu^{ff'} = \pm\,\psi_\mu^{ff'}, \quad \sigma_h\psi_\mu^{ff'} = \pm\,\psi_\mu^{ff'}$$

(indices $f$, $f'$ in the wave function show the number of the excited state in which a molecule of one type or the other happens to be). If we wish to determine only the selection rules and polarization of the transitions, there is no need to know the coefficients in the expressions for functions $\psi_1^{ff'}$, $\psi_2^{ff'}$, $\psi_3^{ff'}$, and $\psi_4^{ff'}$. With the aid of the character tables for the irreducible representations of group $C_{2h}^5$ (see Table 1) we can obtain the following wave-function transformation properties, for example, for $\psi_1^{ff'}$:

$$\left.\begin{aligned}
E\psi_1^{A_gA_g} &= \psi_1^{A_gA_g} \\
c_2\psi_1^{A_gA_g} &= \pm\,\psi_1^{A_gA_g} \\
\sigma_h\psi_1^{A_gA_g} &= \pm\,\psi_1^{A_gA_g} \\
i\psi_1^{A_gA_g} &= \psi_1^{A_gA_g}
\end{aligned}\right\} \quad \begin{aligned}\text{Symmetry} \quad A_g \\ \text{(or } B_g\text{)}.\end{aligned}$$

$$\left.\begin{aligned}
E\psi_1^{A_uA_u} &= \psi_1^{A_uA_u} \\
c_2\psi_1^{A_uA_u} &= \pm\,\psi_1^{A_uA_u} \\
\sigma_h\psi_1^{A_uA_u} &= \pm\,\psi_1^{A_uA_u} \\
i\psi_1^{A_uA_u} &= -\,\psi_1^{A_uA_u}
\end{aligned}\right\} \quad \begin{aligned}\text{Symmetry} \quad A_u \\ \text{(or } B_u\text{)}.\end{aligned}$$

On acting with the operator i on wave function $\psi_1^{A_gA_u}$ we find that

$$i\psi_1^{A_gA_u} \neq \pm\,\psi_1^{A_gA_u},$$

so that excited states of the $\psi_1^{A_gA_u}$ type are not realized in the crystal, since this function does not transform according to irreducible representations of the crystal symmetry group.

Similar calculations may also be made for functions $\psi_2^{ff'}$, $\psi_3^{ff'}$, $\psi_4^{ff'}$. The results obtained are shown in Table 4, which indicates the crystallographic directions in which the corresponding transitions are polarized.

TABLE 4

| State of the molecule | Crystal wave function | Irreducible representations of the crystal func. | Transition |
|---|---|---|---|
| $A_g A_g$ | $\psi_1 {}^{A_g A_g}$ | $A_g(B_g)$ | Forbidden |
| | $\psi_2 {}^{A_g A_g}$ | $B_g(A_g)$ | » |
| | $\psi_3 {}^{A_g A_g}$ | $B_g(A_g)$ | » |
| | $\psi_4 {}^{A_g A_g}$ | $A_g(B_g)$ | » |
| $A_u A_u$ | $\psi_1 {}^{A_u A_u}$ | $A_u(B_u)$ | $\parallel b(\perp b)$ |
| | $\psi_2 {}^{A_u A_u}$ | $B_u(A_u)$ | $\perp b(\parallel b)$ |
| | $\psi_3 {}^{A_u A_u}$ | $B_u(A_u)$ | $\perp b(\parallel b)$ |
| | $\psi_4 {}^{A_u A_u}$ | $A_u(B_u)$ | $\parallel b(\perp b)$ |

It is seen from this table that for each permitted transition in the molecule there are four transitions in the crystal, two being polarized along and two perpendicular to the b axis.

## §4. Free and Localized Excitons

All that we have said so far relates to immovable molecules, firmly fixed in the lattice. Davydov also developed a theory allowing for the possible motion of the molecules in the lattice.

Two situations may arise in the transfer of electronic excitation energy in the crystal from molecule to molecule.

1) Excitation may occur so rapidly that the molecules may be unable to move into their new equilibrium positions (free excitons).

2) While the excitation is being transferred, the molecules may succeed in reaching their new equilibrium positions, and local deformation will move about the crystal in company with the excitation (localized excitons).

Let us denote the time for the molecules to move from their old equilibrium positions to new ones by $\tau_c$, and the time for the transfer of excitation by $\tau_{nb}$. The first case corresponds to the inequality

$$\tau_c > \tau_{nb}, \tag{I.14}$$

and the second to

$$\tau_c < \tau_{nb}. \tag{I.15}$$

In molecular crystals there is possible a transition of free excitons into localized and vice versa. Clearly, when the excitation energy transfer time $\tau_{nb}$ from one molecule to another is much less than the time $\tau_c$ required for the creation of local crystal deformations around the molecule, the assumption that the molecules are firmly fixed at the lattice nodes is justified.

But since this is just the condition for the excitation of free excitons, the results of the foregoing theory (Davydov splitting, selection rules, and polarization of the components) are valid if free excitons arise during the excitation.

The physical meaning of the results following from the theory regarding the polarization of transitions may thus be formulated. Transitions corresponding to free excitons are polarized with respect to the crystallographic axes because a free exciton belongs to the whole crystal, not to a single molecule. The exciton is determined by the symmetry of the crystal lattice as a whole, and the polarization of the corresponding transitions is linked with this symmetry, which naturally leads to polarization with respect to the crystal axes. When localized excitons are produced, however, the crystal as a whole has little influence on the excited state, and the corresponding transition does not reflect the crystal structure, its polarization being determined by the orientation of the molecules in the crystal lattice, thus corresponding to the oriented gas model.

Small values of $\tau_c$, leading to the formation of localized excitons, characterize crystals with molecules possessing small mass and small moments of inertia, and also those crystals in which the wave functions determining the optical behavior of the molecules are greatly dependent on the distance between the molecules. On the other hand, in crystals of large dimensions with wave functions only slightly dependent on the intermolecular distances, the time $\tau_c$ is large, and conditions are right for the formation of free excitons. Examples of the latter type are organic crystals, with molecules containing multiple bonds, in particular crystals of aromatic compounds. In the molecules of these compounds, the intramolecular bonds are created mainly by $\sigma$-electrons, and the optical behavior is governed by $\pi$-electrons. The CH, $CH_3$, and other groups situated on the periphery of these molecules and joined only by $\sigma$-bonds in a certain sense screen the $\pi$-electrons from external perturbations.

Davydov solved equations (I.8) and (I.10) in the zero approximation for the particular case of crystals of the monoclinic system with space group symmetry $C_{2h}^5$, containing two molecules in the elementary cell. The results confirmed the conclusions regarding selection rules and transition polarization obtained by group theory for firmly fixed molecules. But the solution made these conclusions such more precise. If a dipole transition is permitted in a molecule, then in general the transition is permitted in the crystal. In the absorption of light, however, transition may be made not to any point of the energy band $\varepsilon_\mu^f(\mathbf{k})$, but only to those points corresponding to values of wave number $\mathbf{k}$ satisfying the following condition:

$$\mathbf{Q} - \mathbf{k} = \begin{cases} 0 \\ \pm 2\pi\mathbf{a}_l^{-1}, \end{cases} \quad l = 1, 2, 3, \tag{I.16}$$

where $\mathbf{Q}$ is the wave vector of the light, and $\mathbf{a}_l^{-1}$ is the reciprocal lattice vector. These conditions are the result of the translational symmetry of the crystal. For visible and ultraviolet light we may neglect the wave vector of the light wave ($\mathbf{Q} \approx 0$).

Then the selection rules take the form

$$\mathbf{k} = \begin{cases} 0 \\ \pm 2\pi\mathbf{a}_l^{-1} \end{cases}. \tag{I.17}$$

The solution confirmed that one of the Davydov splitting components was polarized along and the other perpendicular to the b-axis. For a crystal plate cut paralled to plane ab, one component will be polarized along the b-axis and the other along the a-axis. Moreover, Davydov's theory leads to the conclusion that the ratio of the intensities of the components polarized along the b- and a-axes equals the ratio of the squares of the sums of the electric moment vector components of the transitions in oriented molecules for the same axes. If, as is usually done, we call the intensity ratio of the components polarized along mutually perpendicular crystallographic axis the "polarization ratio," this result may be formulated as follows: the polarization ratio of the intensities of the Davydov splitting components equals the polarization ratio calculated from the oriented gas model.

Craig [16-18] showed in his papers that this conclusion is only valid for an approximation which only considers the interaction of molecules with identical energy levels, i.e., speaking of the molecules in an actual crystal, the interaction of molecules existing in one and the same excited state. Craig showed that if one considers the interaction of levels of different excited states, the magnitude of the splitting changes, and the polarization ratio differs from that derived for the oriented gas model. Furthermore, second-order perturbation may be very considerable if the second transition (influencing the transition whose splitting is being considered) is the stronger, and if the band systems corresponding to these transitions are spectrally close.

Application of Craig's results to the degree of polarization for the absorption bands of anthracene crystals gives the results presented below. Calculation was made for bands corresponding to transition $A_g \rightarrow B_{2u}$ in the molecule, polarized with respect to the transverse axis of the molecule (experiments supporting this will be discussed in Chapter II). For three bands the following respective degrees of absorption polarization were obtained from the oriented gas model and Davydov's theory, counting only the first approximation: 75, 77, 78%.

From Craig's theory [17], making allowance for the second-order effect, the corresponding values were 42, 48, 46%.

The difference is clearly substantial. The splitting also changes considerably.

In Davydov's theory, one further point is important. We refer to the role of the direction of the incident light. Both components of the molecular term splitting appear only when the projections of the radiating oscillators (i.e., the electric moment vectors of the transitions) of differently oriented molecules make angles differing in sign and magnitude with a definite crystallographic axis (for example, if the light falls perpendicularly to the ab plane of naphthalene or anthracene). If, however, these projections of the molecular oscillators on a plane perpendicular to the incident light make identical or nearly identical angles with a definite crystal axis (for example, the a-axis), only one of the Davydov splitting components will be absorbed. It thus appears as if there were only one type of molecular orientation rather than two. The same thing happens, for example, with the ac plane of naphthalene.

Calculation of the excitation energy of the crystal for the case in which localized excitons are formed leads to the following expression:

$$\Delta E_{\mu}^{f} (\mathbf{k})_{loc} = \Delta E_{m\mu}^{f} + \vartheta^{f_0} - \vartheta^{0} + \Delta \varepsilon_{\mu}^{f} (\mathbf{k}), \tag{I.18}$$

where $\Delta E_{m\mu}^{f}$ is the excitation energy of one molecule, and $(\vartheta^{f_0} - \vartheta^{0})$ is the change in lattice energy associated with local crystal deformation near the excited molecule. Energy $\Delta \varepsilon_{\mu}^{f}(\mathbf{k})$ is connected with the displacement of the excitation together with the local deformation. Since index $\mu$ takes $\sigma$ values, the excited state of the crystal will be marked by $\sigma$ energy bands as in the case of free excitons. Both the band widths and the distances between bands will, however, be far less than in the case of free excitons.

It is extremely interesting to compare the excitation energy of the crystal on the formation of a localized and free exciton respectively, in other words, to determine the relative positions of the transitions in the spectrum. For this we must compare expressions (I.6) and (I.18):

$$\Delta E_{\mu}^{f} (\mathbf{k})_{free} = \Delta E_{m\mu}^{f} + D_{m\beta}^{f_0} + \varepsilon_{\mu}^{f} (\mathbf{k}), \tag{I.6}$$

$$\Delta E_{\mu}^{f} (\mathbf{k})_{loc} = \Delta E_{m\beta}^{f} + \vartheta^{f_0} - \vartheta^{0} + \Delta \varepsilon_{\mu}^{f} (\mathbf{k})_{loc} \approx \Delta E_{m\beta}^{f} + \vartheta^{f_0} - \vartheta^{0}. \tag{I.18}$$

The difference in excitation energy between the free and the localized exciton is

$$\Delta E_{\mu}^{f} (\mathbf{k})_{free} - \Delta E_{\mu}^{f} (\mathbf{k})_{loc} = D_{m\beta}^{f_0} - (\vartheta^{f_0} - \vartheta^{0}) + \varepsilon_{\mu}^{f} (\mathbf{k}). \tag{I.19}$$

This difference may, generally speaking, be either positive or negative. The sign depends on the type of crystal, the excited molecular states, the bands of excited crystalline states, and the wave vectors of the free excitons.

As mentioned earlier,

$$| \varepsilon_{\mu}^{f} (\mathbf{k})_{free} | \gg | \Delta \varepsilon_{\mu}^{f} (\mathbf{k}_{loc}) | \approx 0. \tag{I.20}$$

It may also be confirmed that

$$D_{m\beta}^{f_0} \geqslant \vartheta^{f_0} - \vartheta^{0}, \tag{I.21}$$

since $D_{m\beta}^{f_0}$ gives the difference between the interaction energies of excited and normal molecules with their surroundings, while $(\vartheta^{f_0} - \vartheta^{0})$ gives the same difference for the new equilibrium configuration, including the lattice vibration energy, and the transition into the new equilibrium state cannot be accompanied by a rise in energy.

The case in which the excited level of a free exciton lies above the level of the localized exciton is indicated schematically in Fig. 2a (section I corresponds to the localized and section II to the free exciton). Since condition (I.21) always holds, this requires that $\varepsilon_{\mu}^{f}(\mathbf{k}) \geq 0$. If, however, $\varepsilon_{\mu}^{f}(\mathbf{k}) < 0$ and fairly large in absolute magnitude, then the excited level of the free exciton will be lower than that of the localized exciton (Fig. 2b).

Thus both cases are a priori possible. This gives great interest to the possibility of determining the actual state of affairs experimentally. We shall return to this later (see Chapter V).

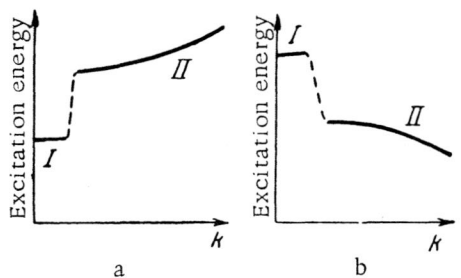

Fig. 2. Relative diposition of exciton excitation levels (schematic). I) Localized exciton; II) free exciton.

When localized excitons are generated, the excitation of the molecule is accompanied by the appearance of a few phonons. Hence, when localized excitons are generated, we must expect wide absorption bands, and, since the separate bands are close to one another, one complex band will result.

## §5. Deforming Excitations

In theoretical papers of Rashba [19,20], the Davydov theory of excitons in molecular crystals is developed for the case of a strong link between molecular excitations and phonons. A general expression is derived for the energy surfaces of an excited molecular crystal. The minima of these surfaces correspond to states with undeformed lattice and a free exciton. For a certain lattice deformation, an additional minimum may appear near the bottom of the energy band. This will correspond to a state with a deformed lattice, and the excitation associated with this deformation will migrate among the molecules in a finite region of the crystal. The author calls excited states of this type "deforming excitations." Such states constitute a peculiar intermediate case between free and localized excitons.

Generally speaking, an energy level corresponding to a deforming excitation may occur either above or below the bottom of the exciton band, depending on the specific parameters of the particular crystal and electron transition. In optical phenomena, however, it is more likely that we shall have cases in which the level of the deforming excitation lies below the bottom of the exciton band, since only in this case will radiationless transitions from the free exciton state to that of the deforming excitation have significant probability. The stronger the exciton-photon link, the lower will be the potential barrier between the minima, and the more strongly will the band and deforming states "fuse." Rashba also showed that, in three-dimensional crystals, the deforming states only arise if the change in the interaction energy of a molecule with its surroundings on excitation is fairly large. This condition is not required in molecular chains (of the polymer type), and only to a lesser degree in two-dimensional lattices. Thus planar or chain structure in crystals (i.e., structure in which the interaction of the molecules in any one or two directions is much stronger than in others) facilitates the appearance of deforming excitations.

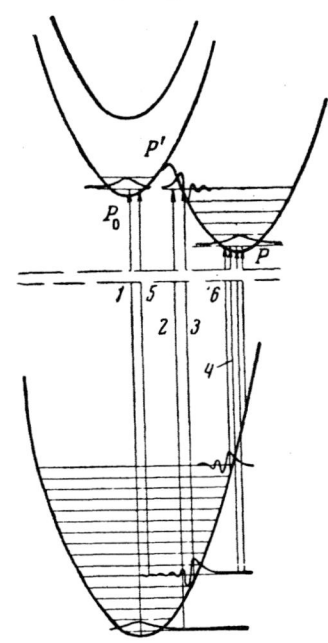

Fig. 3. Potential curve scheme of the ground and excited exciton and deforming states, and course of the $\psi$ -functions (Rashba's calculations).

The existence of deforming excitation levels must have a considerable influence on the spectrum of crystals. Figure 3 shows a scheme for the potential curves of the ground and excited exciton and deforming states, and the run of the $\psi$-functions according to Rashba's calculations. It is clear from this figure that at low temperatures the most intense transitions will be those into the excited state of the free exciton (transition 1), while transitions into the state of the deforming excitation will be weak. Moreover, among the latter, transitions (2) to high vibrational levels of the deforming excitation will predominate. On raising the temperature, there will be a strengthening of type (3) transitions from high vibrational levels of the lower state to high vibrational levels of the deforming excitation. The intensity of transitions (4) to the lowest vibrational levels of the deforming excitation also increases, but transitions (5) to the free exciton state weaken. On still further increase in temperature the probability of transitions (6) rises, and the absorption spectrum extends still further into the long-wave region. This potential curve scheme may be also used to explain features of the luminescence spectra of molecular crystals (see Chapter V).

An important consequence of Rashba's theory [20] is the conclusion that the polarization of the luminescence from the states of the deforming excitation differs from that based on the oriented gas model, because it appears essentially dependent on the wave function of the exciton.

## §6. Luminescence of Molecular Crystals

The excitation energy of a crystal which has absorbed light may pass over into the energy of photochemical reactions, into heat, or once more be radiated in the form of luminescence. In molecular crystals the first process, as a rule, is absent or not very effective; in what follows we shall be interested in the relation between the last two. Transition into heat means the transfer of excitation energy by the intramolecular vibrations of the atoms (intramolecular conversion) or vibrations of the molecules in the lattice. If the probability of intramolecular deactivation is large, then the luminescence output of the substance will, as a rule, be small, both in the crystalline and rarefied states (for example, in solution). It is thus natural to expect intense luminescence only for molecular crystals, the molecules of which luminesce well in the rarefied state. If, when light is absorbed in a crystal, free excitons are formed, then, according to Frenkel [8] and Seitz [21], there first occurs a process of localization of the exciton (as a result of the giving up of kinetic energy to the lattice vibrations and loss of velocity). The energy of the localized excitons then passes over into heat. Thus the localization of an exciton is a necessary intermediate stage in the process whereby excitation energy is transformed into heat. These authors, however, failed to take into consideration the fact that the fall in the energy of an exciton is not always associated with a fall in its wave number and velocity, but, on the contrary, may sometimes lead to an increase in these (Fig. 2b). The principle described is only applicable in the case of Fig. 2a (but, as noted earlier in discussing Rashba's work, this is also the most important for optical phenomena). The probability of a direct transition of free exciton energy into the energy of lattice vibrations is insignificant.

Davydov's theory enables us to determine the probability of the transition of a free exciton into a localized one. The wave function of an excited crystal with one free exciton takes the form

$$\psi^{j\mu}_{1\,\text{free}} = \frac{1}{\sqrt{\sigma N}} A^0_0 (R) \sum_n e^{ikn} \left( \sum_{\alpha=1}^{\sigma} B^{\mu}_{\alpha} \chi^f_{n\alpha} \right).$$

The wave function of the excited crystal with one localized exciton is

$$\psi^{j\mu}_{1\,\text{loc}} = \frac{1}{\sqrt{\sigma N}} \sum_n e^{ikn} \left[ \sum_{\alpha} B^{\mu}_{\alpha} A^f_1 (n\alpha) \chi^f_{n\alpha} \right].$$

Between these two states spontaneous transitions may take place. The probability of this per second is

$$P_n = \frac{2\pi}{h} \left| \int \psi^{*j_1\mu_1}_{1\,\text{free}} (T_R + H_0) \psi^{j\mu}_{1\,\text{loc}} d\tau \, dR \right|^2 \delta \left( U_{\text{loc}} - U_{\text{free}} \right).$$

It follows from this that the transition probability differs from zero if the energy of the system as a whole is conserved: $U_{\text{loc}} = U_{\text{free}}$. After the formation of the localized exciton (immediately on the absorption of light or as a result of a transition from the state of a free exciton), for some time its energy will pass over into heat or be radiated in the form of luminescence. The properties of this luminescence (for example, polarization), as we might expect, should be more or less linked with the oriented gas model. If the radiation takes place not from localized but from deforming excited states, we must expect a deviation from this model. Still greater deviations must be expected for radiation from free exciton states.

Speaking of the mechanism of molecular crystal luminescence, we must bear the following in mind. As already indicated, as a result of the translational symmetry of the crystal, only those transitions are permitted for which condition $k = Q$ (I.16) or apprioximately $k = 0$ (I.17) are satisfied. Theoretical investigations of the structure of the exciton band [22,23,24] show that this structure is complex. The point $k = 0$ may be both an extremal and a saddle point of surface $E(k)$, or different exciton bands may meet at this point.

Exact calculations of the exciton bands in crystals of specific classes of symmetry have àt present not been carried out, so that the position of point $k = 0$ on the energy surface is not known exactly, but there is enough data to indicate that, as a rule, the bottom of the band corresponds to a point $k \neq 0$, and the point $k = 0$ is situated some distance off.

At low temperatures (small $kT$) the main mass of excitons will be found at the bottom of the exciton band, and the sparse filling of $k = 0$ states associated with this will make luminescence from the exciton states more difficult. If transitions from points close to the bottom of the band and satisfying selection rules (I.17) take place at all, the corresponding intensity will be weak. The polarization of these transitions must be determined by the exciton rules, i.e., they must be polarized along one or other of the crystallographic axes. The difficulty of exciton luminescence mentioned, however, is mainly significant at low temperatures only. On raising the temperature to values where $kT$ becomes of the order of the band width, the exciton luminescence intensity grows. We must bear in mind, however, that this effect is neutralized by temperature quenching.

On increasing the temperature, the potential barrier between the exciton band and the state of deforming excitations (Fig. 3) may also be surmounted. At the moment of radiation, the greater part of the excitons pass over into this state. The luminescence spectrum will be wide, shifted towards the red, and its polarization will differ from that of the exciton. Transitions corresponding to radiation from a free exciton state may take place only at the short-wave edge of the luminescence spectrum (transition 1 in Fig. 3) if they correspond to a point near to $k = 0$. Instead of the deforming excitations, an analogous part may be played by localized exciton states if their level lies below that of the free excitons.

Davydov further indicated the important role of crystal lattice defects [25]. A free exciton in a crystal, if it does not pass over into a localized exciton state (or into deforming excitation), will move about the crystal until radiation is produced. The difficulties associated with radiation for the reasons discussed above may be removed by simultaneous emission or absorption of a phonon ($k = Q \pm q$, where $q$ is the wave vector of the phonon) or by radiation at crystal lattice defects, where the excess quasi momentum $\hbar(k-Q)$ is transferred to the defect. In this, by defects we understand any breaking of the translational symmetry of the crystal (surface, block boundaries, cracks, vacant lattice sites, extraneous molecules, thermal flucuations of density, etc.).

A number of experimental data obtained by Prikhot'ko and her colleagues [22], of which more details will be given later, leave no doubt as to the large part played by lattice defects in the luminescence of molecular crystals.

Arganovich [26] approached the question of the part played by defects in crystal luminescence from very general standpoints.

In the theory of the excited exciton states in crystals, their energy levels are calculated by means of the Hamiltonian of the electrons and nuclei of the crystal, no account being taken of the delayed interaction between the electrons. In other words, the interaction between charged particles in the crystal is regarded as being of the coulomb type. But in fact the interactions are not instantaneous, but propagate at a finite velocity; the results of the theory are therefore approximate. On making allowance for the delayed interaction in the zero approximation, elementary excitations constituting a "mixture" of excitons and transverse photons arise. These elementary excitations have received the name of "polaritons" [27] or normal waves in crystals. The interaction of a crystal with a light wave may with certain reservations be considered as follows. The incident photon may produce an exciton in the crystal; in view of the conservation laws, the energy and quasi momentum of this must equal those of the photon. The exciton so produced may in its turn be converted into a photon, and so on. If the energy and the quasi momentum are not given up to the lattice, then this process of successive transformations from photon to exciton and vice versa will also form an ordinary optical wave in the medium, obeying the macroscopic equations of Maxwell and their associated laws of refraction, reflection, etc. If, however, the exciton interacts with the lattice and changes its own energy or momentum, then the resultant photon will not be coherent with respect to the incident photon. In this case there will be absorption of light in the crystal.

Only those excitons with energy and quasi momentum corresponding to the intersection of the branches of exciton energy and transverse photons on the energy-dispersion graph (Fig. 4a) are able to transform into transverse photons. For excitons which substantially alter their quasi momentum on interacting with the lattice,

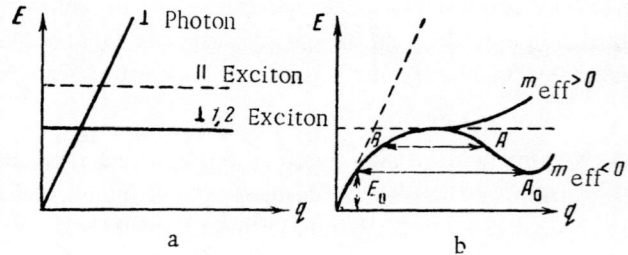

Fig. 4. Energy dispersion of elementary excitations in crystals. a) Without allowance for retardation and space dispersion; b) with allowance for retardation and space dispersion.

transformation into a photon is impossible. This is just the impediment to exciton luminescence which was mentioned earlier. Luminescence may be engendered when the ideal crystal lattice is distorted by defects. The role of the defects may moreover be twofold. First they may lead to the creation of local electron excitation levels. The role of these levels in luminescence may be considerable, even when the defect concentration is small, since their energy may be supplemented by excitons. In this case the part played by the defects would be identical with that of impurity molecules in mixed crystals (for example, anthracene with naphthacene), for which Arganovich [28] showed that the transfer of excitation energy from the base material to the impurity molecules is brought about by free excitons. On the other hand, it is possible to have a mechanism for the transformation of excitons into photons at defects which is not accompanied by the localization of excitation energy.

Figure 4b shows the energy dispersion of elementary excitations with due allowance for retardation and space dispersion. The lower exciton branch has a minimum at point $A_0$. An exciton situated at this "momentum" potential well cannot transform into a transverse photon, since its quasi momentum is not equal to that of the latter. By giving up the surplus quasi momentum to a defect, however, the exciton may achieve an intraband transition of the $A \rightarrow B$ type at constant energy, after which transformation into a photon becomes possible. This process may have considerable significance in the case in which the region of small quasi momenta does not correspond to a stable minimum of exciton energy, i.e., when exciton luminescence is impeded. In this mechanism, the crystal lattice defects act, in effect, as catalysts for the exciton luminescence.

The excess momentum from the exciton may be transferred not only to a lattice defect but also to a phonon originating during the transformation from exciton to photon. In this,

$$\mathbf{k} - \mathbf{Q} - \mathbf{q} = 0,$$

where $\mathbf{q}$ is the wave vector of the phonon.

This may be called the "de-excitation" of excitons at thermal fluctuations, just as the earlier-described process may be called the "de-excitation" of excitons at defects.

Popov and Selivanenko [29] calculated the probability w of this process, using perturbation theory methods analogous to that used for calculating "indirect" transitions in semiconductors. The value obtained was $w \sim 10^7 \, sec^{-1}$. Evidently the influence of impurities (defects) on the luminescence of free excitons becomes important when

$$w \approx vn\sigma,$$

where $v$ is the velocity of the exciton, $n$ the number of defects per $cm^3$, $\sigma$ the interaction cross section of the exciton with the impurity. The authors deduced from this that at $T \sim 300°K$ the role of the defects becomes substantial for $n\sigma \sim 1 \, cm^{-1}$.

§7. Experimental Study of Molecular Crystal Absorption Spectra

Outstanding among treatments of the spectroscopy of molecular crystals both in diversity and extent is a large group of investigations begun by I. V. Obreimov and developed by a number of investigators working in Kiev under the direction of A. F. Prikhot'ko.

Let us first of all turn to investigations into the absorption spectrum of crystals which have provided confirmation of Davydov's theory.

Prikhot'ko [30,31] studied the absorption spectrum of naphthalene single crystals at liquid hydrogen temperature and compared this with the absorption spectrum of naphthalene vapor. We should expect that, neglecting the interaction of molecules in the crystal, there would be the following differences: displacement of lines, disappearance of the rotational structure of the absorption bands of the vapor, and appearance of frequencies corresponding to molecular vibrations in the crystal lattice. As it turned out, part of the crystal spectrum was in keeping with this. However, the crystal spectrum also revealed series of lines and single lines and bands having no analogy in the spectrum of the vapor. Of special interest are electron transitions in the molecule with frequencies 32,455 and 36,398 $cm^{-1}$. In crystals, the first is represented by two lines: one at 29,944 $cm^{-1}$, an intense line polarized along the b-axis, and one at 29,931 $cm^{-1}$, weaker and polarized along the a-axis. Corresponding to the second transition we have the doublet 31,620 $cm^{-1}$ (polarized along b) and 31,474 $cm^{-1}$ (polarized along a). Apart from the long wave shift on passing from the vapor to the crystal spectrum, an important feature here is the experimental observation of Davydov splitting, with polarization in accordance with the foregoing theory.

In a paper by Shpak and Sheka [32], a study was made of the luminescence of naphthalene crystals specially purified by zone melting. At a temperature of 20.4°K, a weak luminescent band was observed, polarized along the a-axis and with a frequency maximum of 31,480 $cm^{-1}$. This band of luminescence practically coincides with an absorption band also polarized along a. On increasing the temperature to 77°K, a band of frequency 31,623 $cm^{-1}$ appears in the luminescence spectrum; this is polarized along the b-axis and coincides with a similarly polarized absorption band. These bands may be ascribed to the radiation of free excitons.

The absorption and luminescence spectra of anthracene have been the subjects of quite a large number of investigations. The absorption spectra at a temperature of liquid hydrogen were studied by Obreimov and Prikhot'ko [33,34]. Besides absorption bands agreeing in a- and b-components, associated with localized excitons, these workers observed a series of bands at the long-wave end of the spectrum, having a b-component and caused by the excitation of free excitons in the crystal. These were bands with frequencies 24,760, 24,961, and 25,109 $cm^{-1}$. It is interesting to note, moreover, that only one of the two possible types of exciton ($\mu = 1, 2$) arises in anthracene.

Craig and Hobbins [18] experimentally investigated the long-wave absorption band system of anthracene crystals (approximately 3800 A) at room and low temperatures (–140° and –250°C). Davydov band-splitting was observed at low temperatures, and this was analyzed using the second approximation to the theory developed by Craig, as discussed earlier. This analysis led to the conclusion that the system at 3800 A corrseponded to a transition of symmetry $A_g - B_{2u}$ in the molecule, polarized along its short axis. The absorption band system at 2500 A, however, belonged to molecular transition $A_g - B_{3u}$ polarized with respect to the long axis [16].

Lyons carried out experiments with the vacuum spectrograph [35], studying the splitting of this short-wave band system. Bands were observed at 2200 and 1890 A (theory would predict a b-component at 2680 and an a-component at 1900 A). Interpretation of these results is made difficult not only by the considerable difference between theoretical and experimental values of splitting, but also by the fact that both components appear to be so polarized that the electric vector component along the a-axis is greater than along the b-axis. The only assertion in holding good is that this transition corresponds to the molecular transition $A_g - B_{3u}$, polarized along the long axis of the molecule.

The existence of anthracene absorption bands polarized along the b-axis and caused by free excitons was confirmed by later investigations of Prikhot'ko and Brodin [36] carried out by the interference method at temperature 20.4°K. However, this conclusion relates to strong absorption extending from 25,300 $cm^{-1}$ into the higher frequency region. Weak absorption in the 24,900 to 24,680 $cm^{-1}$ range is observed only in the b-component, and only for crystal thickness less than 1 $\mu$. At larger crystal thicknesses, these lines are also observed in the a-component [37,38]. These lines are polarized mainly with respect to b, but the degree of polarization differs. The intensity of the lines varies from sample to sample. The lines agree quite well in absorption and luminescence. The authors come to the conclusion that the absorption in this range does not have an exciton nature, but, just as the luminescence, is produced by local levels arising in the crystal as a result of various lattice defects. This point of view regarding the role of defects in the process of luminescence will be further discussed separately.

The absorption spectrum of anthracene crystals at the temperature of liquid helium was studied by Sidman [39]. Sidman did not observe strong bands polarized along the b-axis in the 25,400 to 3000 $cm^{-1}$ range, but found that the absroption was polarized along the b-axis in the 25,060 to 25,400 $cm^{-1}$ range. The earlier paper [37], however, casts some doubt on Sidman's results on acount of the methods which he used, since Sidman's crystals were less than 1 $\mu$ thick and were pressed between two quartz plates. Under these circumstances, considerable mechanical stress arises in the crystals during cooling, and this may substantially affect the polarization in the spectra.

The absorption spectra of naphthacene and phenanthrene were studied by Éichis [40]. It was established for naphthacene that the crystal spectrum was displaced by approximately 2600 $cm^{-1}$ relative to that of the vapor. Moreover, each absorption band was split into two components, polarized along the a- and b-axes. These data indicate that free excitons are formed on absorption of light. The magnitude of the splitting is considerable, of the order of 400 $cm^{-1}$, which is caused by the comparatively large oscillator strength of the transition.

In phenanthrene, on the contrary, the low oscillator strength of the corresponding transition in the molecule leads to a low value of splitting (order of 45 $cm^{-1}$). The component polarized along the b-axis begins in the spectrum from frequency 28,649 $cm^{-1}$, and that polarized perpendicular to the b-axis from frequency 28,604 $cm^{-1}$.

The laws in the absorption spectrum of benzene crystals appear to be quite complicated. The theory of absorption in this case was also worked out by Davydov [41]. Crystalline benzene belongs to the orthorhombic system and space group symmetry $V_h^{15}$ with four molecules in the elementary cells. It was shown by group theory that the forbidden molecular transition to level $B_{1u}$ becomes permitted and polarized along the b-axis in the crystal, the forbidden molecular transition to level $B_{2u}$ corresponds in the crystal to a doublet, one of its components polarized along the c-axis of the crystal and the other along the a-axis, while the permitted twice-degenerate molecular term splits into three components in the crystal, these being polarized along the three crystallographic axes a, b, and c.

The absorption spectra of benzene crystals were studied experimentally by Broude, Medvedev, and Prikhot'ko at temperatures of liquid nitrogen [42,43] and liquid hydrogen [44]. In the first of these only the ac plane of the crystal was investigated, and in the second ab and bc as well. At the temperature of liquid nitrogen, a "sharply polarized series of bands" (as the authors called bands polarized along the crystallographic axes) was observed for the a- and c-directions only, and not for the b-direction. Furthermore, the sharply polarized lines had no analogs in the benzene vapor spectrum. The investigation with liquid hydrogen gave a more complicated picture. In general, the spectrum may be divided into two series of bands. The series of sharply polarized bands the authors called the K series, and the series of bands analogous to those of benzene vapor they called the M series. Series K was quite complicated, but in general was characterized by the appearance of split triplets with components polarized along the a-, b-, and c-axes. The amount of splitting differed in various parts of the series. The M-series bands were "weakly polarized" (as the authors called bands not polarized along any one of the crystallographic axes, but having components along two axes). The M series is associated with localized absorption, and the K series with the formation of free excitions. The main features of the experimental results agree with the aforementioned theory.

The existence of "crystalline" absorption bands in crystals of benzene and its derivatives was ingeniously used by Broude [45] to study polymorphic crystal modifications. The essence of the method is based on the fact that the "crystalline" lines are different in different crystal modifications formed from the same molecules.

In [46-48] fellow-workers of the same Kiev school studied the absorption spectra of stilbene at temperatures of liquid nitrogen and liquid hydrogen. Analysis of the spectra was made difficult by such circumstances as the considerable width of the bands superimposed on an intense background of continuous absorption, the strong dependence of the bands on the individual characteristics of the sample or its treatment, deformation of the substrate, etc. Analysis was somewhat helped by comparing the spectra of pure stilbene with those of crystalline solutions of stilbene in dibenzyl- and diphenylacetylene; the comparison helped to explain the structure of the electron-vibrational excited states of the molecules, since the spectra of these solutions at $20°K$ have a discrete character [49,50].

Despite all the difficulties, these various investigations have certainly established the existence of a few absorption bands polarized with respect to definite crystallographic axes of the crystal.

Thus our review of experiments on the absorption spectra of molecular crystals shows without doubt that these confirm Davydov's theory and establish the existence of Davydov splitting and the formation of free excitons on the absorption of light by molecular crystals in definite spectra ranges for a whole series of substances.

## §8. Experimental Study of Molecular Crystal Luminescence

Experimental data on the luminescence of molecular crystals are quite abundant, especially in relation to aromatic compounds, many of which, in the crystalline state, possess high luminescent yields.

In general, the laws linking the state of aggregation of molecules with their luminescence are complex and not very well known [51].

Polycyclic hydrocarbons (benzene, naphthalene, anthracene) in crystals give a high yield of luminescence. This fact by itself is very interesting and enigmatic. The point is that (like many others) these substances in solution show concentration quenching of luminescence. The transition into the crystalline state, where the molecular concentration per unit volume becomes a maximum, nevertheless leads to an increase in the luminescent yield.

This law is, however, by no means always obeyed.

Polycyclical hydrocarbons with a larger number of rings than anthracene, as a rule, give low luminescent yields in crystals. True, even here there are exceptions: coronene, rubrene (tetraphenylnaphthacene), and others. Any strict rules relating luminescent power and properties to molecular structure are still less established for crystals than for solutions, where the properties of the individual molecules appear with greater clarity.

The majority of dyes, which constitute a classic example of bright and well-studied luminescence in solution, luminesce very poorly or not at all in crystals, even at low temperatures.

For studying various laws in the luminescence of crystals, it is usual to choose substances with large yields, the range of which is quite extensive. In many of these, the luminescent output is extremely high. Galanin and Chizhikova [52] made quantitative measurements of the absolute quantum yield of luminescence for a series of molecular crystals. For anthracene this was 0.7, for stilbene 0.5, for naphthalene 0.16, for paradiphenylbenzene 0.82, and so on. In order of magnitude, these results agree with the data of Wright [53]. It should be noted that these data were obtained for polycrystalline powders, where multiple reflection and scattering, leading to the reabsorption of luminescence (impossible to allow for exactly), lower the true yield value. Measurement of the yield from anthracene single crystals carried out in the same investigation by Galanin and Chizhikova, after introducing the necessary corrections for refraction of the luminescent light at the boundaries of the crystal, allowing for the anisotropy of the latter, gave a yield value close to unity.

Naturally, the polarization of luminescence as a method of studying molecular crystals has attracted the attention of scientists. A medium formed from molecules oriented in a definite way and optically anisotropic must clearly always give polarized luminescence, independently of the means of excitation. Study of this polarization may give information both of the anisotropy of the molecules themselves and of the anisotropy of the whole crystal.

We may cite a group of papers by Indian authors devoted to this question, in their interpretation mainly linked with the oriented gas model.

Krishnan and Seshan [54] correlated the polarization of luminescence with x-ray structural analysis for chrysene and came to the conclusion that the radiation oscillator lay in the plane of the molecule. Analogous results were obtained by the same authors for some other crystals: 1,2,5,6-dibenzanthracene, and also anthracene and chrysene with naphthacene impurity.

Later, experiments were carried out by Ganguli and Chaudhuri [58] with single crystals of pure anthracene, and these also led the authors to the conclusion that the radiation oscillator lay in the plane of the molecule.

In other investigations made by these authors [59,60], measurements of the polarization of luminescence for anthracene, phenanthrene, benzyl, and pyrene crystals were used to determine the orientation of the radiation oscillators of molecules relative to the crystallographic axes. The measurements, it is true, were carried out in a single direction, and the orientation was determined only in one projection. The results agreed qualitatively with x-ray structural analysis.

Chaudhuri [61] made especially careful measurements of the polarization of luminescence for thin single crystals of anthracene. The polarization was measured in separate spectral fluorescence bands, introducing corrections for reabsorption, etc.

The author noted, as before, qualitative agreement between the measured degree of polarization of the luminescence and that calculated from the oriented gas model. At the same time, there was a substantial difference in the polarization ratios for luminescence and absorption in the first band (the latter substantially less than the former), leading to the conclusion that the luminescence took place in those places where the lattice symmetry was broken, i.e., at defects. The reduced polarization ratio in absorption, also observed by Craig, may be explained as a result of the fusing of electronic states under the action of the crystal field. In places where the field is distorted (defects), this fusing is disturbed. At the same time, the defects stimulate luminescence. Thus, these results show that migration of energy takes place from the lattice to the defect sites, where de-excitation occurs. Moreover, the polarization ratio of the luminescence more or less agrees with the oriented gas model.

In a recent paper [62], Chaudhuri and Ganguli obtained analogous results for naphthacene in anthracene.

We must also note a group of investigations by French authors. First Brodersen [63] and later Pesteil [64-66], measuring the polarization ratio of luminescence for various crystallographic planes, determined the orientation of the radiation oscillators in molecules, linking the direction of the oscillators with one or another of the molecular axes. Brodersen did this for naphthalene and Pesteil for stilbene, naphthalene, diphenyl, acenaphthene, fluorene, and various other compounds. Furthermore, Pesteil studied the vibration structure of the luminescence spectra at low temperatures (liquid nitrogen and liquid hydrogen), and also showed that the degree of polarization of the integral luminescence did not vary with temperature. It is true that this last important and interesting observation has rather a qualitative character.

The largest number of investigations on the polarization of luminescence have fallen to the share of anthracene, one of the most brightly luminescing molecular crystals. Apart from the papers already mentioned, anthracene was studied by Canadian authors [67], who measured the angular distribution of luminescent intensity and the polarization ratio in crystals grown from the melt and obtained by sublimation and crystallization from solution. It was shown that the polarization ratio was determined not so much by the means of growing the crystal as by the perfection of its surface.

In a paper mentioned earlier [39], Sidman studied the fluorescence spectra, as well as the absorption of anthracene crystals at 4°K. The spectrum consisted of wide bands possessing structure. The spectra were taken in components polarized along the b- and a-axes, which gave a qualitative picture of the spectral distribution of luminescence polarization. Component b was stronger than a over the whole spectrum, increasing towards the short-wave end. Quantitative investigations were unfortunately not made. In order of magnitude, the polarization ratio agreed with other authors. On the basis of vibrational analysis of the spectrum, there emerged the conclusion that the luminescence was emitted by the anthracene itself and not by impurities. Sidman came to the conclusion that the luminescence proceeded form localized exciton states. The main argument for this was the marked spectral shift between the strong absorption edge and the luminescence.

The same author studied impure crystals: anthracene in naphthalene and phenanthrene [68], and also naphthacene in naphthalene and anthracene [69]. This revealed the interesting fact that the fluorescence of anthracene in the first case and naphthacene in the second was partly polarized at room temperature but appeared unpolarized at 20°K.

This result, however, was not confirmed by Ferguson and Schneider [70], who also studied the polarization of luminescence for crystals of anthracene with naphthacene. Besides this, the authors studied pure anthracene and found that for single crystals obtained by sublimation the polarization ratio $I_b/I_a$ for luminescence was higher than for absorption, and equal approximately to $5:1$. On reducing the temperature to $-196°C$, the ratio hardly altered. The degree of polarization in luminescence, which is higher than that found in absorption, approximating that given by the oriented gas model, is interpreted as a proof of the fact that fluorescence arises mainly in those parts of the crystal where the lattice symmetry is disturbed, i.e., near defects or dislocations. This agrees with the results of the investigation mentioned [61].

Lyons and White recently studied the temperature variation of luminescence in anthracene crystals over a wide temperature range, 4 to 300°K [71]. The changes in the spectrum are of a complex kind. Apart from the spectral shift of the bands, the intensities of different bands change in different ways with temperature; some increase, others decrease, and a few are relatively unchanged. The author draws for interpretation purposes on various quite trivial possibilities: transitions from different levels at different temperatures, the Franck-Condon principle, reabsorption, and temperature quenching, which gave him a qualitative explanation for the experimental results.

The other bands of the luminescence spectrum, which also appear in the absorption spectrum (from 25,370 to 24,670 cm$^{-1}$) are associated by the author with localized excitons which radiate as a result of interaction with phonons. As will be shown later, other investigators [37,48] associate the production of these bands with energy levels disposed in the lattice near defects.

In other very recent papers [72,73], Lyons and his colleagues established the existence of a very weak band, polarized along the b-axis, at 25,040 cm$^{-1}$ in very carefully purified anthracene crystals at 4°K. This band was ascribed to radiation of free excitons. More intense, longer wave bands situated near the former were associated with impurities or lattice defects.

A weak band at 25,100 cm$^{-1}$ in the fluorescence spectrum of anthracene, polarized along the b-axis, was also observed by Wolf [74]. Wolf, however, associated all the luminescence of the crystals with lattice defects on the basis of such data as the variability of position and intensity of the bands in spectra of different samples, and the same sample treated in different ways. This conclusion agrees with those of the Kiev school, of which mention will be made later.

Results analogous to Sidman's were obtained by McClure [75] for phenanthrene. The electric moment of the transition is directed along the perpendicular axis of the molecule. Strong absorption in the crystals leads to the formation of free excitons (there is splitting of the bands at 20°K with $\Delta \nu = 27$ cm$^{-1}$).

The large shift of the luminescence spectrum with respect to the strong absorption leads the author to the conclusion that the levels from which radiation arises correspond to localized excitons associated with lattice faults and deformations.

The role of lattice defects in the luminescence process of molecular crystals forms the main theme of a large group of investigations by the Kiev school carried out in recent years [22].

The variability of the weak, long-wave absorption and luminescence lines at low temperatures in anthracene, stilbene, and other crystals, as well as a number of semiconductors, was established in [37,48], etc. This variability in spectral position and intensity appeared from sample to sample, and also depended on the heat treatment of the samples.

From this collection of results, we may well draw the conclusion that the luminescence of molecular crystals takes place from local levels arising in the crystal near various lattice defects. Disturbance of the ideal crystal lattice, of course, always exists in any crystal. The cause may be, for example, vacancies (elementary lattice nodes not occupied by molecules), interstitial molecules, microcracks in the crystal, and so forth.

Summarizing the results of the survey which we have just conducted, we may say that the absorption and luminescence of molecular crystals are of a complex nature. To some extent, the oriented gas model can be applied to explain these processes, but often it is quite inadequate. The substantial influence of exciton processes on these phenomena in crystals, and also the important part played by lattice defects, have been established beyond doubt.

The relation between all these factors is still far from clear. To what extent these results and conclusions contradict or support one another, and what the true nature and mechanism of these optical phenomena in molecular crystals is, are questions which to a large degree still remain unanswered.

It would seem to us that in the study of molecular crystals very poor use has so far been made of one powerful method, which indeed suggests itself from the very nature of the subjects of investigation and problem posed: this is the quantitative and diversified study of the polarization of luminescence, with its rich store of characteristics and methods developed for isotropic media (solutions and cubic crystals), chiefly by the Soviet school of luminescence founded by S. I. Vavilov [1,3].

Of course, the case of anisotropic crystals will have many peculiar features, and the extension of these methods to the new field will require certain changes involving the development and remodeling of the old methods as well as the working out of new ones.

## § 9. Methods of Polarized Luminescence and Their Possibilities

The state of polarization is one of the main characteristics of radiation, together with its intensity and spectral distribution. Study of the polarization of radiation gives valuable information on the structure and properties of the radiating systems themselves (atoms, molecules, crystals), and also on their interaction with the surrounding medium (for example, molecules in solution).

Already the electromagnetic theory of light, having enabled us to construct the simplest model of an elementary radiator (linear oscillator dipole, identified with quasi-elastically bound electrons), further enables us to inquire into the polarization arising in the elementary radiative act. It follows from this simplest model that the light radiated in the elementary act is linearly polarized. The superposition of coherent dipoles leads to cases of more complex elementary radiative systems, to circular and elliptic oscillators, and also to higher multipoles, quadrupoles, octupoles, and so on.

Later, the quantum theory of radiation confirmed the position of classical physics on the total anisotropy of the radiation emitted in the elementary act. The state of polarization of radiated light is determined by the anisotropy of the matrix element of the transition from the excited to the ground state. Calculation of the matrix element components of the transition dipole in a rectangular coordinate system shows that only the component along an axis coinciding with that of the dipole differs from zero.

Hence only linearly polarized light is emitted and absorbed. Transitions with $\Delta m = \pm 1$ correspond to emission or absorption of circularly polarized light.

It is, however, only possible to calculate the wave functions and matrix elements of the dipole moment of a transition for atoms and diatomic molecules. For complex molecules, the corresponding wave functions are unknown. In such cases we shall have to be satisfied with classical representations, which are graphic but not strict. The simple model of the linear oscillator, however, turns out to be very fruitful when applied to complex molecules. This model provides a formal description for all phenomena associated with the polarized luminescence of complex molecules in solutions.

In this, as Levshin [76] showed, it proves sufficient to work with the model of the linear oscillator. Use of models of elliptic [77] or three-dimensional oscillators with incoherent components [78] leads to nothing much new in the interpretation of the phenomenon.

Although a strict quantum-mechanical theory cannot be constructed for complex molecules, the oscillator model may to some degree be substantiated with the aid of the symmetry properties of molecular states [3]. The symmetry properties of the states form important characteristics in studying the optical properties of molecules. Although the eigenfunctions for complex molecules cannot be calculated, the polarization and selection rules can be determined even without knowing them. Knowledge of the molecular symmetry enables us to judge the symmetry properties of the wave functions as regards symmetry operations. The selectrion rules are determined by whether the terms have corresponding symmetry properties. The same properties determine the direction of the dipole moments of the transitions, i.e., the polarization rules.

The probability of an electron transition between states m and n is given by the matrix element of the transition

$$D_{mn} = \int \psi_m^* D \psi_n d\tau,$$

where $\psi_m^*$ and $\psi_n$ are wave functions of the excited and ground states, and D is the operator of the molecular dipole moment. In order to find the selection rules, we must find under what circumstances this integral vanishes. In order to establish the polarization rules, we must find the components $D_{mn}$ in definite directions. It may be shown that the integral differs from zero only when the product of wave functions $\psi_m \psi_n$ contains at least one term with the same transformation properties as D. In its turn, D has the same symmetry properties as the translation T. The latter is obvious, since the components of D have the form

$$D_x = \sum e_i x_i;$$

and so on. The direction of translation determines the direction of the dipole moment of the transition. Hence, if the symmetry of the states is known, we can in principle determine the direction of the dipole moment of the transition, i.e., the direction of the molecular oscillator.

Feofilov applied this conception to the case of linear symmetric molecules of symmetry $D_{\alpha h}$. In this, he used data on the symmetry of products of eigenfunctions of states with different symmetries obtained by Spooner and Teller [79]. Having considered all combinations corresponding to permitted transitions, he established that the dipole moment of the transition is oriented along the molecular axis if $\Delta \Lambda = 0$, and perpendicular to the axis if $\Delta \Lambda = \pm 1$, where $\Lambda$ is the molecular quantum number characterizing the projection of the sum of the angular momenta of the constituent atoms on the molecular axis.

The majority of complex luminescent molecules have linear chains of conjugate bonds. Evidently this is not by chance. A molecule will not be able to luminesce if there is a high probability of internal conversion, i.e., if the potential surfaces of the ground and excited states approach one another closely or intersect. This will be the more likely, the more strongly bound the optical electron is in the molecule [80]. On the other hand, for the "quasi-free" $\pi$-electron, the potential curves are similar to one another and their branches nearly parallel. We can thus understand why fluorescent materials are chiefly found among compounds with conjugate bonds.

But if we approximate a molecule by a linear chain of conjugate bonds, then we may use the abovementioned polarization rules. One possible scheme of levels, for example, is that in which the oscillators of long-wave absorption and radiation are parallel to the molecular axis, and the oscillator corresponding to the second absorption band is oriented perpendicularly to the axis. Analogous principles apply to molecules of other classes of symmetry.

Thus the absorption and luminescent processes in complex molecules may be modeled by means of linear oscillators, oriented in a definite way with respect to the molecular axes. Hence complex molecules are optically anisotropic, and the radiation arising in elementary acts is always polarized. The luminescence of macroscopic media comprising a large number of molecules may be either polarized or unpolarized, depending on two factors, namely, whether the molecules are ordered or disposed chaotically, and whether the excitation is anisotropic.

If the molecules are chaotically disposed, orientation being absent, then the medium as a whole will be isotropic. Examples of this are the liquid solutions of fluorescent materials. Such media give unpolarized luminescence if the excitation is isotropic—for example, if the excitation is caused by natural light in the same direction as that in which the luminescence is viewed. If the excitation comes from linearly polarized light, however, then from all the chaotically distributed molecules those with absorption oscillators parallel to the electric vector in the incident light will in the main be selectively excited. If this partial anisotropy created on excitation is not lost while the excited state exists, then the luminescence will be partially polarized.

A certain small anisotropy is also created by natural exciting illumination if its direction differs from that in which the luminescence is observed (especially if these directions are at right angles). Apart from solutions, examples of such media include cubic crystals, in which the radiating oscillators are oriented, not chaotically, but uniformly along three mutually perpendicular axes, as a result of which the system as a whole is isotropic.

If, however, a medium is macroscopically anisotropic, i.e., if the molecules are completely or even partially oriented, then that medium will show polarized luminescence even for completely isotropic excitation. The molecular crystals forming the theme of the present treatment are just such media. Other examples of such media are cellophane or polymer films dyed by organic substances, and mechanically drawn [81-86] fibers, as well as a few natural organic fibers [87,88]. Partial anisotropy may also be created in isotropic bodies by electric and magnetic fields [89], and also by means of the Maxwell effect (orientation effect in flux) [90].

All these examples are fairly rare in nature, and made artificially only with difficulty, except in the case of molecular crystals, which present a noteworthy subject for studying polarized luminescence.

It was evidently no mere chance that the phenomenon of polarized luminescence was first discovered quite long ago in anisotropic crystals of platinicyanic salts [91]. Later, however, scientists turned their main attention to isotropic media. The polarization of fluorescence from an isotropic solution under polarized excitation was first observed by Weigert [92]. Vavilov and Levshin [93] established experimentally that the fluorescence of a solution may be partly polarized even with natural excitation, if the direction of this does not coincide with that in which the fluorescence is observed. These authors also theoretically calculated the degree of polarization of the fluorescence under linearly polarized excitation. This proved to be 50%.

Limiting Degree of Polarization. Experimental data always fall short of this theoretical value of 50%. The cause of this discrepancy is connected both with the structure of the molecules themselves and with outside conditions (temperature, viscosity, concentration).

First of all, the Brownian rotational motion of the molecules constitutes a depolarizing factor. In the course of the excited state, the molecule turns through a certain angle, as a result of which the anisotropy of the solution created at the moment of excitation is partly or wholly lost. The rotation angle determining the depolarization of the fluorescence depends on the temperature T and viscosity coefficient $\eta$ of the solution, the volume V of the molecules, and the duration $\tau$ of the excited state. Levshin [76,94] and later Perrin [95] carried out calculations with the aid of Brownian motion theory and obtained the formula

$$\frac{1}{P} = \frac{1}{P_0} + \left(\frac{1}{P_0} - \frac{1}{3}\right) \frac{RT}{V\eta}\, \tau,$$

where $P_0$ is the so-called limiting degree of polarization, corresponding to the case $T \to 0$, $\eta \to \infty$. The linear form of the function $1/P = f(T/\eta)$ is duly confirmed by many experimental data. This formula gives a convenient experimental means of determining the limiting polarization from the variation of polarization with temperature and viscosity of the solution. Clearly, experiment always demands the use of fairly viscous solvents (glycerine, oil, etc.). It turns out, however, that even the limiting degree of polarization for the luminescence of solutions is always less than 50%.

Here we must note that things are very different as regards this depolarizing factor in crystals. Rotational thermal vibrations of molecules fixed in the lattice do take place even in this case, but their effect on the polarization of the luminescence differs essentially from that in solutions. More will be said of this in Chapter III.

One reason for the limiting degree of polarization in solutions being less than the theoretical value may be the symmetry of the molecule. For example, the plane benzene molecule, with a sixth-order symmetry axis perpendicular to the plane of the molecule, must absorb and radiate light with equal probability when the electric vector lies in six directions in the plane of the molecule. Simple calculations show that if the order of symmetry of the axis exceeds two, the degree of polarization will be approximately 14%. Other forms of molecular symmetry also [96] also lead to a reduction in the degree of polarization. This was shown experimentally for benzene and its derivatives, and also for porphyrins in [97].

An important depolarizing factor is concentration depolarization. This phenomenon was observed by several authors almost simultaneously [98-100] in viscous fluorescing solutions. The phenomenon is as follows:

over a considerable range of small concentrations, the polarization remains constant; then, at concentrations of the order of $10^{-4}$ to $10^{-3}$ g/cm$^3$ it starts falling rapidly, and practically vanishes at concentrations of the order of $10^{-2}$ g/cm$^3$. For such small concentrations, this effect cannot be explained as the result of collisions. Concentration depolarization of the fluorescence of solutions is one of the most important proofs of the theory of resonance migration of excitation energy given by Vavilov, Calanin, and Förster [1,101,102]. If there is a certain interaction between the molecules, and their absorption and radiation spectra overlap, then there is a certain probability, dependent on the distance between the molecules, of the occurrence of the transfer of electron excitation energy from an excited molecule to an unexcited one. Clearly, this must lead to depolarization of the fluorescence, since the transfer takes place from an excited molecule, the orientation of which is determined by the direction of the electric vector in the exciting light, to a molecule with random orientation. Finally, we must also always bear in mind a trivial depolarizing factor, that is, secondary fluorescence excited as a result of reabsorption. This may be excluded by working with very thin fluorescing layers. By this means, Galanin [103] obtained degrees of polarization quite close to 50% for a number of dyes.

Up to the present, the long-wave absorption band, as well as the luminescence band, has been considered as one unit and modeled by a single linear oscillator. Strictly speaking, this is inaccurate. Even in 1924, Levshin showed [104] that, on reducing the wavelength of the exciting light within the limits of the first absorption band, the polarization of the fluorescence fell slightly. Sevchenko and colleagues [105-107] observed an analogous relation for increasing wave length of luminescence. For molecules in which the law of mirror symmetry of the absorption and emission spectra is fulfilled, these polarization relations are also symmetrical. Polarization increases on approaching the point of intersection of the spectra, corresponding to the frequency of a purely electronic transition, and near this reaches values very close to the theoretical 50%. These results show that the orientations of the dipole moments of the direct and inverse transitions corresponding to different vibrational levels are somewhat different. Thus the absorption or emission band must be described by means of a certain cone of oscillators with a definite aperture angle. A transition between a single pair of levels, either direct or inverse (for example, a purely electronic transition), may, however, be described by a single linear oscillator.

For molecules with unsymmetrical spectra, the polarization relations are also unsymmetrical, but the increase in polarization on approaching the intersection of the spectra still holds.

Polarization Diagrams. An important and interesting characteristic of luminescent systems is provided by polarization diagrams describing the spatial distribution of polarization and allowing the nature (multipolarity) of the radiator to be determined.

Electric dipoles are not the only possible form of elementary molecular oscillators, although they are the most widespread. There may also be multipoles of higher order, beginning with electric quadrupoles and magnetic dipoles.

One certain indication of the multipolarity of a radiator is the anisotropy of the spatial distribution of radiation. For example, for the electric dipole, the amplitude of the emitted wave in a direction at an angle of $\alpha$ to the dipole axis is proportional to $\sin \alpha$, and for a quadrupole to $\sin 2\alpha$. Thus for the dipole there is only an absence of radiation along the axis, whereas for the quadrupole this holds both along and perpendicular to the axis.

Vavilov [108] proposed the wide-angle interference method for determining the multipolarity of radiators. This method, however, only gives information on the multipolarity of the emitting system.

In order to determine the multipolarity of both emitting and absorbing systems in solutions, Vavilov proposed the method of photoluminescence polarization diagrams [109]. This term signified the variation of the polarization of the luminescence with the direction of observation and the direction of the electric vector in the exciting illumination. To fix ideas, the following parameters are introduced: $\chi$, the angle between the direction of the exciting radiation and the direction of observation, and $\eta$, the angle between the exciting electric vector and the vertical. Vavilov calculated the polarization diagrams for various combinations of emitting and absorbing electric dipoles and quadrupoles on the condition that the limiting degree of polarization was $P_0 = 0.5$. Feofilov [110] made calculations for magnetic dipoles. Polarization diagrams for arbitrary values of $P_0$ were calculated in later papers [111,112]. Polarization diagrams have specific forms for different combinations of radiators, and may thus serve to determine the multipolarity of systems. This method was used, for example, to prove the electric dipole character of luminescence from a number of organic compounds of uranium glasses. With certain changes, the method was applied by Feofilov to study the multipolarity of elementary radiators in cubic crystals [113].

Polarized Luminescence of Cubic Crystals. Feofilov [3] studied the polarization of the luminescence of cubic crystals $CaF_2$, NaF, LiF, and certain others containing color centers. The luminescence was polarized on excitation by polarized light. Experimental examination of the azimuthal variation of polarization (i.e., variation with position of the electric exciting vector relative to the crystallographic axes), and comparison with the corresponding calculated variations, showed that the anisotropic luminescent centers were oriented along a fourth-order axis of symmetry in $CaF_2$ and a second-order axis of symmetry in NaF and LiF.

Thus these luminescent cubic crystals form a "quasi-isotropic" radiating medium, in which the oscillators, in contrast to truly isotropic solutions, are disposed not chaotically, but oriented with respect to the three mutually perpendicular axes. On excitation by natural light along one of these axes and observation along the same axis, the luminescence will of course be unpolarized. Thus, just as in isotropic solutions, the polarization of the luminescence is here created as a result of the anisotropy of the excitation. In this it is extremely important that there should be no migration of excitation energy between centers of different orientation, which would lead to depolarization of the luminescence (analogous to concentration depolarization in solutions). The very existence of polarization shows that there is no migration of energy between luminescence centers in these crystals. Clearly, P = 50% is not here the theoretical maximum degree of polarization. On excitation with linearly polarized light, the electric vector of which coincides with one of the oscillators, the degree of polarization must be 100%. Experiments under these conditions in fact gave an extremely high polarization, over 90%. It is also evident that the medium is only "quasi-isotropic" for directions coinciding with the directions of the oscillators, and is not strictly so for other directions. Thus, for example, in a plate cut parallel to the (110) plane and excited by natural light, partially polarized luminescence is produced, its intensity, moreover, depending on the azimuth.

In order to determine the multipolarity of radiators in these crystals, Feofilov also used the method of polarization diagrams in somewhat altered form [114]. He calculated the variation of polarization and luminescent intensity with the angle between the exciting electric vector and the vertical, for a definite orientation of the crystal, for cases of various multipolarities and orientation of the oscillators along axes of various symmetry. Then he compared the experimental diagrams with the calculated ones, and since the latter had specific forms for cases of different multipolarity, arrived at definite conclusions regarding the nature of the radiators. For example, it emerges that for $CaF_2$, on excitation in the first absorption band, the processes of absorption and radiation can be described by linear electric dipoles. On excitation in the second absorption band, one of these must correspond to a circular electric oscillator (rotator). Special interest attaches to the $Eu^{+++}$ ions in the crystal lattice of fluorspar [115]. The elementary radiators are in this case oriented along third-order symmetry axes. In the luminescence spectrum of $CaF_2$-Eu there are lines corresponding to all possible dipole radiators, among them a magnetic dipole and a magnetic rotator.

Polarization Spectra. "Polarization spectrum" of a fluorescent solution is the accepted term for the variation of the degree of polarization with wavelength of the exciting light. It was Levshin, as mentioned earlier [104], who first observed a slight change in the polarization of the fluorescence of certain dyes on changing the wavelength of the exciting light within the limits of the first absorption band. Later Frohlich [116] observed marked changes in polarization on changing the wavelength of the exciting light for a number of other dyes. Vavilov [117] carried out measurements over a wide ultraviolet region of the spectrum and found sharp changes in the degree of polarization, in particular noting that in some cases the polarization became negative.

The concept of a polarization spectrum implies the fulfilment of two conditions: 1) the independence of the emission spectrum on the wavelength of the exciting light; 2) the constancy of the degree of polarization within the limits of the emission band.

Numerous experiments confirm that the first condition is usually fulfilled. The second condition, however, as already indicated, is not strictly fulfilled. In the case of complex molecules with continuous spectra, however, the variation of the degree of polarization with the wavelength of the luminescence is insignificant. The variation has no great effect on the form of the polarization spectra, and in no case can it alter the qualitative conclusions resulting from these.

Only in the particular cases of the porphyrins [118], pheophytin, and chlorophyll [119] is there a strong variation of polarization with the wavelength of the luminescence. This is a proof of the fact that the luminescence proceeds from different electronic levels, corresponding to two radiation oscillators: linear and planar. The polarization spectrum concept defined above is clearly inapplicable to such cases, and they must be regarded as exceptions.

Feofilov [120] measured polarization spectra of various dyes, and showed that there was a correspondence between these and the absorption spectra. Each band in the absorption spectrum, generally speaking, corresponded to some definite value of the degree of polarization of the luminescence. The simpler the structure of the absorption spectrum, the more clearly did this relationship appear.

It follows, from the nonvariation of the fluorescence spectrum with wavelength of the exciting light, that a transition accompanied by fluorescence always takes place from one and the same excited level, whatever the level to which the molecule passed during absorption, i.e., the radiation is always associated with one and the same elementary oscillator. The polarization spectra indicate that different elementary oscillators are responsible for different absorption bands, these making different angles with the radiation oscillator.

Study of polarization spectra provides a method of determining the mutual orientations of absorption and radiation oscillators in a molecule.

Polarization spectra are especially valuable because, by comparing them with absorption spectra and taking in other data, one can draw definite conclusions regarding the orientation of oscillators in the molecule in relation to the elements of its chemical structure.

It is just this possibility which excites our greatest interest at this moment, since, before proceeding to a study of the anisotropic properties of crystals, we must understand the anisotropy of the molecules themselves and determine the orientation of the radiation oscillators with respect to the molecular axes. Clearly this is a first and essential step in studying crystals.

For this reason, we shall also make use of the method of polarization spectra.

CHAPTER II

# ANISOTROPY OF MOLECULES

DETERMINATION OF THE ORIENTATION OF RADIATING OSCILLATORS

WITH RESPECT TO THE AXES OF THE MOLECULES BY MEANS

OF THE POLARIZATION SPECTRA OF THE LUMINESCENCE OF SOLUTIONS

## §1.  Possibilities of the Methods

As we have already discussed, the phenomenon of polarized luminescence is quite satisfactorily described by means of the linear oscillator model.  This conclusion agrees with classical and quantum theory for atoms and diatomic molecules.  For complex molecules there is no strict quantum-mechanical theory, but the oscillator model may be substantiated with the aid of the symmetry properties of molecular states.

The polarization spectra, the concept of which was formulated at the end of the previous chapter, show that for excitation of luminescence in various regions of the absorption spectrum the polarization of the luminescence varies.  If the degree of polarization differs substantially from 50% (for example, if it takes a negative value), we must come to the conclusion that the absorbing and emitting oscillators have different space orientation.  Here two hypotheses are possible.  According to the first, one and the same oscillator absorbs and emits, rotating through some angle $\alpha$ in the course of the excited state, which exists for time $\tau$.  According to the second hypothesis, absorption and emission are effected by different oscillators making an angle of $\alpha$ with each other.  Clearly these hypotheses are mathematically identical, though they have different physical meanings.

The first hypothesis was set out and mathematically developed by V. L. Levshin [76,94], and the second by F. Perrin [95].  These calculations are also set out in detail in [86].

Both variants of the calculation lead to an identical result:

$$P = \frac{3\cos^2\alpha - 1}{\cos^2\alpha + 3} .$$

(II.1)

This expression is called the Levshin-Perrin formula.  Let us consider two extreme cases:  (1) the oscillators are coincident, $\alpha = 0$; and (2) the oscillators are mutually perpendicular, $\alpha = 90°$.  In the first case P = +50%, and in the second, P ~ −33%.  Actually, the degrees of polarization observed in experiment for the fluorescence of isotropic solutions nowhere attain these limits.

In order to decide which of the two physically differing hypotheses is really true, isolated data on the fluorescence of isotropic solutions are insufficient.  Additional experiments are needed to choose between them.  Such experiments are carried out with anisotropic films colored with fluorescent materials.

Kautsky and Hirsch [81] first prepared cellophane films colored in aqueous solutions of dyes, and studied their fluorescence and phosphorescence.

Yablonskii [82] and Pringsheim [83] studied the anisotropic properties of such films caused by the fact that the dye molecules adsorbed by the ordered and directed cellophane fibers were partly oriented in a particular direction.  These workers established that the fluorescence of the films was partly polarized when excited by natural or linearly polarized light, for observation in the direction of excitation. In isotropic media under these conditions, of course, the fluorescence is unpolarized. In the present case the fluorescence in the films is partly polarized (so called spontaneous polarization), the degree of polarization being independent of the wavelength of the exciting light. The same authors also studied the spectral behavior of the dichroism of films for a number of dyes.

Later Feofilov [84,85] studied polarization spectra for polarized excitation, and also the dichroism spectra and spontaneous polarization of cellophane films colored by various dyes. Feofilov also showed that the spontaneous polarization was independent of the excitation wavelength, while the polarization of the fluorescence of the same films excited by polarized light varied with wavelength in just the same way as in isotropic solutions. This phenomenon may only be explained on the assumption that the degree of orientation of the radiation oscillators is independent of the excitation wavelength, being determined by the orientations of the molecules, i.e., the radiating oscillator is firmly fixed in the molecule. The polarization spectrum moreover shows that different absorption oscillators are oriented differently. This fact already serves as a proof of the hypothesis of separate oscillators. Data on the film dichroism spectra lend weight to the same hypothesis.

The dichroism (the difference in the absorption of polarized light for different orientations of its electric vector) may be quantitatively characterized by the quantity

$$D = \frac{\varkappa_1 - \varkappa_2}{\varkappa_1 + \varkappa_2}, \qquad (II.2)$$

where $\varkappa_1$ and $\varkappa_2$ are the absorption coefficients of light polarized respectively parallel and perpendicular to the orientation axis of the film. Clearly the dichroism of films is caused by preferential orientation of the absorption oscillators of the molecules in a certain direction. The variation of D with wavelength (the dichroism spectrum) shows that the oscillators corresponding to different absorption bands are oriented differently. These facts support the hypothesis of separate oscillators, the absorption of light in various bands being caused by different oscillators, while emission always comes from the same oscillator, close to that of the long-wave absorption. All these oscillators have certain fixed positions in the molecule.

In a number of dyes (trypaflavine, benzoflavine, eosine, and acridine orange), there is a qualitative similarity between the polarization and dichroism spectra. This is only natural, since both phenomena are connected with and determined by one and the same anisotropy of the molecules coloring the film. This is not a trivial matter, since it indicates how the molecules are oriented in the film. If the radiation oscillator of an oriented molecule is directed along the axis of orientation of the film, then approximately the same direction pertains to the long-wave absorption band oscillator. Thus, for excitation polarized along the film axis in this band, the polarization of the fluorescence will have a large positive value, and the dichroism will also be positive ($\varkappa_1 > \varkappa_2$). In the region of excitation in which the absorption oscillator makes a considerable angle with the axis, the polarization takes on a negative value; the dichroism also here becomes negative ($\varkappa_1 < \varkappa_2$). Thus, qualitative coincidence of the polarization and dichroism spectra proves that the radiation oscillator coincides with the axis of orientation of the film, or is close to this axis. We cannot expect quantitative agreement in this, as it is clear from formula (II.1) that the polarization may take values $-(1/3) \le P \le (1/2)$, and the dichroism values $-1 \le D \le 1$, as follows from (II.2).

For such an orientation of the molecules, clearly, the film must have yet another characteristic: its spontaneous polarization must be positive. In the case of the dyes mentioned, the spontaneous polarization is indeed positive, of the order of 10-20% for different substances. Its value also of course depends on the degree of orientation.

We may assume that other molecules may be oriented in some other fashion. This is determined, on the one hand, by the anisotropy properties of the cellophane itself, and on the other by the structure of the oriented molecules. If, for example, a short-wave absorption oscillator, making a considerable angle (almost a right angle) with the long-wave absorption oscillator (and the radiation oscillator), is oriented along the film axis, then the dichroism in the long-wave region will be negative and that in the short-wave region positive, that is, its spectral distribution will be inverse to that of the polarization. The spontaneous polarization in this case must be negative. Such an orientation was found experimentally, for example, for acrichin in a cellophane film.

Thus, in different regions of the spectrum, absorption arises from oscillators having a definite orientation in the molecule and making different angles $\alpha$ with the radiation oscillator. This latter is also fixed in the molecule in a definite direction coinciding with, or at any rate close to, the direction of the first long-wave absorption band oscillator. This is indicated by the spectral proximity of the long-wave absorption band and the emission band. These correspond to direct and reverse transitions between the ground and first excited electron levels. The degree of polarization of the luminescence, which, in the long-wave absorption band region is always positive and closest to 50%, also agrees with this.

From formula (II.1) we may estimate the angles made by various absorption oscillators with the radiation oscillator. Exact evaluation of the angles $\alpha$ is only possible after removing all depolarizing factors, but even a simple evaluation of the angles may give important results.

If, for example, the polarization in some region of the polarization spectrum is negative, then this means that the oscillator of the corresponding absorption band makes very nearly a right angle with the radiation oscillator. Thus, from the polarization spectra, we may determine the orientation of the absorption and radiation oscillators relative to each other, and construct the oscillator model of the molecule.

But this is only the first step in studying the molecule by this method. The oscillator model is still not attached to the molecule. The next step must be to determine the orientation of this oscillators with respect to the skeleton of the molecule and the elements of its chemical structure. The question is one of agreement between the physical representation of the elementary oscillators and the chemical representation of the molecular structure.

In complicated molecules it is impossible to calculate the eigenfunctions and matrix elements of the transition dipole moments exactly. Nevertheless, even qualitative considerations often lead to interesting and important results.

Clearly, simple comparison of the absorption and polarization spectra is not enought to solve the problem in hand, and supplementary data and considerations must be brought in.

It was indicated earlier that in the luminescence of complex molecules a fundamental part was played by the conjugated chains of $\pi$-electrons. These chains evidently constitute the cause of chromaticity in organic compounds, i.e., the presence of a strong long-wave absorption band.

The approximation of a molecule with a linear chain of conjugate bonds, and the application of considerations based on the symmetry properties of molecular states to this model, leads, as indicated earlier, to the conclusion that the long-wave absorption and radiation oscillators are directed along the chain of conjugate bonds.

In many dyes, the chain of conjugate bonds is terminated on both sides by auxochromes ($NH_2$, OH, etc.). For this reason Feofilov suggested that the long-wave absorption oscillator in dyes is oriented along the direction joining the auxochromes [84]. This suggestion found support in theoretical work of Laffitte [121] based on the "metallic model" of the molecule, which, as is well known, enables us to calculate the optical properties of molecules with conjugate chains of bonds to a good approximation.

On the basis of this proposition we may consider that in dyes of the

type the orientation of the oscillators is as follows: the long-wave oscillator is directed along the line $AA^+$; the short-wave oscillator (250 to 270 m$\mu$) is connected with the presence of the benzene ring ,and may be considered parallel to the fundamental oscillator, since the degrees of polarization of the luminescence on excitation in either the long-or short-wave absorption band are close together.

In the case of triphenylmethane dyes of the

type, it may be shown by means of the polarization spectrum and formula (II.1) that the angle between the oscillators responsible for absorption in different regions of the spectrum is close to 60°, in agreement with the orientation of oscillators along the lines joining the auxochromes.

This is one example of the way in which supplementary data help analyze absorption spectra together with polarization spectra and determine the orientation of the oscillators with respect to the molecular axes. Other indirect data may also be brought in, for example, by studying homologous series of substances and considering how regular changes of molecular structure affect the polarization and absorption spectra.

Besides the dyes which we have considered, the literature contains data on the study of polarization spectra of porphyrins, chlorophylls, and others [122,123,124]. We used this method to study several polycyclic hydrocarbons, their derivatives, and some other compounds. Our purpose was to determine the orientation of the radiation oscillators in the molecules as a first and necessary step before studying the polarized luminescence of crystals formed from such molecules.

## §2. Anthracene and Its Derivatives [86,125,126]

The orientation of the oscillators in the anthracene molecule is of special interest to us. We measured the polarization of the blue fluorescence of anthracene solutions in a photoelectric polarization system with a modulating biquartz. *

Twice-distilled glycerine was used as viscous solvent. The anthracene was introduced into the glycerine solution through alcohol. The anthracene was preliminarily dissolved in a small quantity of ethyl alcohol, then this solution was mixed with glycerine, heated to 70 or 80°C, and stirred carefully. The alcohol evaporated almost entirely, leaving a solution in glycerine. A concentration of $1 \cdot 10^{-5}$ g/cm$^3$ was prepared in order to avoid passing into the region of concentration depolarization.

The variation of the polarization of the luminescence of the solution was measured as a function of temperature and viscosity with the aid of a thermostat. At the same time, the temperature dependence of viscosity was measured with an Ostwald viscometer in the same thermostat. This relation is shown in Fig. 5. A graph of $1/P$ vs. $T/\eta$ is given in Fig. 6. Our data are in good agreement with results of other authors [127,128]. The limiting polarization $P_0$ obtained from this graph equals 26.3% (excitation by mercury line, wavelength 365 m$\mu$).

At room temperature and the same wavelength excitation, the degree of polarization is 23%. We see from this that the thermal depolarization is slight in view of the large viscosity of the anhydrous glycerine. For this reason the polarization spectra of anthracene and other substances were measured at room temperature.

The excitation was derived from the light of various spectral lines from a mercury lamp (254 to 405 m$\mu$), analyzed by means of a quartz monochromator. The absorption spectrum was measured with the spectrophotometer SF-4. The absorption and polarization spectra of anthracene are presented in Fig. 7.

The main information derivable from these spectra amounts to the following: the polarization of the luminescence on excitation in the region of the first long-wave absorption band ($\sim$400 m$\mu$) is positive (+23%); on excitation in the short-wave band ($\sim$250 m$\mu$) it is negative ($-7$%). It follows from this that the absorbing oscillators corresponding to these transitions are mutually perpendicular, or, at all events, make an angle very close to a right angle with each other.

---

* A description of all the experimental methods used appears in Chapter VII.

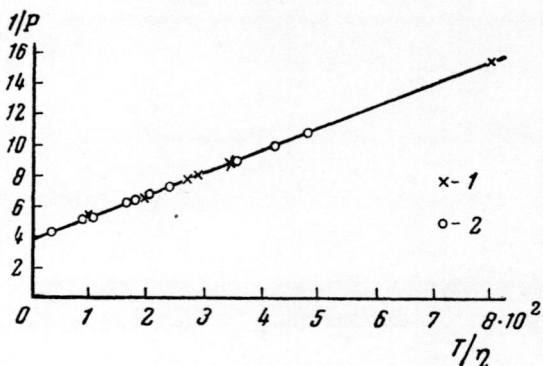

Fig. 6. Variation of $1/P$ with $T/\eta$ for anthracene. 1) From data of Shishlovskii and Spektorov; 2) author's data.

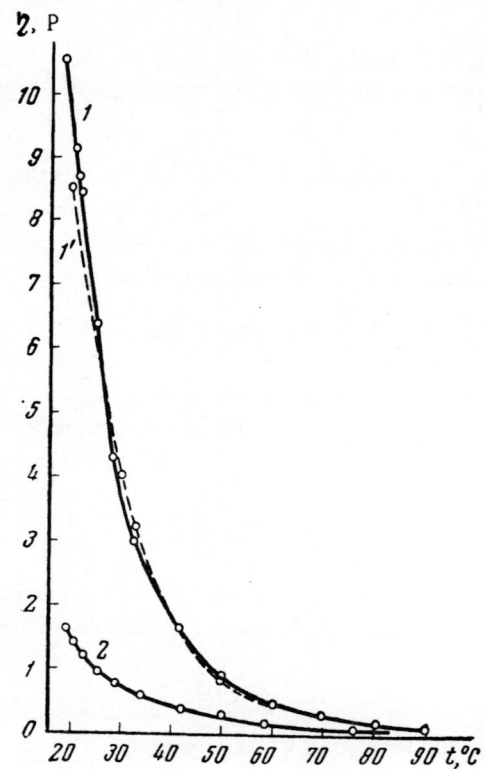

Fig. 5. Temperature dependence of viscosity coefficient $\eta$ (in poise). 1) For glycerine; 1') from data of M. D. Galanin; 2) for vaseline oil.

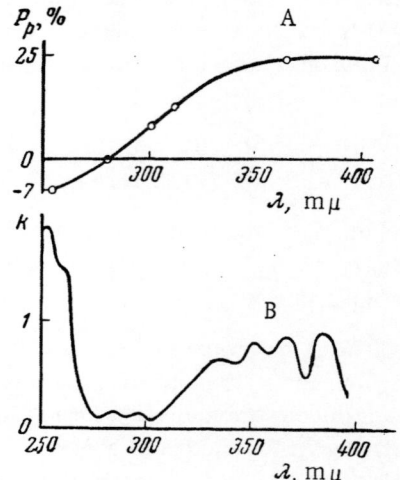

Fig. 7. A) Polarization; B) absorption spectrum of anthracene.

In order to make an attempt at answering the question of the orientation of oscillators in the molecule, let us consider the following additional data. By the method of molecular orbits, Coulson [129] showed theoretically that the long-wave absorption band in molecules of an aromatic series (naphthalene, anthracene, etc.) arises as a result of a transition between levels of symmetry $A_{1g} \rightarrow B_{2u}$, and is polarized along the transverse axis of the molecule. The second, short-wave band is caused by transition $A_{1g} \rightarrow B_{1u}$ polarized with respect to the longitudinal axis of the molecule. An analogous result was obtained in theoretical work of Vol'kenshtein and Borovinskii [130], in which the benzene ring was considered as a one-dimensional, closed potential box, and in the naphthalene and anthracene molecules the symmetry properties of the wave functions were determined from the continuity condition of the wave functions $\psi$ at the mesh points of the molecule (where the benzene rings come together), and the continuity of the derivative $\partial \psi / \partial r$, written in the form of Kirchhoff's law for the mesh points in an electric current network. These results are supported by experimental data of Craig [18] on the absorption of anthracene crystals, as already mentioned.

Immediate confirmation of these conclusions for solutions comes from experiments of Jones [13], who studied absorption spectra of the solutions of a large number of anthracene derivatives with substitute groups introduced at various places in the anthracene framework. It transpired that different substitute positions had different effects on the two main absorption bands around 400 and 250 m$\mu$. On substituting OCH, CN, NO$_2$, phenyl, naphthyl, and other groups at vertices 9 and 10,* the long-wave band shifts considerably more than the short-wave band. The introduction of substitutes in positions 2 and 3, on the other hand, strongly influences the short-wave band. It follows from this that the long-wave band is associated with the transverse axis of the molecule, and the short wave band with the longitudinal axis. Analogous results are given by examining the absorption spectra of arylamine derivatives of anthracene [132].

Only one paper gives results disagreeing with what has been said. Obreimov and Prikhot'ko [133], studying the dispersion of anthracene crystals, came to the conclusion that both transitions were polarized along the transverse axis of the molecule. It is true that the chief aspect of the question interesting us is also confirmed in this case, that is, the long-wave transition is polarized along the transverse axis of the molecule. From what has been said we may conclude that in the anthracene molecule the radiation oscillator is directed along the transverse axis.

It was interesting, for comparison with anthracene, to study the polarization spectra of its derivatives. We did this for the group of 9,10-diaryldiaminoanthracenes, which have a bright-green luminescence. This last circumstance enabled us to carry out measurements by means of a Cornu polarimeter. The base of this group of compounds is 9,10-diphenyldiaminoanthracene:

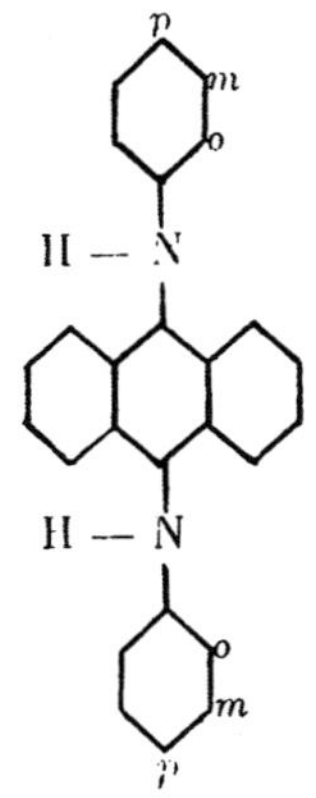

in which $CH_3$ and Cl are substituted into various positions (ortho, meta, and para). Besides this there are the $\alpha$- and $\beta$-9,10-dinaphthyldiaminoanthracenes:

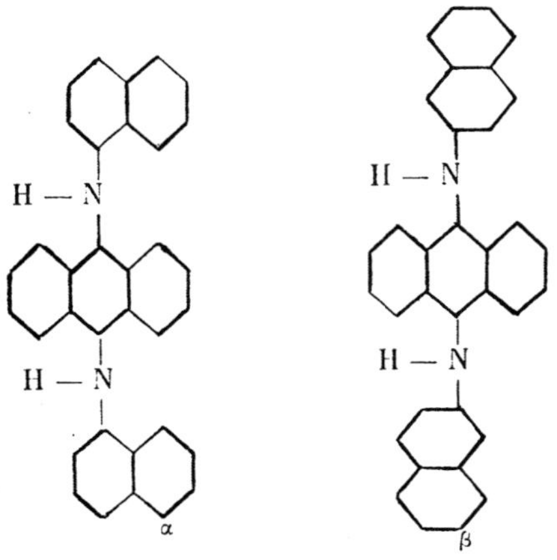

In the absorption spectra of the derivatives with one phenyl ring in the side-chain, the short-wave maximum scarcely shifts from compound to compound, and is very little displaced from anthracene. At the same time, the long-wave maximum is considerably displaced from that of anthracene, and varies in different compounds. In the naphthyl derivatives, however, both long- and short-wave maxima are displaced.

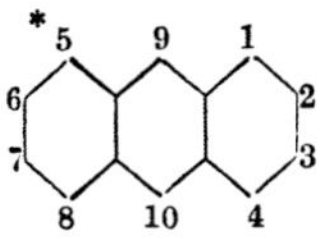

37

The polarization spectra of this group of anthracene derivatives were measured in glycerine solutions (dissolution via alcohol) of concentration $2 \cdot 10^{-5}$ g/cm$^3$.

The polarization spectra of all the compounds with one phenyl ring in the side-chain have more or less the same general features as that of anthracene itself, namely, positive polarization (order of 40%) in the long-wave absorption band region, and negative (around −10%) in the short-wave band. In more recent work by Williams [134], polarization measurements were made with only two excitation wavelengths, $\lambda_{\text{locc.}} = 365$ and 254 m$\mu$, for solutions of a number of substances, among them anthracene and some of its derivatives. These results agree well with ours.

By way of example, Fig. 8 shows our measured polarization and absorption spectra for the ortho- and para-9,10-dichlorophenyldiaminoanthracenes. This form of the polarization spectra once more bears witness to the orthogonality of the long- and short-wave oscillators, and, taking note of the absorption data, we arrive at the conclusion that in these derivatives, just as in anthracene itself, the radiation oscillator is oriented in the same way as the long-wave absorption oscillator, being directed along the transverse axis of the molecule joining points 9 and 10.

The polarization spectra of the naphthyl derivatives are rather different. With these, the polarization does not change sign in the neighborhood of the short-wave absorption band, but remains positive (order of +10%). Figure 9 shows the spectra of 9,10-di-$\alpha$-naphthyldiaminoanthracene.

From this we must conclude that the angle between the two absorption oscillators is considerably less than a right angle, and the long-wave oscillator has components along both longitudinal and transverse axes of the molecule. This agrees with absorption spectrum data. It may be that this orientation of the oscillators is caused by the unsymmetrical disposition of the substitute naphthyl groups with respect to the longitudinal axis of the molecule.

In order to confirm these conclusions, we carried out experiments with anisotropic films colored by arylamine derivatives of anthracene. The cellophane films could not be colored by these substances, as they were insoluble in water. After trying numerous substances and solvents, we prepared polystyrene films in benzene. Polystyrene was diluted in a benzene solution of the material under investigation, poured out on to glass, and dried at 40°C. The resultant film was again put to swell in benzene, and then drawn out in an arbitrary direction on a special frame, by which means it acquired partial anisotropy in the drawing direction. With due care, a fourfold extension could be achieved. It is easy to check the anisotropy of the film by placing it between crossed Nicols. For these films we measured the polarization of the fluorescence as a function of the wavelength of the exciting light for natural and polarized illumination, and also the dichroism spectra.

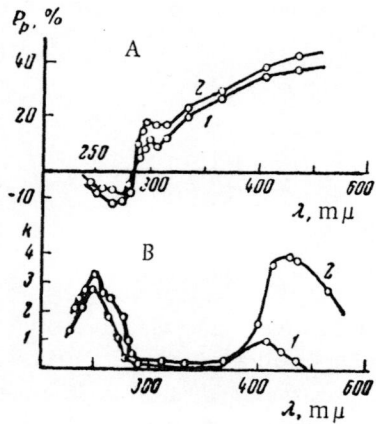

Fig. 8. A) Polarization; B) absorption spectra of the dichlorphenyldiaminoanthracenes. 1) Ortho-; 2) para-.

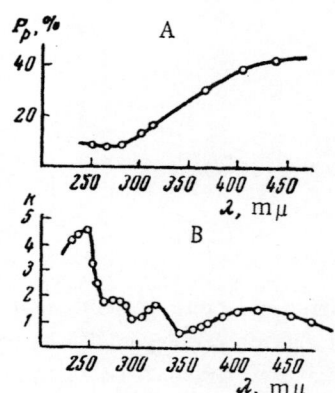

Fig. 9. A) Polarization and B) absorption spectra of 9,10-di-$\alpha$-naphthyldiaminoanthracene.

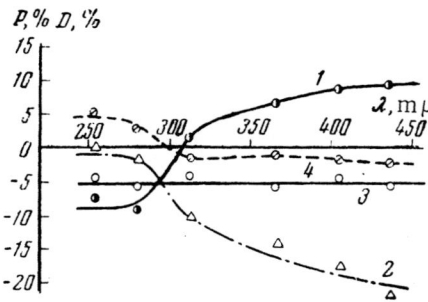

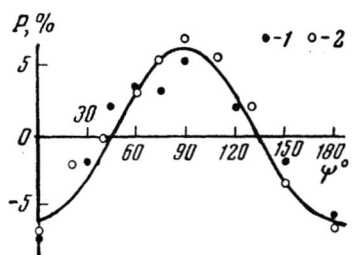

Fig. 10. Polarization and dichroism spectra of an anisotropic polystyrene film with 9,10-dimetachlorphenyldiaminoanthracene. 1) Excitation polarized along the orientation axis; 2) excitation polarized perpendicular to the orientation axis; 3) natural excitation (spontaneous polarization); 4) dichroism.

Fig. 11. Spontaneous polarization as a function of film azimuth. 1) Para-9,10-ditolyldiaminoanthracene; 2) meta-9,10-dischlorphenyl-diaminoanthracene.

The results of the measurements for one of the substances are shown in Fig. 10. The spontaneous polarization (for natural excitation) is independent of wavelength, as in the case of the dyes, and has a negative value, approximately 6%.

The polarization spectrum (on excitation by polarized light) caused by nonoriented molecules has the same character as in an isotropic medium. However, the presence of oriented molecules giving spontaneous polarization leads to the polarization spectrum being displaced with respect to the axis of ordinates and undergoing some deformation.

It is clear that in isotropic solutions the polarization spectra for two mutually perpendicular polarized excitations must be symmetrical with respect to the axis of abscissas. In anisotropic films, their displacement must be such that the spontaneous-polarization straight line becomes a line of symmetry. Figure 10 shows that this is also found by experiment. Thus measurement of these polarization spectra is an indirect, independent measure of the same spontaneous polarization. The existence of two methods is important, since the spontaneous polarization is quite small, and relative errors may be substantial. The experimental values of spontaneous polarization agree farily well with the mean values of corresponding points in the polarization spectra.

The negative value of the spontaneous polarization indicates that the radiation oscillator makes a considerable angle with the orientation axis. The polarization spectra have the same character as in the glycerine solutions, which confirms our conclusion regarding the orthogonality of the long-and short-wave absorption oscillators. Hence we may assert qualitatively that the molecules orient themselves with the longitudinal direction of the benzene rings along the film axis, and the angle between the transverse axis of the molecules and the film axis is approximately a right angle.

This is also supported by the azimuthal variation of the spontaneous polarization of the film (Fig. 11). The continuous line in this figure represents the calculated azimuthal variation for oscillators oriented perpenducularly to the film axis, normalized with respect to the maximum experimental value of the spontaneous polarization. The experimental points for various azimuth angles of the film agrees satisfactorily with this.

Figure 10 also gives the dichroism spectrum. The small degree of polarization results in a small value of the dichroism, but qualitatively the behavior of the spectrum is clearly visible; in the long-wave region the dichroism is negative (−2 to −3%), and in the short-wave region positive (+5%). Qualitatively, these data completely agree with the results of all the other experiments: the oscillators of long- and short-wave absorption are mutually perpendicular, and the latter is directed along the film axis. Thus the experiments with the anisotropic films qualitatively confirm and supplement the experiments with solutions.

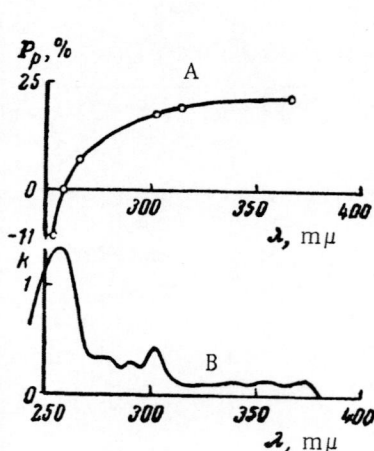

Fig. 12. A) Polarization and B) absorption spectrum of phenanthrene.

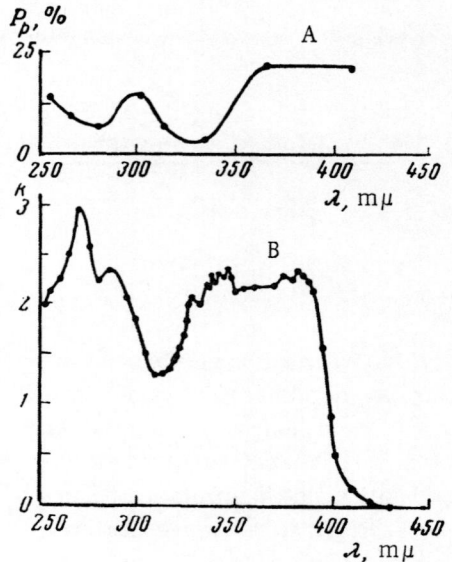

Fig. 13. A) Polarization and B) absorption spectrum of dixanthylene.

## §3. Phenanthrene, Dixanthylene, and Certain Other Compounds [126]

A glycerine solution of phenanthrene was also obtained by preliminary dilution in alcohol. As in the case of other substances described in this paragraph, the measurements were carried out in a photoelectric polarization apparatus with a modulating biquartz. The polarization and absorption spectra of phenanthrene are shown in Fig. 12. In general, the character of the polarization spectrum is the same as in anthracene: positive polarization in the long-wave region and negative in the short. It follows that in phenanthrene also the angle between the oscillators corresponding to these absorption bands is nearly a right angle.

Interesting results are obtained for dixanthylene

For this substance, we used vaseline oil as solvent, since it was insoluble in glycerine.

Vaseline oil has quite high viscosity, but is difficult to purify and always possesses its own luminescence, which must either be separated out experimentally or allowed for as a correction. The polarization and absorption spectra of dixanthylene appear in Fig. 13. In these spectra an important fact is that the degree of polarization is positive, and approximately the same for long- and short-wave excitations.

This is good ground for considering that the long-wave absorption oscillator (and hence also the radiation oscillator) is directed along the C = C bond linking the transverse axes of analogous parts of the molecule. As regards the short-wave oscillator, it follows from the form of the polarization spectrum that this cannot be regarded as directed along the chain of benzene rings as in anthracene. It may well be that in the dixanthylene

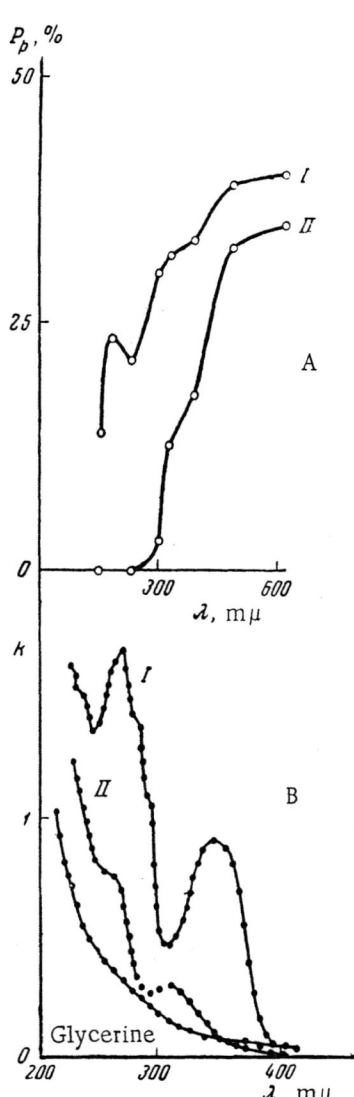

Fig. 14. A) Polarization and B) absorption spectra. I) 2-(oxy-phenyl)benzoxazole; II) 3-ami-nophthalimide.

structure the effect of the oxygen atom substantially changes the configuration of the electron cloud in the anthracene framework, as a result of which the short-wave oscillator is oriented in a direction close to the transverse axis of the molecule.

We also studied the polarization spectra of some other complex molecules. Figure 14 shows, for example, the spectra of glycerine solutions of 2-(oxyphenyl)benzoxazole

and 3-aminophthalimide

The structural formulas of these have analogous features. In both cases the benzene ring adjoins a five-term ring. The structural similarity is reflected in certain similarities of the absorption and polarization spectra. The angle between the oscillators responsible for absorption in the long- and short-wave parts of the spectrum is negligible, being greater in the 3-aminophthalimide than in 2-(oxphenyl)benzoxazole.

Unfortunately, we were unable to measure the polarization spectrum of stilbene, tolan, and diphenylbenzene on account of the very small intensity of the violet luminescence from these substances, which could not reliably be separated from the fluorescence of the solvents.

§4. Naphthalene Derivatives [135]

We were unable to measure the polarization spectrum of naphthalene itself, as we could not prepare viscous solutions of naphthalene owing to its poor solubility.

We studied a series of naphthalene derivatives, namely, certain naphthylamines:

phenyl-α-naphthylamine

phenyl-β-naphthylamine

paratolyl-β-naphthylamine CH₃

All these substances dissolved in glycerine (introduced via alcohol). The fluorescence of the solutions was bright blue. Measurements of the polarization of the luminescence from the naphthylamines and polyenes were

41

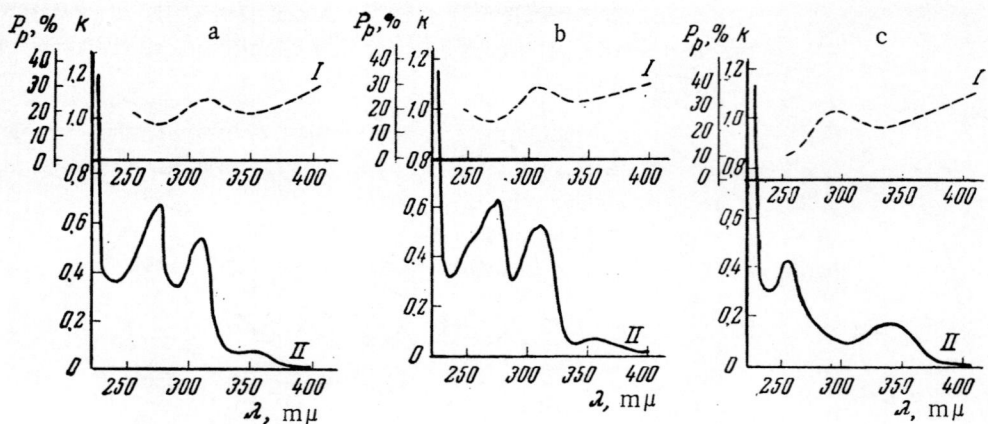

Fig. 15. I) Polarization and II) absorption spectra of the naphthylamines. a) Phenyl-β-naphthylamine; b) paratolyl-β-naphthylamine; c) phenyl-α-naphthylamine.

carried out by means of the Kavraiskii polariscope with compensator (see Chapter VII). The polarization and absorption spectra of the naphthylamines are shown in Fig. 15.

The fact that the sign of the polarization remains unchanged over the whole spectral range studied leads unequivocally to the following qualitative conclusions: the oscillators responsible for the second and third absorption bands are directed roughly parallel to the first long-wave oscillator. In this respect there is an essential difference from anthracene and its derivatives.

Brodersen [63], studying the polarization of the fluorescence from naphthalene single crystals, and taking into consideration x-ray structural data on the orientation of the naphthalene molecules in the lattice, came to the conclusion that the radiation oscillator was directed along the transverse axis of the molecule. If we assume that this also holds for naphthalene derivatives, it follows that in the naphthylamine all the oscillators are directed roughly along the transverse axis of the molecules.

## §5.  Polyenes [135]

We had at our disposal a limited number of polyenes, mainly derivatives of butadiene:

1) 1,1,2-triphenylbutadiene $(C_6H_5)_2C = C(C_6H_5) - CH = CH_2$;
2) 1,1,4,4-tetraphenylbutadiene $(C_6H_5)_2C = CH - CH = C(C_6H_5)_2$;
3) 1,1,4,4-tetraphenyl-2,3-dimethylbutadiene $(C_6H_5)_2C = C(CH_3) - C(CH_3) = C(C_6H_5)_2$;
4) tetraphenylallene $(C_6H_5)_2C = C = C(C_6H_5)_2$;
5) tetraphenylethylene $(C_6H_5)_2C = C(C_6H_5)_2$.

(In the last we were unable to measure the spectrum in the short-wave region on account of the weak luminescence.)

These substances were all insoluble in glycerine, and were studied in solid solutions in polystyrene. The absorption and polarization spectra are shown in Fig. 16.

The absorption spectra have no distinct structure. The general features are as follows: there is a wide, diffuse long-wave absorption band; on moving into the short-wave region, there is a certain fall in absorption, followed by a very marked rise in short-wave absorption, the maximum being outside the measuring range.

The polarization spectra of the polyenes have certain general common features: a considerable positive degree of polarization in the long-wave region, then a certain fall as one passes into the short-wave region, followed by a renewed rise. The structure of the spectra arising in some of the compounds is hard to explain.

In order to arrive at an explanation, low temperature measurements are needed. We can draw qualitative conclusions at once, however, e.g., that the absorption-band oscillator in the short-wave region is approximately parallel to the long-wave oscillator.

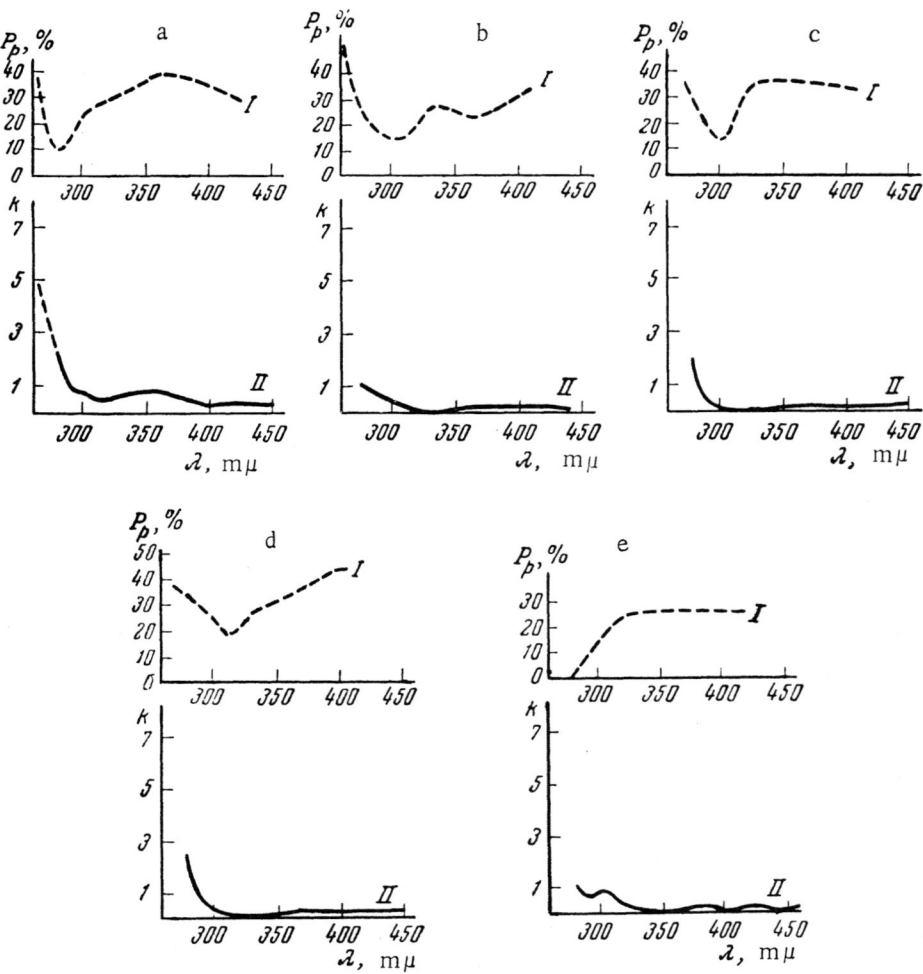

Fig. 16. I) Polarization and II) absorption spectra of the polyenes. a) Tetraphenyl-butadiene; b) 1,1,2-triphenylbutadiene; c) 1,1,4,4-tetraphenyl-2,3-dimethylbuta-diene; d) tetraphenylallene; e) tetraphenylethylene.

Polyene spectra have been theoretically calculated by various methods. In particular, W. Kuhn [136] made a calculation of the spectrum of the polyene chain from classical theory, considering the vibrations of the electrons in various links of the chain as being coupled. Kuhn obtained good experimental agreement for the dependence of $\lambda_{max}$ of the first absorption band on the number of links in the polyene chain. Kuhn also determined the polarization of the electron transition, showing that this was polarized along the polyene chain. In other words, the long-wave absorption oscillator is directed along the long axis of the polyene molecule. It follows then from the polarization spectra that the short-wave oscillator (which is approximately parallel to the long-wave oscillator) is also directed along the long axis of the polyene molecule.

These have been a few examples of the study of polarization spectra of various substances in solutions. They show that in a number of cases it is possible to determine the orientation of the radiating oscillators with respect to the axes of the molecule and the elements of its chemical structure. This is extremely important when we pass on to the study of molecular crystals.

The importance of this circumstance is obvious, since, from the point of view of the optical behavior of crystals, we shall always be mainly interested in the orientation of the molecular oscillators in the lattice rather than the molecules themselves. Thus we need to know how to determine the orientation of the oscillators in the molecule by independent means. The polarization spectra of solutions may be regarded as one of such means.

# CHAPTER III

# THE ORIENTED GAS MODEL AND ITS APPLICATION TO MOLECULAR CRYSTALS

## POLARIZATION DIAGRAMS OF LUMINESCENCE

### §1. Essence of the Model and Method of Investigation

The oriented gas model is the simplest representation of the molecular crystal lattice. Within the framework of this model, the interactions of the molecules are neglected, and the crystal is considered as an accumulation of periodically disposed and regularly oriented isolated molecules.

It follows from the close-packing principle discussed in Chapter I that, in general, the molecules in the crystal cannot all be oriented similarly; there are in fact several kinds of orientation for the molecules in the lattice. For example, in crystals of the monoclinic system, the number of different kinds of orientation is often two.

Of course, the oriented gas model is an extremely simplified representation of a crystal, since it neglects intermolecular interactions, without which the crystal could not exist at all. The validity of the model up to a certain approximation is based on the fact that the forces of interaction between the molecules are considerably less than the intramolecular interaction forces.

The problem for experimental investigation consists largely of finding to what extent the model is valid, and on what occasions and in what phenomena deviations arise.

The polarization of the luminescence from a crystal on excitation by natural light in the same direction as that in which the luminescence is viewed must wholly and completely be determined by the orientation of the molecules, if the oriented gas model is correct. The degree of polarization for observation in any direction is clearly determined by the orientation of the projection of the molecular oscillators on a plane perpendicular to this direction. In order to study the orientation of the oscillators in space, it is necessary to measure the degree of polarization of the luminescence for different directions relative to the crystal axes. In other words, we must study the space distribution of the polarization of the luminescence, i.e., the polarization diagrams.

At this point we must clear up the terminology. The term "polarization diagrams" introduced by Vavilov for solutions embraces two kinds of angular dependence, namely, the way in which the degree of polarization of the luminescence depends on the angles determining the direction of observation and the direction of the electric vector in the exciting light; these are $P(\chi)$ and $P(\eta)$, respectively. In this, $\chi$ is the angle between the direction of the exciting beam and the direction of observation, while $\eta$ is the angle between the electric vector of the exciting light and the vertical.

The second type of angular dependence $P(\eta)$ for molecular crystals will be discussed in detail in the following chapter. At the moment we merely note that, as shown by experiment, the polarization of the luminescence from molecular crystals is independent of the direction of the electric vector in the exciting light and of the direction of the exciting beam itself relative to the crystal axis. This is only important to us just now from the point of view of the experimental determination of the relationship $P(\chi)$. If this independence did not exist, then on changing the angle $\chi$ we should not be able to regard the excitation as isotropic, and this would have to be borne in mind when carrying out the experiments.

We shall only give the name "polarization diagrams," in the case of molecular crystals, to the variation of the degree of polarization of the luminescence $P = (I_1 - I_2)/(I_1 + I_2)$ on the angle $\chi$ between the direction of observation and a definite crystal axis, for a definite orientation of the crystal relative to the main axes of the polarization measuring apparatus, along which components $I_1$ and $I_2$, respectively, are polarized. It is plain that the polarization diagrams will be different for different orientations of the crystal. Experimental methods of measuring polarization diagrams are described in detail in Chapter VII.

As already mentioned, the polarization diagrams of fluorescing solutions are of particular interest in that they enable us to determine the multipolarity of the elementary radiators experimentally, being different for the absorption and radiation of light by different kinds of multipole (dipoles, quadrupoles, etc.). The case is somewhat different in crystals. If two oriented media (crystals) consist of radiators possessing different space distributions of radiation but identical polarization distributions, then the polarization diagrams of these crystals will be identical, whereas the polarization diagrams of solutions in an analogous case would be different.

For example, the electric dipole and quadrupole have differing radiation diagrams. The intensity in a direction making an angle $\alpha$ with the multipole axis is proportional to $\sin^2 \alpha$ for a dipole, and to $\sin^2 2\alpha$ for a quadrupole; the space distribution of the polarization of the radiation in these is, however, identical. Crystals formed from similarly oriented dipoles or quadrupoles will also have identical polarization diagrams (in the expression for the degree of polarization $P = (I_1 - I_2)/(I_1 + I_2)$ the terms $\sin^2 \alpha$ or $\sin^2 2\alpha$ respectively have been contracted). There is no difference in principle in the case of a crystal with any number of differently oriented dipoles or quadrupoles, the respective terms then being $(\sin^2 \alpha_1 + \sin^2 \alpha_2 + \sin^2 \alpha_3 + \ldots)$ or $(\sin^2 2\alpha_1 + \sin^2 2\alpha_2 + \sin^2 2\alpha_3 + \ldots)$.

In isotropic solutions, the dipole and quadrupole cases give different polarization diagrams, since the fluorescence can only be polarized on anisotropic excitation, and with this dipoles and quadrupoles of a definite orientation will be excited differently, giving different contributions to the total polarization of the radiation. There is thus an important role here for the difference between the radiation diagrams of dipoles and quadrupoles, although this is immaterial in crystals under isotropic excitation.

Clearly, crystals formed from radiators with different space distributions of the polarization of the radiation will have different polarization diagrams (for example, in the case of an electric dipole and a magnetic dipole). Thus, as a method of determining the multipolarity of radiators, polarization diagrams have less generality in crystals than in isotropic media. However, they do enable us to solve the problem of determining the orientation of the molecules in the lattice in crystals, or, in other words, to study the applicability of the oriented gas model.

Generally speaking, in the case of crystals characterized by a spatial distribution of radiation from individual elementary radiators, in order to determine the multipolarity of the latter we may in principle use, instead of polarization diagrams, the variation in luminescent intensity as a function of the direction of observation and the direction of the electric vector in the exciting light. However, the specificity of the multipoles will be greatly weakened by this owing to the existence of several forms of orientation of the radiators, as well as of the migration of energy between these (more details of this will appear in Chapter IV).

§2. Polarization Diagrams of Anthracene Single Crystals [137]

As an object for study it is appropriate first to select a substance for which we know (a) the orientations of the radiation oscillator in the molecule, (b) the optical anisotropy of the crystal (its indicatrix, i.e., the refractive index as a function of direction in the crystal), in order to be able to introduce corresponding corrections into the measured polarization of fluorescence from the crystal, and (c) the orientation of the molecules in the crystal lattice, so as to be able to compare the orientation deduced from the fluorescence polarization measurements with these data.

The first such object selected was anthracene, for which all the data mentioned above are available. The radiation oscillator in the anthracene molecule lies in the plane of the molecule and is directed along its transverse axis. This follows both from theoretical calculations [129,130] and from experimental data, especially from a study of absorption spectra [131] and polarization spectra of solutions of anthracene [126] and its derivatives [86,125]. These we discussed in detail in Chapter II.

The indicatrix of the anthracene crystal was investigated by Obreimov and his colleagues [138]. These authors measured the refractive index for various angles in the (010) plane and obtained the corresponding section of the indicatrix.

X-ray structural analysis of crystalline anthracene was carried out by Robertson in 1933 [11] and later repeated by Robertson and colleagues in 1950. The crystal lattice of anthracene belongs to the monoclinic system. Robertson obtained the following values for the unit cell:

$$a = 8.56 \text{ A}, \ b = 6.04 \text{ A}, \ c = 11.16 \text{ A}, \ \beta = 124°42'.$$

The elementary cell contains two molecules. The molecules are set at the lattice nodes and in the centers of the (001) faces (Fig. 17). The orientations of the molecules vary. Below we present the angles between the molecular and crystal axes. These angles determine the orientation of molecules at the lattice nodes according to Robertson's latest data, which differ little from the older data published in 1933.

|    | L      | M      | N      |
|----|--------|--------|--------|
| a  | 119°,7 | 71°,3  | 36°,2  |
| b  | 97     | 26,6   | 115,5  |
| c' | 30,6   | 71,8   | 66,2   |

Here L, M, N are the axes of the anthracene molecule (L = longitudinal, M = transverse, N = normal to the plane of the molecule), a, b, c' are the crystal axes (a and b = crystallographic axes, c' = perpendicular to the plane ab).

The position of the molecules in the centers of the (001) faces is obtained from that of the molecules at the nodes by means of reflection in the (010) plane, or rotation around the b-axis and displacement by a/2, b/2. Since the radiation oscillator is directed along the transverse axis of the molecules, we shall be most interested in the orientation of just these molecular axes in the lattice. Figure 18 shows the oscillator scheme for the elementary cell of the anthracene crystal.

Radiation oscillators $00_1$, $0'0''$ (transverse axes of the anthracene molecules situated at the node and in the center of the face) make angles $\varphi = 26°,6$; $\psi = 71°,3$ with the b- and a-axes. The two oscillators lie in a plane

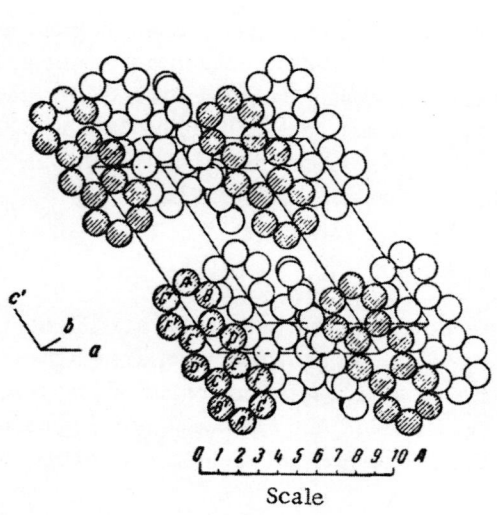

Fig. 17. Position of the anthracene molecules in the elementary cell.

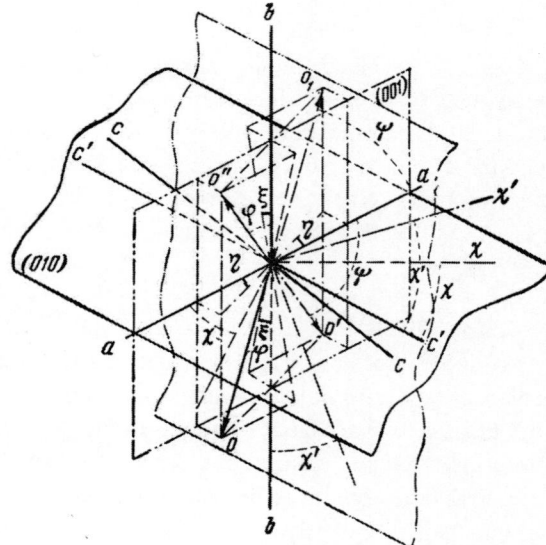

Fig. 18. Oscillator scheme of the elementary cell in the anthracene crystal.

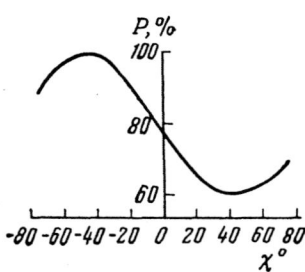

Fig. 19. "Positive" polarization diagram calculated from x-ray structural data.

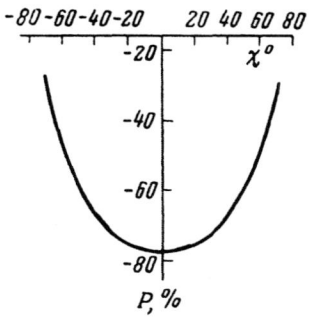

Fig. 20. "Negative" polarization diagram calculated from x-ray structural data.

passing through the b-axis and making an angle $\eta$ with the ab-plane. The angle between the b-axis and the projection of each of the oscillators on the bc'-plane equals $\xi$. The angles $\eta$ and $\xi$ are easily calculated:

$$\eta = 44°, \ \xi = 19°,3.$$

We may further calculate, for this scheme of oscillators, the variation of the degree of polarization of the radiation $P = (I_1 - I_2)/(I_1 + I_2)$ with the direction of observation, that is, with the angle $\chi$, reckoned from the c'-axis in the ac'-plane or in the bc'-plane. For the first case (in which angle $\chi$ lies in the ac'-plane):

$$P(\pm \chi) = \frac{\cos^2 \varphi - \sin^2 \varphi \cdot \cos^2 (\eta \pm \chi)}{\cos^2 \varphi + \sin^2 \varphi \cdot \cos^2 (\eta \pm \chi)} . \tag{III.1}$$

For the second case (angle $\chi$ in the bc'-plane):

$$P(\pm \chi) = \frac{2 \cos^2 \psi - [\sin^2 \psi \cdot \cos^2 (\xi \mp \chi) + \sin^2 \psi \cdot \cos^2 (\xi \pm \chi)]}{2 \cos^2 \psi + [\sin^2 \psi \cdot \cos^2 (\xi \mp \chi) + \sin^2 \psi \cdot \cos^2 (\xi \pm \chi)]} . \tag{III.2}$$

Substituting the values of $\varphi, \eta, \psi$, and $\xi$ given above in formulas (III.1) and (III.2), we may calculate the polarization diagram of the anthracene crystal for the cases in which the crystal is oriented so that the b-axis is parallel to the principal direction of the polarimeter (III.1), or so that it is parallel to the a-axis (III.2). The results obtained are shown in Figs. 19 and 20. In the first case the polarization is positive over a wide range of angles; in the second it is negative. We shall arbitrarily call these the "positive" and "negative" polarization diagrams.

The single crystals were obtained by sublimation of purified anthracene in porcelain crucibles. The crystals were obtained in the form of very thin plates (thickness of order $10\mu$), which by means of needles could be set in optical contact on a glass of quartz substrate. The plate surface was formed by the (001) plane. The best of a number of single crystals formed by sublimation were selected; these were strictly single, homogeneous, and without cracks or defects. It was desirable to diaphragm out the edges of the single crystal in order to remove their depolarizing effect.

The single crystal on the substrate was set up on the vertical axis of the goniometer in such a way that the axis lay in its plane. Along the graduated circle of the goniometer moved a Cornu polarimeter, so placed that its optical axis passed through the center of rotation. This made it possible to measure the polarization of the fluorescence for various directions as given by angles $\chi$. Further, the crystal was fixed in a special holder with a vertical circle, so that it could be rotated around a horizontal axis passing through the center of the crystal. This was needed in order to vary the orientation of the single crystal.

Excitation of fluorescence was achieved by means of the light of a mercury lamp in the region of the 365 m$\mu$ line. The direction and polarization of the exciting light play no part, since the polarization of the fluorescence from a single crystal is independent of the anisotropy of the excitation. In the majority of cases the fluorescence was excited by natural light along the normal to the crystal surface from the side opposite to

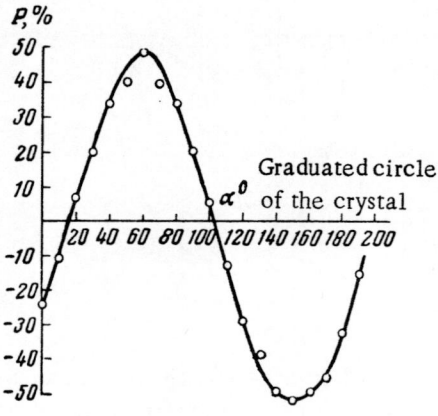

Fig. 21. Azimuthal variation of the polarization of fluorescence from an anthracene single crystal.

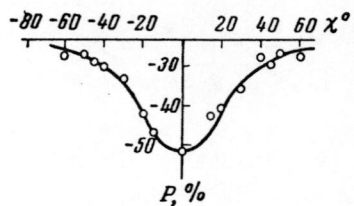

Fig. 22. Experimental "negative"
polarization diagram for an anthracene single crystal.

the viewer. The polarization of the fluorescence from the crystal clearly depends on its orientation. In measurements along the normal to the crystal surface, the polarization of the fluorescence was determined by the
angles between the projections of the oscillators on this surface, i.e., on the (001) plane. As seen from the
oscillator scheme (Fig. 18), these projections are symmetrically disposed relative to crystallographic axes a
and b, the a-axis being the bisector of the obtuse angle between the projections, and the b-axis being that of
the acute angle between these. If we rotate the crystal around the horizontal axis, the degree of polarization
$P = (I_1 - I_2)/(I_1 + I_2)$ of the fluorescence along the normal to the crystal will change smoothly ($I_1$ = component
parallel to the principal direction of the polarimeter, $I_2$ = component perpendicular to this). We shall call this
relationship the azimuthal variation. The positive extremum of the azimuthal variation of polarization corresponds to the orientation for which the b-axis is vertical, and the negative extremum to that in which the
a-axis is vertical.

A characteristic azimuthal variation of the polarization of fluorescence from one of the anthracene single
crystals studied is shown in Fig. 21.

The azimuths corresponding to the extrema determine the positions of the crystal for which the "positive"
and "negative" polarization diagrams are to be measured. Clearly, for each kind of diagram there are two such
positions, differing by 180° from each other. In order to increase accuracy and remove chance errors, the diagrams were measured for all four azimuths.

In this way we studied a considerable number (several tens) of anthracene single crystals. The results obtained were as follows.

The negative polarization diagrams in all the single crystals have identical form and agree qualitatively
with the negative diagram constructed on the basis of x-ray structural data. A typical experimental diagram is
shown in Fig. 22. In absolute magnitude, the measured polarization is less than theoretical, but this discrepancy
was not the same in different crystals. This may be explained by different depolarizing factors; in particular,
the difference in the absolute values of the polarization for different samples may be caused by the presence of
unnoticed cracks, surface irregularities, and so forth, in some of these.

The difference in the degree of polarization of the luminescence from individual samples was 5 to 10%.
This is considerably less than the difference between the mean measured value and that calculated theoretically,
which was around 30%.

The positive polarization diagrams for anthracene single crystals in the majority of cases had the form
shown in Fig. 23. Here again, the polarization is less than the theoretical value, and furthermore the behavior
of the curves is no longer in qualitative agreement with those deduced from x-ray structural data. In some
single crystals, the absolute value of the polarization for $\chi = 0$ reaches 60%, while in other cases it is rather less,

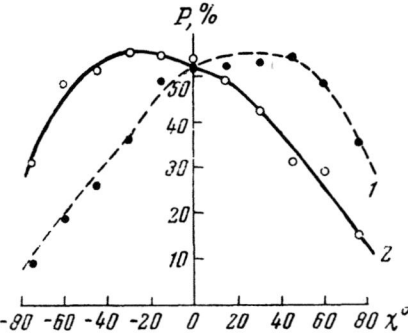

Fig. 23. Experimental "positive" polarization diagrams of the first type for anthracene single crystals. Curves 1 and 2 relate to two azimuths of crystal, differing by 180°.

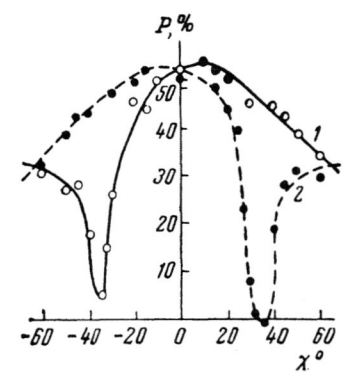

Fig. 24. "Positive" polarization diagrams of the second type for anthracene single crystals. 1, 2 as in Fig. 23.

but the form of the polarization diagrams is maintained. Evidently depolarization in these cases is also caused by imperfections in the single crystals grown. The polarization of fluorescence may thus itself serve as an indicator of the homogeneity and regularity of a sample.

Besides the positive polarization diagrams of the type described, in a few specific cases we obtained positive diagrams of another type for anthracene single crystals (Fig. 24). The negative diagrams for these single crystals, however, had the same form as before.

## §3. Effect of Double Refraction of the Luminescence, Calculation of Corrections

Before comparing the experimental data with theory, we must make allowance for certain important corrections.

In the calculated polarization diagrams we have in no way taken account of the optical anisotropy of the crystals and its effect on the light propagated through them. The calculation was made on the basis of a scheme of oscillators oriented in a certain manner, but situated in an isotropic medium.

The real fluorescing molecular crystal, however, is not only an anisotropic radiator but also an anisotropic, doubly refracting medium for light passing through it, in particular for its own fluorescence.

Anthracene crystals belong to the monoclinic system and are biaxial. In monoclinic crystals the crystallographic b-axis is perpendicular to the other two (a and c), the angle between the latter differing from a right angle. One of the principal axes of the indicatrix coincides with the b-axis, and the other two lie in the ac-plane but do not coincide with either the a- or c-axes, their position being peculiar to the particular substance. The optical axes are perpendicular to the circular sections of the indicatrix and lie in the plane of its major and minor principal axes.

The light of the fluorescence radiated by molecular oscillators of definite orientation and linearly polarized in a given azimuth will be propagated through the crystal in all directions, experiencing double refraction. In every direction this light will be propagated in just the same way as light similarly polarized and directed falling on the crystal from outside at the point at which the oscillator is radiating. For every direction of the wave normal of such incident light, there will be two directions in the crystal along which light will be propagated with different polarizations. In the biaxial crystal of present interest, the polarization of the light (direction in which the induction vector D vibrates) in these directions is determined by the well-known theorem of Fresnel [139].

The directions of the vibrations lie in planes forming the bisectors of the angles between the planes passing through the direction of the wave normal and the optical axes. A uniaxial crystal is a particular case in which the two optical axes collapse into one; correspondingly, one direction of the vibrations lies in the principal plane (extraordinary ray) and the second perpendicular to this (ordinary ray).

The ratio of the intensities of polarized light in these two directions is determined by the orientation of the electric vector of the incident light relative to these particular directions in the crystal in which the vibrations take place. The intensities are equal if the electric vector of the incident light makes an angle of 45° with these directions, and also for natural incident light.

If, however, the electric vector of the incident linearly polarized light coincides with one of the directions of vibration in the crystal (the so-called uniradial azimuths) [140], then only one ray will be propagated in the crystal, and the intensity of the second will vanish.

Thus, either incident or emitted light propagating through the crystal is split into two rays, which fall on the exit face at different angles. The linear separation of the rays will be negligible on account of the small thickness of the crystal (for a thickness of 10 $\mu$ and an observation angle $\chi = 60°$, the linear deviation between the rays in anthracene is 1.3 $\mu$). These rays will possess different refractive indices. After refraction, they emerge into the air parallel, but the ratio between their intensities will differ from that in the crystal, since they experience different losses in reflection. Hence the degree of polarization of the light emerging from the crystal in a given direction will differ from that of the light emitted in the corresponding direction. In order to determine this true polarization, we must introduce corrections for the reflection of the rays with different polarizations.

Let us use Fresnel's formula for the reflection and refraction of light at the boundary between two media. These were derived for isotropic media, but we shall use them for anisotropic media, individually for the rays with different polarizations, taking account of the difference in the refractive indices for the different directions and for differently polarized light. Fresnel's formulas connecting the amplitude of the incident light $A^E$ with that of the refracted light $A^D$ for cases where the electric vector lies in the plane of incidence ($A_p$) and perpendicular to this ($A_s$) have the following form:

$$A_p^D = \sigma_p A_p^E, \qquad A_s^D = \sigma_s A_s^E,$$

where

$$\sigma_p = \frac{2 \cos \delta \cdot \sin \chi}{\sin (\delta + \chi) \cdot \cos (\delta - \chi)},$$

$$\sigma_s = \frac{2 \cos \delta \cdot \sin \chi}{\sin (\delta + \chi)}.$$

(III.3)

Here $\delta$ is the angle of incidence in the crystal, and $\chi$ is the angle of refraction in air.

If, in the anthracene crystal, the light is propagated in the ac-plane, then we have the case of the positive polarization diagram. The optic axes lie in the ac-plane, since the medium principle axis of the indicatrix coincides with the crystallographic b-axis. Both planes passing through the direction of the wave normal and the optic axes merge into one plane (the ac-plane). Hence one direction of the electric vector vibrations coincides with the b-axis, and the other is perpendicular to this and lies in the ac-plane. The first vector ($A_s$) is perpendicular to the plane of incidence, and the second ($A_p$) parallel ($A_s \parallel b$; $A_p \perp b$). In [138] there appear experimental data on the refractive indices for various directions of light in plane ac and for various directions of the electric vector $\parallel b$ and $\perp b$. The values of the refractive index calculated from these data are shown in Fig. 25.

The principal axes of the indicatrix (except the medium axis) and also the optical axes in the anthracene crystal are subject to dispersion. The data presented relate to wavelength 436 m$\mu$, which falls in the region of the luminescence spectrum of crystalline anthracene near the maximum [141].

Knowing the refractive index values, we may find the angles of incidence $\delta$ in the crystal (different for different directions of vibration of the electric vector) corresponding to the angle of refraction $\chi$ in air for the case of the positive polarization diagram, and hence also the values of coefficients $\sigma_s$ and $\sigma_p$. This enables us to pass from the amplitudes of the refracted light $A^D$, the ratio of which $-\rho = A_p^D / A_s^D$ —may be found from the experimentally measured degree of polarization, to the amplitudes $A^E$ on the incident light in the crystal. Then

50

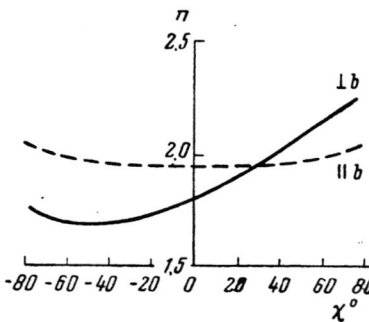

Fig. 25. Refractive index as a function of direction in the anthracene crystal.

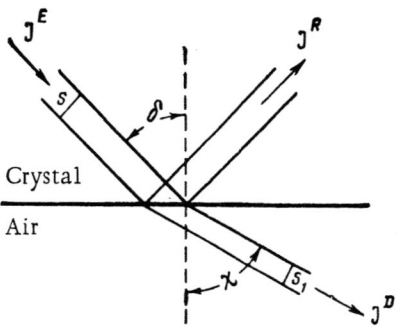

Fig. 26. Scheme for calculating corrections for refraction at the crystal-air interface.

we must pass on from the amplitudes to the intensities of the respective components, the flux of which is measured experimentally.

From the law of conservation of energy, the flux of incident light energy in the absence of absorption equals the sum of the fluxes of reflected and refracted light:

$$I^E = I^D + I^R.$$

We must remember, first, that the intensity is proportional to the square of the amplitude, secondly, that density of light energy for the propagation of light in the medium is proportional to the square of the refractive index ($u \approx A^2 n^2$) [142,143], and, thirdly, we must remember the difference in the cross sections of the incident and refracted beams and the change in the velocity of light on passing from one medium to another.

For the incident flux (in the crystal) we may write:

$$I^E = v s u^E,$$

where v is the velocity of light in the crystal, s is the cross section of the incident beam (Fig. 26), and the energy density is $u^E \approx (A^E)^2 n^2$.

For the refracted flux (in air) correspondingly

$$I^D = c s_1 u^D,$$

where c is the velocity of light in air, $s_1$ is the cross section of the refracted beam, and the energy density is $u^D \approx (A^D)^2$.

The ratio of these fluxes is

$$\frac{I^E}{I^D} = \frac{v s (A^E)^2 n^2}{c s_1 (A^D)^2} .$$

Remembering that c/v = n, $A^D/A^E = \sigma$ and $s/s_1 = \cos \delta / \cos \chi$, we finally obtain the relation between the incident and refracted light intensities $I^E$ and $I^D$ in the form:

$$I^E = I^D \frac{n}{\sigma^2} \frac{\cos \delta}{\cos \chi}. \tag{III.4}$$

For light polarized $\parallel b(I_s)$, and for light polarized $\perp b(I_p)$, the quantities $\sigma, n,$ and $\delta$ have different values: for $I_s - \sigma_s$, $n_{\parallel b}$ and $\delta_{\parallel b}$, and for $I_p - \sigma_p$, $n_{\perp b}$ and $\delta_{\perp b}$.

With the aid of formula (III.4) we calculated corrections for the experimentally measured positive polarization diagrams.

The results of the calculation for the diagram of the first type shown in Fig. 23 are given in Fig. 27. It is seen that the corrections are comparatively small for small angles $\chi$ (roughly up to 30°), but become extremely large for considerable changes in angle $\chi$.

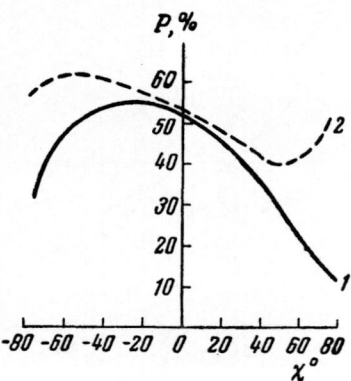

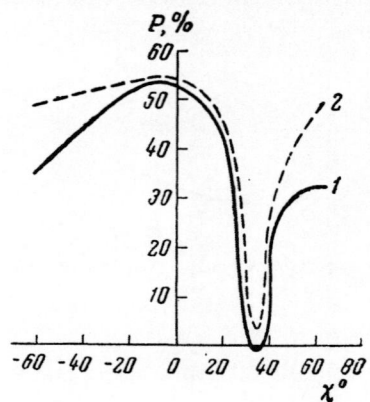

Fig. 27. "Positive" polarization diagrams of the first type. 1) Experimental; 2) with corrections.

Fig. 28. "Positive" polarization diagrams of the second type. 1) Experimental; 2) with corrections.

If we compare the positive polarization diagrams, namely, the experimental diagram with corrections (Fig. 27) and that constructed from x-ray structural data (Fig. 19), we may say that the general shapes of the two curves agree. Quantitatively, the experimental values are lower than those given by the x-ray structural data. Thus for χ = 0 the experimental degree of polarization, with allowance for corrections, will be 53%, and the calculated value 77.5%.

Thus the main experimental result may be formulated as follows: the polarization diagrams of anthracene calculated from the oriented gas model and measured experimentally agree in form, but quantitatively the experimentally measured degree of polarization is always less than the calculated.

We shall return later to the study of this result.

The corrections for the "negative" diagrams were not calculated, since the corresponding refractive indices were not known. However, from the fact that, qualitatively, the form of the negative diagrams agrees with the calculated diagrams without introducing corrections, we may conclude that the corresponding section of the indicatrix in this case is closer to a circular section than in the case of the positive diagrams.

As far as the "positive" polarization diagrams of the second type are concerned (Fig. 28, cf. Fig. 24), the introduction of corrections in this case does not lead to agreement with the calculated diagrams (Fig. 19). Evidently these cases correspond to another structure of the lattice and orientation of the molecules.

One possible suggestion regarding this is that we are here dealing with dianthracene or with a considerable proportion of this present as an impurity in the anthracene. The dianthracene lattice has a structure differing from the anthracene lattice [144].

Davydov showed theoretically [9] that in the absorption of light by crystals the operative factor is the total moment of the elementary cell of the crystal lattice rather than the moments of the transitions of individual molecules. More specifically, for crystals containing two molecules in the unit cell, from one molecular frequency we obtain two absorption frequencies of the crystal, one of which will correspond to an electric moment parallel to the sum of the electric moments of the two molecules, and the other to one parallel to their difference.

It is known in advance whether, in luminescence, the moment of the cell or that of individual molecules will be operative. At first glance it would appear that an unequivocal answer to this question might come from the agreement or lack of agreement of the polarization diagrams with the oriented gas model. However, this is not quite the case.

The point is this, that the vector sum and difference of the dipole moments of the molecule are in general not obliged to be directed along the crystallographic axes. In other words, the dipole moment of the cell

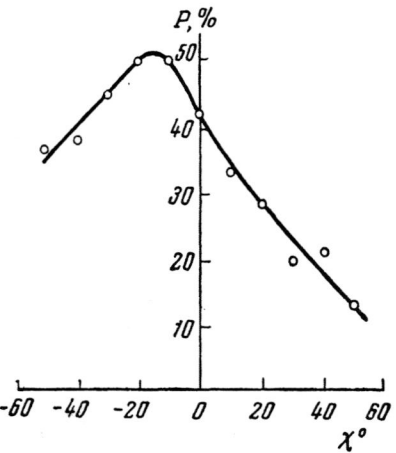

Fig. 29. "Positive" polarization diagram of naphthacene in anthracene.

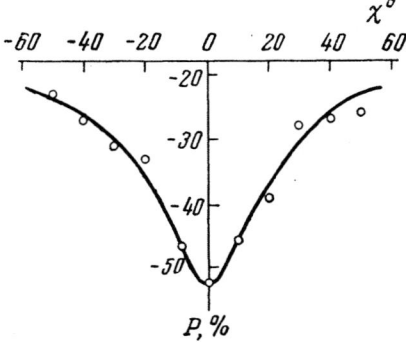

Fig. 30. "Negative" polarization diagram of napthacene in anthracene.

is not essentially directed along one of the crystallographic axes, as is the moment of a transition for which a free exciton is excited.

When we are concerned with the moment of the cell, everything depends on the orientation of the molecules, or more precisely on their characteristic moments. For example, in anthracene the molecular oscillators lie in a plane not coinciding with the ab-plane, being oriented symmetrically with respect to the b-axis. It follows from this that the vector sum of the molecular oscillators will be directed along the b-axis, but their difference will not be directed along the a-axis.

Clearly, the space distribution of the polarization of the radiation will not depend on whether the separate molecular oscillators or the moments of the cell (sum and difference of the molecular moments) radiate. In particular, the asymmetry of the positive polarization diagram (measured at various angles in the ac-plane) relative to the b-axis must be the same in both cases. In other words, the result in both cases will agree with the oriented gas model.

There is, nevertheless, one possibility of setting up an experiment so as to be able to know in advance which polarization diagrams give the individual molecular moments. For this experiment we used mixed crystals of anthracene with naphthacene. It has been established by various workers [145-147] that in such crystals, even with very small concentrations of naphthacene ($10^{-5}$ to $10^{-4}$ g/g), on absorption in the base substance (the anthracene), the majority of the luminescence comes from the impurity (the naphthacene). This is a direct proof of the migration of energy from the anthracene to the naphthacene. But at the moment we are interested in this from another point of view. First, for such concentrations, the naphthacene molecules are found at mean distances of 10 to 20 lattice constants from one another. Secondly, they are oriented roughly in the same way as anthracene molecules themselves. This is confirmed by measuring the polarization of luminescence from naphthacene and anthracene in the same samples (more details of this will appear in Chapter V). Hence we know that the polarization diagrams of naphthacene luminescence must correspond to isolated, individual molecular oscillators.

Figures 29 and 30 show the "positive" and "negative" polarization diagrams of the naphthacene luminescence in anthracene crystals (concentration $10^{-4}$ g/g). The results are interesting because the form of the diagrams is exactly the same as for pure anthracene. The corrections for double refraction of the luminescence will be approximately the same, since the main mass of the crystal is as before anthracene.

For napthacene in anthracene, therefore, the polarization diagrams, on the one hand, are known to correspond to separate, isolated molecular oscillators, and on the other hand they have the same form as for pure anthracene. This constitutes an indirect indication that the luminescence of pure anthracene crystals is also caused by individual molecular oscillators.

§4.  Experiments with Spherical Single Crystals of Stilbene and Tolan [148,149]

Thus the study of polarization diagrams for single-crystal laminas of anthracene and anthracene with naphthacene has shown that, if we take account of the corrections for double refraction of the luminescence, the experimental diagrams qualitatively agree with the diagrams calculated on the basis of x-ray structural data. The introduction of corrections greatly complicates matters, involving quite heavy calculations, and requiring a knowledge of the optical anisotropy of the crystal, which itself constitutes a rather tedious experimental problem.

However, for directions close to the crystal surface normal (up to values χ = ± 20°), the corrections are very slight, only becoming serious for large angles between the normal and the viewing direction.

It is not hard to understand why the depolarization of the luminescence associated with the different reflection of the split components at the surface of the crystal is considerably less for normal incidence than for oblique. For the same divergence angle of the components, the difference between the angles at which they fall on the surface is independent of the angle of incidence. However, the loss in reflection at oblique incidence will be correspondingly greater in the two components than for near-normal incidence. Hence the difference between the reflection losses of the components (which determines the depolarization) will also be greater.

This leads to the idea of changing the experimental conditions in such a way that for all directions the light from the luminescence falls on the crystal surface normally or nearly so. These conditions can be achieved by using spherical crystals and exciting the luminescence at the center.

We managed to prepare good samples, quite large and homogeneous and suitable for making spheres, from stilbene and tolan.

In order to facilitate excitation of the luminescence at the center only, the spheres were cut into hemispheres along definite crystallographic planes, and the plane of the cut was masked so as to excite only a very small part of the crystal (diameter 1 to 2 mm) at the center of the sphere. The spheres themselves were quite large, some 5 to 10 mm in radius. Excitation was effected by the natural light of a mercury line with λ = 365 mμ from the plane side, while the polarization of the luminescence was measured from the spherical side. The polarization diagrams were measured on the same apparatus, the main parts of which comprised the goniometer and Cornu polarimeter.

The hemispherical samples required for measurement were prepared from single crystals of stilbene grown in a sealed test tube by the Obreimov-Shubnikov method and single crystals of tolan grown in a thermostatic bath by the Kiropulos method. The large single crystal was sawn up by means of a special saw with a silk thread (see Chapter VII) into sections of the required size, which were then given spherical form with the aid of a rotating tube and emery; the sphere so obtained was polished by rotation in the same tube, or in a sphecial rotating chamber, with a soft cloth moistened in kerosene. After this treatment, perfectly transparent spheres with smooth, polished surfaces were obtained.

The next problem was to orient the spheres, i.e., to determine the direction of the crystallographic axes. Stilbene and tolan are biaxial crystals. The spherical form of the sample makes it possible, from the conoscopical picture, to determine and fix the emergent points of the optic axes on the surface of the sphere, using the method proposed by Shubnikov [150].

Knowledge of the directions of the optic axes in the crystals is fundamental for the orientation of the samples. The oriented spheres were cut by saw or ground along a definite plane to hemispheres. The cut was polished with a cloth moistened in kerosene. Hemispheres cut along the following planes were perepared: 1) along the plane of the optic axes, 2) perpendicular to the bisector of the acute angle between the optic axes, 3) perpendicular to the bisector of the obtuse angle between the optic axes. The lattices of stilbene and tolan belong to the monoclinic system. As shown by x-ray structural data, they are very similar to one another, and the orientation of the molecules in them is almost the same ([6], pp. 401-406, [14], [151]).

Below we give these data on the lattice constants and the angles between the crystallographic and molecular axes.

For Stilbene:

Elementary Cell

$$a = 12.35 \text{ A}, \ b = 5.70 \text{ A},$$
$$c = 15.92 \text{ A}, \ \beta = 114°.0'.$$

Orientation of Molecule

| | | L | | M | | N | |
|---|---|---|---|---|---|---|---|
| | | I | II | I | II | I | II |
| (φ) | $a$ | 33°,4 | 31°,4 | 114°,8 | 107°,9 | 110°,9 | 114°,9 |
| (ψ) | $b$ | 79,9 | 88,4 | 35,0 | 33,6 | 123,2 | 123,5 |
| (ω) | $c'$ | 58,5 | 58,6 | 66,8 | 62,6 | 40,8 | 44,0 |

For Tolan:

Elementary Cell

$$a = 12.75 \text{ A}, b = 5.73 \text{ A},$$
$$c = 15.67 \text{ A}, \beta = 115°,2.$$

Orientation of Molecule

|  | | L | | M | | N | |
|---|---|---|---|---|---|---|---|
|  | | I | II | I | II | I | II |
| ($\varphi$) | $a$ | 35°,9 | 34°,5 | 117°,5 | 110°,2 | 111°,2 | 114°,7 |
| ($\psi$) | $b$ | 75,3 | 84,5 | 33,8 | 29,8 | 119,7 | 119,9 |
| ($\omega$) | $c'$ | 58,0 | 56,1 | 72 | 69 | 37,7 | 41,6 |

Figures 31 and 32 give the disposition of the molecules in the cells of the stilbene and tolan lattices.

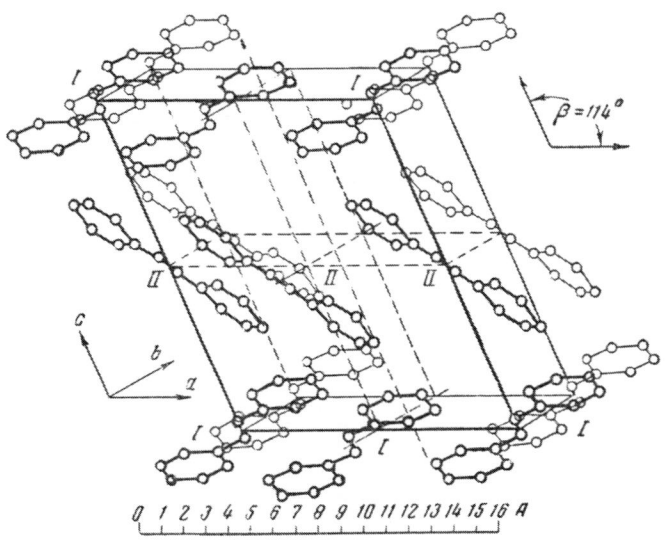

Fig. 31. Disposition of the molecules in the stilbene cell.

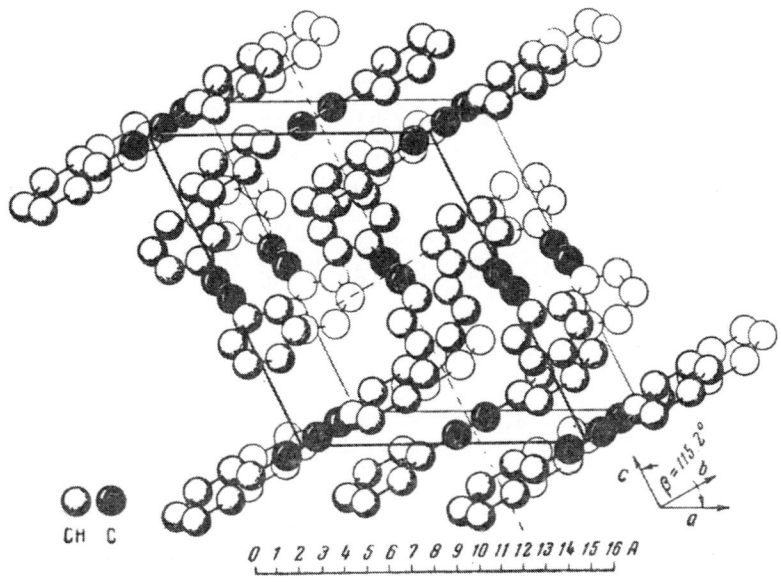

Fig. 32. Disposition of the molecules in the tolan cell.

In both crystals, the elementary cell contains four molecules, but there are only two different forms of orientation. The structure consists of two symmetrically independent layers. Two molecules of one layer transform one into the other as a result of transposition and rotation through 180° around an axis lying in a plane parallel to that of the benzene rings and intersecting the double bond in the middle. Hence the orientation of the axes of these two molecules is identical.

In view of the closeness—almost identity,—of the stilbene and tolan lattices, all questions connected with the orientations of the crystals themselves and the molecules in their lattices, and so forth, are solved for these materials in analogous ways.

Considerable help in the orientation of the samples may be obtained by determining the cleavage planes from the appearance of a series of cracks formed on cooling the sample by flooding it with a rapidly evaporating substance (for example, dichloroethane). In the substances studied, the series of cracks was parallel to the plane of the optic axes. It follows from x-ray structural data that the greatest lattice constant corresponds to the c-axis; this means that we may assume this to have the weakest bond. Then the cleavage plane will be the ab-plane (this assumption is confirmed by experiment).

Hence the a- and b-axes lie in the plane of the optic axes corresponding to the mean wavelength of the luminescence spectrum. In crystals of the monoclinic system, the b-axis must coincide with one of the principal axes of the indicatrix, i.e., in our case, either with the obtuse or acute bisector of the optic axes. In future we shall call the bisectors of the obtuse and acute angles between the optic axes the "obtuse and acute bisectors of the optic axes," for the sake of brevity, as usually accepted in crystallographic literature. We can decide which bisector the b-axis coincides with by studying the azimuthal variation of the degree of polarization. By "azimuthal variation" we here mean, as before, the variation of the degree of polarization with the orientation of the crystal relative to the principal axis of the polarimeter, i.e., the variation of the degree of polarization with the angle of rotation of the crystal around a horizontal axis coinciding with the axis of the apparatus, from which in future the angles in the polarization diagrams are reckoned.

This azimuthal variation may be calculated for any cut of the crystal from the oriented gas model, since the orientation of the molecules in the lattice is known from x-ray structural data. Unfortunately, the orientation of the radiation oscillator in the stilbene and tolan molecules is not known (for certain reasons mentioned in Chapter II, we were unable to measure the polarization spectra of solutions of these substances). In view of this we have to make a definite assumption regarding the orientation of the oscillator in the molecule, and then check whether this agrees with experimental data.

Let us assume that the radiation oscillator is directed along the longitudinal axis of the stilbene molecule. After projecting the axes of the two molecules with different orientations on the ab-plane, by means of simple trigonometry it is easy to show that the projections of the oscillators on this plane are situated close to the a-axis, making angles of 11.5 and 2° with it. From this, without any quantitative calculation, it is clear that the positive polarization maximum approximately corresponds to the position of the crystal for which the a-axis is vertical, and the negative maximum to that position for which the b-axis is vertical. Experiment shows that the negative maximum holds when the obtuse bisector is vertical, and the positive when the acute bisector is vertical. Hence the b-axis coincides with the obtuse bisector, and the a-axis with the acute.

Thus we may determine the orientation of the crystallographic axes in all the hemispherical samples under investigation as follows: 1) for the cut along ab, the c'-axis coincides with the axis of the apparatus; 2) for the cut along bc', a coincides with the axis of the apparatus; 3) for the cut along ac', b coincides with the axis of the apparatus.

It is experimentally convenient always to measure the polarization diagrams in one and the same plane (the horizontal) for various orientations of the sample being measured. In our investigations, we always measured the diagrams for the two orientations corresponding to the maximum positive degree of polarization ("positive" diagrams) and the maximum negative degree of polarization ("negative" diagrams). We remember that the degree of polarization is determined by the relation $P = (I_1 - I_2)/(I_1 + I_2)$, where $I_1$ and $I_2$ are the vertical and horizontal components of the intensity of the light vibrations, coinciding with the principal directions of the polarimeter.

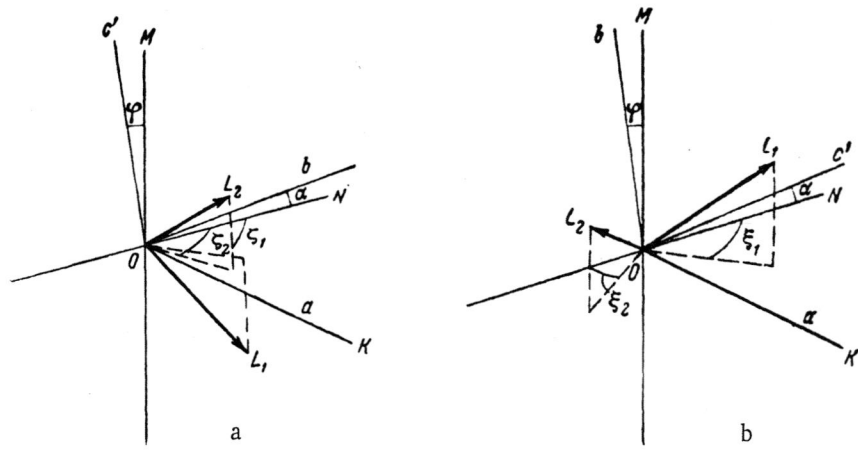

Fig. 33. Relating to the calculation of polarization diagrams.

The positions of the crystal corresponding to the positive and negative maxima were determined by preliminary measurements of the azimuthal variation of polarization for each sample.

Our problem was to compare the experimental diagrams with the diagrams calculated on the assumption that the polarization of the luminescence was caused by the orientation of the radiation oscillators of individual molecules in the lattice. For calculation we used x-ray structural data regarding the orientation of the molecules, i.e., data on the angles between the axes of the molecules and the crystallographic axes. If we know the direction of the crystallographic axes of the given sample and the orientation of the molecular axes with respect to these, then calculation of the polarization diagrams and the azimuthal variation is not too complicated, though it still constitutes quite a heavy geometrical problem. It is now only necessary to know the orientation of the radiation oscillators in the molecule. We shall not show the calculations for all cuts and orientations of the crystals. These are all analogous to one another.

By way of example, let us carry out a calculation of the tolan polarization diagrams for a cut through the bc'-plane (Fig. 33).

As already discussed, we assume that the radiation oscillator is directed along the longitudonal axis L of the molecule. There are two forms of molecular orientation in the lattice. The angles between the oscillators $L_1$ and $L_2$ of the molecules of the two forms and the crystallographic axes a, b, and c' are $\varphi_1$, $\varphi_2$, $\psi_1$, $\psi_2$, $\omega_1$, and $\omega_2$; from x-ray data we take: $\varphi_1 = 35°54'$, $\varphi = 34°30'$, $\psi_1 = 75°18'$, $\psi_2 = 84°30'$, $\omega_1 = 58°$, $\omega_2 = 56°6'$.

Projecting the oscillators on the bc'-plane, it is easy to calculate the degree of polarization for any orientation of the crystal on viewing along the a-axis (azimuthal variation). The positive maximum corresponds to the case shown in Fig. 33A (where the b- and c'-axes make angles of $\alpha = 7°30'$ with the horizontal and vertical, respectively). For this case, we must calculate the "positive" polarization diagram, i.e., the variation of the degree of polarization with the angle $\chi$ between the direction of viewing in the horizontal plane and the a-axis. For these calculations we must first of all transform from the angles with the b- and c'-axes to the angles with axes ON and OM (we denote these angles by primes: $\psi'_1$, $\psi'_2$; $\omega'_1$, $\omega'_2$) by means of the formulas for the rotation of coordinate axes. We thus obtain the following values: $\psi'_1 = 79°20'$; $\psi'_2 = 88°40'$; $\omega'_1 = 56°$; $\omega'_2 = 55°30'$. The angles $\varphi_{1,2}$ do not change. The degree of polarization along any direction in the horizontal plane may easily be obtained in the following form:

$$P(\pm \chi) = \frac{\cos^2 \omega'_1 + \cos^2 \omega'_2 - [\sin^2 \omega'_1 \cdot \cos^2 (\zeta_1 \pm \chi) + \sin^2 \omega'_2 \cdot \cos^2 (\zeta_2 \pm \chi)]}{\cos^2 \omega'_1 + \cos^2 \omega'_2 + [\sin^2 \omega'_1 \cdot \cos^2 (\zeta_1 \pm \chi) + \sin^2 \omega'_2 \cdot \cos^2 (\zeta_2 \pm \chi)]}, \qquad (III.5)$$

where $\zeta_1$ and $\zeta_2$ are the angles between the projections of the oscillators on the horizontal plane and the ON-axis; these are determined by the following expressions:

$$\cos \zeta_1 = \frac{\cos \psi'_1}{\sin \omega_1}; \qquad \cos \zeta_2 = \frac{\cos \psi'_2}{\sin \omega_2}.$$

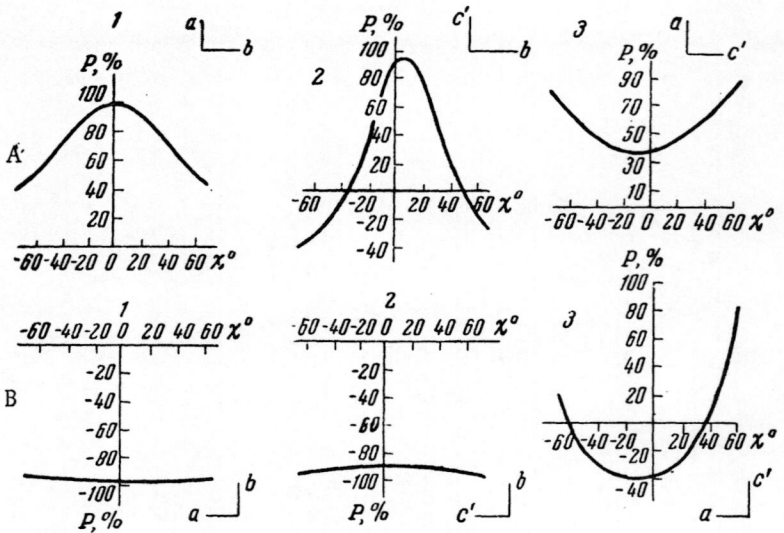

Fig. 34. Calculated (A) "positive" and (B) "negative" polarization diagrams of a stilbene crystal. Cuts through planes: 1) ab; 2) bc'; 3) ac'.

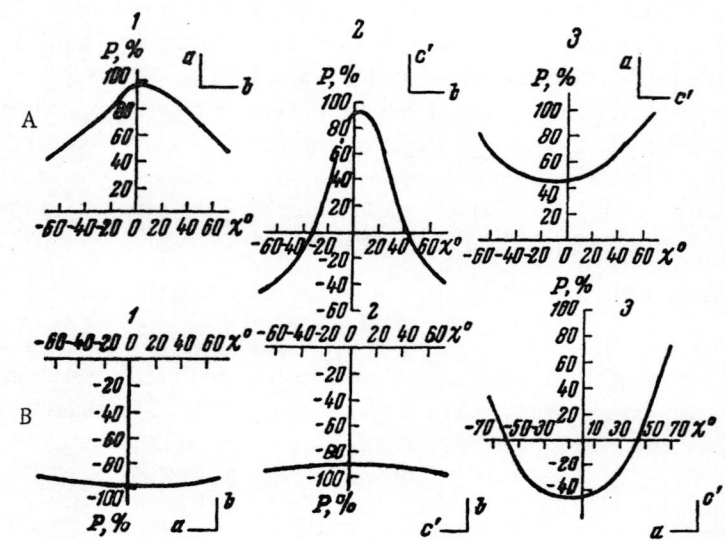

Fig. 35. Calculated (A) "positive" and (B) "negative" polarization diagrams for a tolan crystal. Notation of planes as in Fig. 34.

Analogous calculation of the "negative" polarization diagram for an orientation of the crystal obtained by rotating the crystal 90° around the a-axis (Fig. 33 B) leads to the following result:

$$P\,(\pm\chi) = \frac{\cos^2\psi_1' + \cos^2\psi_2' - [\sin^2\psi_1'\cdot\cos^2(\xi_1\pm\chi) + \sin^2\psi_2'\cdot\cos^2(\xi_2\mp\chi)]}{\cos^2\psi_1' + \cos^2\psi_2' + [\sin^2\psi_1'\cdot\cos^2(\xi_1\pm\chi) + \sin^2\psi_2'\cdot\cos^2(\xi_2\mp\chi)]},\qquad\text{(III.6)}$$

where $\xi_1$ and $\xi_2$ are determined by the expressions

$$\cos\xi_1 = \frac{\cos\omega_1'}{\sin\psi_1'}; \qquad \cos\xi_2 = \frac{\cos\omega_2'}{\sin\psi_2'}.$$

Calculations for stilbene and tolan cuts along other crystallographic planes are analogous to these.

Figures 34 and 35 show the "positive" and "negative" polarization diagrams calculated for stilbene and tolan respectively in the three cuts mentioned: 1) in the plane of the optic axes ab, 2) in a plane perpendicular

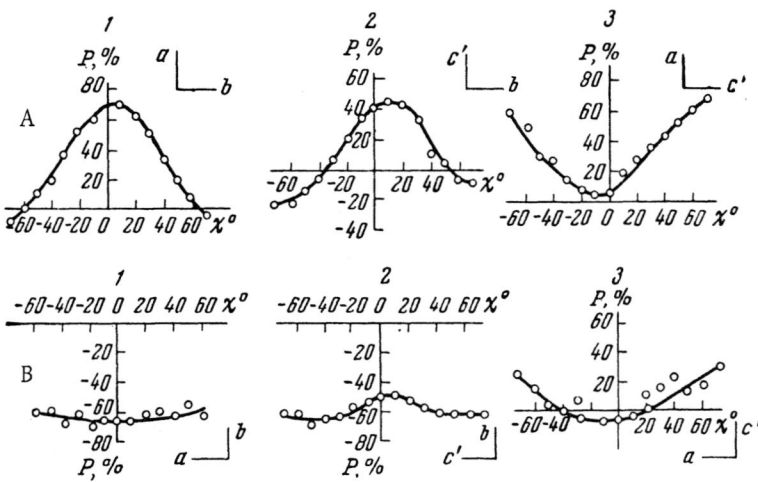

Fig. 36. Experimental (A) "positive" and (B) "negative" polarization diagrams for a stilbene crystal. Notation of planes as in Fig. 34.

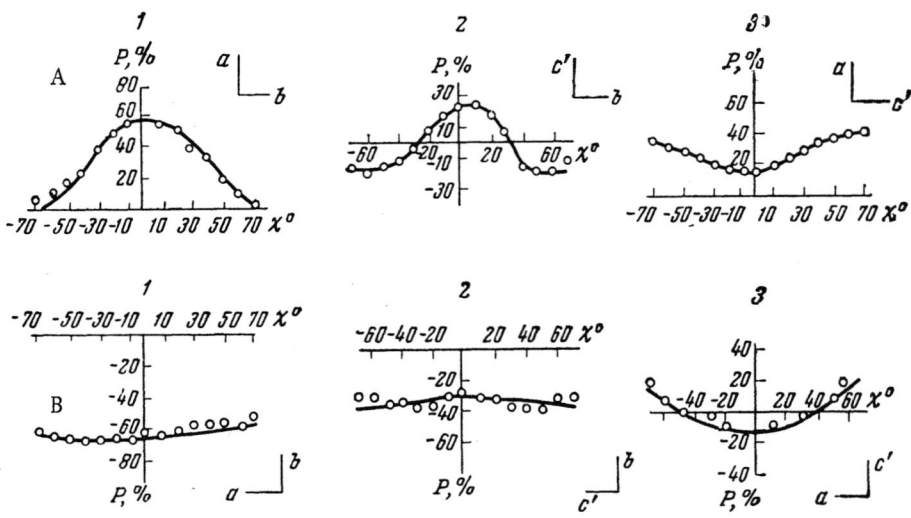

Fig. 37. Experimental (A) "positive" and (B) "negative" polarization diagrams for a tolan crystal. Notation of planes as in Fig. 34.

to the acute bisector bc', and 3) in a plane perpendicular to the obtuse bisector ac'. The corresponding stilbene and tolan diagrams are very much alike, which is natural in view of the fact that their structures are extremely similar.

Figures 36 and 37 show the experimental polarization diagrams for stilbene and tolan with the same orientations of the hemispheres. We see that for all cases measured there is good qualitative agreement between the calculated and experimental diagrams, without the introduction of any double refraction corrections. Above all, this agreement indicates the validity of the assumption that the oscillator is directed along the longitudinal axis of the molecule.

For comparison, we made analogous calculations for a few cuts and orientations of stilbene on the assumption that the oscillator was directed along the transverse axis of the molecule. The results for cuts along the ab- and bc'-planes are shown in Fig. 38. If we compare these calculated polarization diagrams with experiment, we find no qualitative agreement whatsoever between them (especially for the "negative" diagrams).

There are certain indications in the literature [50] that tolan is extremely difficult to purify from stilbene impurities, and it has been suggested that pure tolan does not luminesce itself, but that the luminescence comes from the stilbene impurity. There is no doubt that separation of these two substances is difficult, in view of the

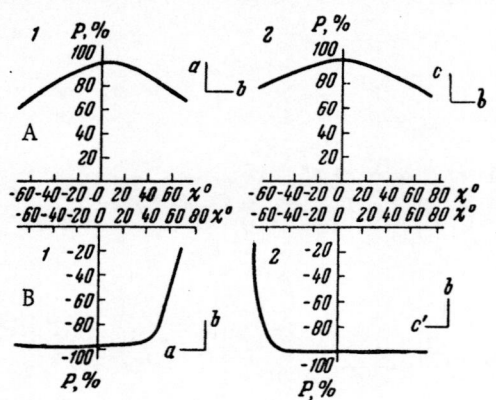

Fig. 38. Calculated diagrams for two orientations of stilbene in the case of an oscillator directed along the transverse axis of the molecule. A) Positive; B) negative; cuts along planes: 1) ab; 2) bc'.

similarity between their structures, molecules, and lattice. If, in fact, it is the stilbene impurity which fluoresces, then the mixed tolan-stilbene crystal is similar to the anthracene-napthacene crystal. From the point of view of the polarization diagrams, this hypothesis cannot be denied. If we examine stilbene molecules set in a tolan lattice and oriented identically with the tolan molecules, then clearly the polarization diagrams must be exactly the same as in pure stilbene. This similarity is actually observed in experiment, just as in the case of pure anthracene and anthracene-naphthacene.

In the case of fluorecence from stilbene existing in the form of a low concentration of impurity in tolan, clearly the radiation comes from individual molecular oscillators. The similarity between the diagrams of pure stilbene and stilbene in tolan may be explained in the same way as that between pure anthracene and napthacene in anthracene. This constitutes an indirect indication that the luminescence of pure stilbene is also caused by individual molecular oscillators.

Thus, comparison between the calculated and experimental polarization diagrams leads to the following conclusion: the qualitative behavior of the two kinds of diagram is in good agreement in all cases. This agreement is the more convincing in view of the fact that the functional behavior of the diagrams is very distinctive for the various cuts and orientations. In laminar crystals, such agreement is only obtained after allowing for double refraction corrections, but in spherical crystals no such corrections are required. Quantitatively, however, the calculated and experimental degrees of polarization disagree in all cases, the measured values being considerably less in absolute magnitude than the calculated.

As in the anthracene laminas, so also in the stilbene spheres there is some variation in polarization from sample to sample. This amounts to 5 to 10% and is evidently due to the imperfection of the samples, including microcracks, insufficiently well polished surfaces, and so forth. The variation is considerably less than the difference between the mean measured and theoretically calculated degrees of polarization. This latter difference is approximately 30 to 40%.

We may thus say with certain reserve that the oriented gas model agrees qualitatively but not quantitatively with the polarization diagrams. Let us now consider what might be the cause of this deviation.

§5. Effect of Temperature on the Polarization of Luminescence from Molecular Crystals

There must certainly be an effect of temperature on the polarization of luminescence from molecules in any state of aggregation on account of the depolarizing action of the rotational vibrations of the molecules around equlibrium positions. Translational vibrations have no effect on the polarization. We have already seen the marked influence of the temperature of the medium on the polarization of luminescence in Chapter II, where we were discussing the luminescence of solutions.

In crystals, also, there must be a depolarizing effect of the rotational oscillations of the molecules around equilibrium positions. There may, however, be an an essential difference between solutions and crystals in this respect. First of all, the effect of temperature on the polarization of luminescence in crystals will depend on whether the luminescence is associated with the moment of the cell as a whole or with separate independent molecular moments. As indicated earlier, there are some grounds (though even these cannot constitute strict proof) for inclining towards the second possibility. We shall therefore first consider what effect temperature will have in the case in which the radiation is associated with the moments of individual molecules rather than the moment of the cells.

In this case, molecular crystals will behave very differently from solutions. The difference lies in that, in solutions, the directions of the predominant oscillations of the radiated light and the radiating oscillators coincide, whereas in general this is not so for crystals. Only in the case in which all the molecules in the lattice are oriented similarly will these directions coincide. But in general, molecular crystals have several (often two) different kinds of orientation of the molecules. In such a case the predominant direction of the oscillations will coincide with neither of the directions of the radiating oscillators. For example, if the two differently oriented oscillators lie in a plane normal to the direction of viewing, then the direction of the predominant oscillations will coincide with the bisector of the acute angle between the oscillators. This is true independently of the state of polarization of the exciting light, since the polarization of the fluorescence of the crystal does not depend on this. We shall be discussing this in more detail in the next chapter.

Let us now consider the difference between the depolarizing action of the rotational vibrations of molecules in solutions and crystals, respectively. When the direction of the oscillator coincides with the preferred direction of the oscillations (Fig. 32 A), any deviation of the oscillator from the equilibrium position will lead to depolarization. In the second case (Fig. 32 B), a deviation of the oscillator from the equilibrium position bringing it closer to the direction of the preferred oscillations will lead to an increased degree of polarization, while a deviation in the opposite sense will reduce this. Clearly, depolarization will be less in the second case than in the former.

Let us carry out the corresponding calculation. We shall confine ourselves to vibrations in one plane (Fig. 32 B); Oy is the direction of the preferred oscillations of the radiated light, OL is the oscillator in the equilibrium position, while OL' and OL" are its extreme positions during vibration.

The components of radiated light intensity $I_1$ and $I_2$ are given by the following expressions:

$$I_1 = \int_0^{2\gamma_0} \cos^2(\omega - \gamma_0 + \varphi)\, d\varphi,$$

$$I_2 = \int_0^{2\gamma_0} \sin^2(\omega - \gamma_0 + \varphi)\, d\varphi.$$

Carrying out the integration and substituting the result in the expression for the degree of polarization, we obtain

$$P = \frac{P_0 \sin 2\gamma_0}{2\gamma_0}, \tag{III.7}$$

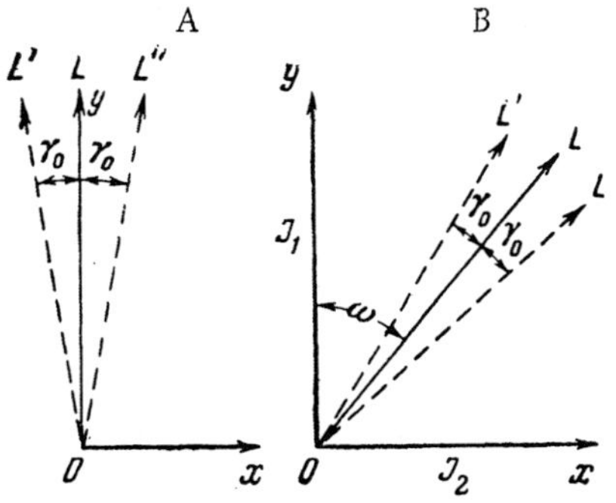

where $P_0 = \cos 2\omega$ is the degree of polarization in the absence of vibrations. The term $(\sin 2\gamma_0 / 2\gamma_0)$ is close to unity for small angles.

The angle $\gamma_0$ is the angular amplitude of the molecular vibrations in the crystal, and may be calculated in the following way.

The kinetic energy of the vibrating molecules equals

$$\frac{J\gamma^2}{2} = \frac{\hbar\omega}{2}\left(\frac{1}{e^{\frac{\hbar\omega}{kT}} - 1} + \frac{1}{2}\right),$$

where J is the moment of inertia of the molecule with respect to an axis perpendicular to the radiation oscillator, $\omega$ is the vibration frequency, and $\gamma = \gamma_0 \cos \omega t$ is the deviation angle. Hence

$$\gamma_0^2 = \frac{2\hbar}{J\omega}\left(\frac{1}{e^{\frac{\hbar\omega}{kT}} - 1} + \frac{1}{2}\right).$$

Fig. 39. Relating to the calculation of depolarization due to thermal vibrations in (A) solutions and (B) crystals. L = direction of radiating oscillator in its equilibrium position; y = direction of preferred oscillations; $\gamma_0$ = amplitude of vibrations.

61

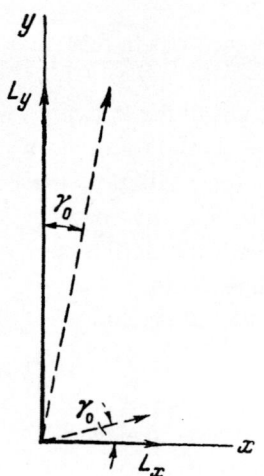

Fig. 40. Relating to the calculation of depolarization for thermal vibrations of the cell moments.

Vibration analysis of the absorption and luminescence spectra of certain molecular crystals at low temperatures, carried out by Obreimov and Prikhot'ko [152] and Pesteil [153], showed the existence of frequencies of the order of $\tilde{v} = 40$ to $50\ cm^{-1}$, characteristic for the crystalline state and ascribed to the rotational vibrations of molecules. For stilbene and tolan these data have not been published, but it is reasonable to suppose that the frequencies for these crystals will have the same order of magnitude. The moments of inertia of the stilbene and tolan molecules with respect to the transverse axis of the molecule (i.e., an axis perpendicular to the radiation oscillator) may be calculated if one knows the geometrical values of the bond lengths and the angles in the molecules, and the masses of the carbon and hydrogen atoms. Calculation gives the following values:

$$J = 2.95 \cdot 10^{-37}\ g \cdot cm^2 \text{ for stilbene;}$$
$$J = 3.3 \cdot 10^{-37}\ g \cdot cm^2 \text{ for tolan.}$$

For $kT \gg \hbar\omega$ we obtain approximately

$$\gamma_0^2 = \frac{\hbar}{J\omega} + \frac{2kT}{J\omega^2}. \tag{III.8}$$

Substituting here the value of $\omega = 2\pi\tilde{v}c$, J, and T = 293°C, we obtain a value of $\gamma_0$ of the order of 4°. Putting this value in (III.7), it is easy to see that the depolarization is negligible to an accuracy of the third decimal place.

We must now consider what will happen in the case in which the luminescence comes from the moments of the cell as a whole. One of these cell moments equals the vector sum of the molecular moments, and the other the vector difference. If the elementary cell had only one natural moment, then this case would correspond to Fig. 39A, i.e., for one cell moment the situation would be exactly the same as for solutions, vibrations only resulting in depolarization. On the other hand, the vibrations of the second, perpendicular moment will lead to a reduction of this depolarization. The angles of rotation of the cell moments (the resultant of the thermal rotational vibrations of molecules in the lattice) will have the same order of magnitude as the angles of rotation of the molecular oscillators.

We may calculate the magnitude of the thermal depolarization from the scheme of cell (Fig. 40). Let the crystal be so oriented that the vector sum of the molecular moments is vertical and their vector difference horizontal. The corresponding intensities will be in the ratio 3:1 (these agree with the observed degree of polarization). Thermal vibrations with angular amplitude 4°, as may easily be shown, change this ratio very little. The corresponding reduction in the degree of polarization is in all approximately 0.5% of its initial value.

The thermal depolarization in anthracene crystals can be calculated in an analogous fashion. From geometrical data on the bond lengths and angles in the anthracene molecule, and from the masses of the carbon and hydrogen atoms, the moment of inertia with respect to an axis perpendicular to the radiation oscillator can be calculated. In anthracene, in contrast to stilbene, this will be the moment of inertia with respect to the longitudinal axis of the molecule. We remember that according to data obtained from polarization spectra of solutions and other investigations the radiation oscillator is directed along the transverse axis in the anthracene molecule, but the longitudinal axis in stilbene.

It is natural that, in view of this, the moment of inertia in the anthracene molecule of interest to us will be considerably less than that in stilbene. Calculation, in fact, gives $J = 2.1 \cdot 10^{-38}\ g \cdot cm^2$.

Correspondingly, the amplitude of the angle of rotation determined by formula (III.8) will be greater for anthracene. Calculation gives $\gamma_0 = 10°$.

In analogy with the stilbene case, we may calculate the kind of depolarization which these vibrations give.

In the first case (if individual molecules radiate), it follows from formula (III.7) that such vibrations may reduce the degree of polarization in absolute magnitude by about 2.5%, i.e., the depolarization is not more than 5% of the original value of $P_0$ in the absence of vibrations. In the second case (if the radiation is associated with the moment of the cell as a whole), calculation according to the scheme of Fig. 40 shows that the fall in the degree of polarization as a result of thermal vibrations is approximately 7% of its original value. Although this is not so negligibly small as in stilbene, it is still considerably less than the observed deviation between theory and experiment.

All that has been said indicates that the temperature depolarization of the luminescence from molecular crystals is very slight, and, in any case, cannot explain the observed difference between the calculated and experimental polarization diagrams.

We notice that in the theoretical treatment of McRae [154] regarding the effect of intramolecular vibrations on the polarization of crystal luminescence, it is shown that these vibrations do not lead to any temperature-dependence of the polarization, whereas they may produce some difference in the polarization of different spectral bands. These considerations are supported by the fact, established experimentally in a number of papers, that the polarization of fluorescence in molecular crystals is independent of temperature or only very nearly so [70, 155].

We ourselves measured the degree of polarization of the luminescence from a number of crystals (anthracene and its derivatives, some phthalimides, stilbene, etc.) at room temperature and the temperature of liquid nitrogen. Within the limits of experimental error, the polarization was unaltered. We shall be speaking of experiments on the polarization of crystal luminescence at low temperatures in more detail in Chapters IV and V.

## §6. Polarization Diagrams and the Oriented Gas Model

Since depolarization as a result of thermal vibrations of the molecules cannot explain the observed quantitative discrepancy between the calculated and experimental polarization diagrams, we must seek other means of explaining this deviation.

One of the causes of the deviation in question may be the incomplete anisotropy of the radiation oscillator in the molecule. For example, in anthracene, the limiting polarization in solutions remain considerably lower than 50% even when all the depolarizing factors are carefully eliminated. Yet the diagrams are calculated for linear oscillators. In this case deviation may take place even if the oriented gas model is completely justified, and hence our hypothesis may not only not contradict the model but, on the contrary, may speak in its favor. However, the cause of the incomplete anisotropy in the molecular oscillator is still not explained.

It is very important to examine the dependence of the degree of polarization on the wavelength of the luminescence, in order to establish whether it varies in certain parts of the spectrum, and whether such variations lead to an observable effect in the degree of polarization of the total (not spectrally resolved) luminescense. This idea arises from the possibility of luminescence arising from states corresponding to free excitons; such radiation must be polarized along the crystal axes, i.e., it may lead to a departure of the degree of polarization of the total luminescence from the oriented gas model. Chapter V will be devoted to a detailed exposition of this. At the moment we shall only say that, although such radiation was in fact observed in certain regions of the spectrum, and although its inclusion should actually reduce the degree of polarization of the total luminescence, the intensity of radiation from free excitons appears to be very weak, and completely inadequate to explain the observed deviation of the experimental polarization diagrams for integrated radiation from those calculated from the oriented gas model. Thus we can partly, but by no means fully, explain the observed discrepancy in this way.

We are thus forced to the conclusion that the discrepancy is due, not to some subsidiary (however important) depolarizing factor, but to a deep-seated cause lying in the very nature of the luminescence of molecular crystals. In this discrepancy we find a limit to the applicability of the oriented gas model to molecular crystals. Strictly speaking, we cannot consider the excited molecules in the crystal as isolated, without interaction among themselves and with the remaining unexcited molecules. Yet it is just this concept which constitutes the basis of the oriented gas model. If they do not correspond to free excitons, we must consider the excited states in the crystal as localized excitons or deforming excitations (see Chapter I). Neither the former nor, in particular, the latter correspond to isolated excited molecules, but rather to excited local regions of the crystal. For the case of deforming excitations, Rashba [20] showed theoretically that the polarization of the integral luminescence will differ from that following from the oriented gas model.

Still, we must not fail to take into account the fact that the qualitative behavior of the experimental polarization diagrams in all cases agrees very well with those calculated on the basis of the oriented gas model. This means that the model gives many of the important features of the structure and properties of the molecular crystal and is true up to a point. Deviations from the model appear when the interaction between molecules begins to take effect. But in order to conclude that the oriented gas model is to a considerable extent justified in describing phenomena in the luminescence of molecular crystals we must have agreement with other aspects of this problem, i.e., with the role of defects in the luminescence of crystals.

As already discussed in Chapter I, the important part played by crystal lattice defects in the process of crystal luminescence has been unequivocally established as a result of a large number of investigations by various authors, in particular those of the Kiev school (Prikhot'ko, Broude, and their colleagues) [22]. In this, defects are understood in the broad sense; they may be vacancies, interstitial molecules, microcracks, impurity molecules, etc. One of the main proofs of this is the experimentally established variability of the weak longwave absorption and luminescence lines in the spectra of crystals at low temperatures. This variability in spectral position and intensity appears on passing from one specific sample of the same substance to another, and also on heat treatment of the samples; it is natural to connect this with defects. However, to be more precise, we should speak not of the luminescence of defects, but of the radiation of molecules near defects. In this, there may not be very great changes in the orientation of the radiating molecules, and the space distribution of the polarization will characterize the true orientation of the molecules in the lattice, in other words will correspond to the oriented gas model. At the same time, the energy terms of the molecules may experience considerable displacement under the influence of the crystal lattice field distorted near the defects. This change in the terms, revealed in the variability of the spectral structure at low temperatures, the displacement of the lines and the changes in their intensities, must clearly be dependent on the nature and character of the defects.

This point of view evidently agrees with all the well-known and widespread data on the luminescence of molecular crystals reviewed in Chapter I. The distortion of the lattice field near defects is probably responsible in some measure for the fact that the experimental polarization diagrams deviate from the theoretical, i.e., the orientation of molecules situated near defects differs somewhat from normal.

These differences are small, however, and cannot alter the general character of the orientation symmetry in the lattice.

The qualitative agreement of the form of the polarization diagrams with independent data on the orientation of molecules in the lattice allows us to use polarization diagrams of the luminescence of molecular crystals as an auxilaiary method of studying crystal structure, which may be useful in crystallography. This naturally touches only the orientation structure, since the polarization of luminescence cannot give any information on translational structure.

Even the fact that the qualitative character of the diagrams allows us to form a quite definite picture of the orientation of molecules in the lattice constitutes a valuable auxiliary source of information, since the x-ray structural analysis of molecular crystals is quite complicated, and a knowledge of the main features of orientation in the structure may make this easier.

As a convincing example, we may quote just the "positive" polarization diagram of anthracene. The asymmetry of the diagram at once leads to the conclusion that the plane in which the molecular oscillators lie does not coincide with the crystallographic ab-plane, but is rotated with respect to this through a certain angle around the b-axis. We may further determine this angle from the position of the experimental maximum in the diagram. We must emphasize that in so doing we use only the qualitative functional form of the diagram, and not the absolute degree of polarization, i.e., the data obtained are fully reliable. If we use the value of the degree of polarization at the extreme minimum of the diagram, we may also determine the angle between the two molecular oscillators with different orientations. This value of angle, however, will be inexact, being larger than the true value, since the polarization is smaller than would follow from exact data on the orientation of the oscillators.

Even such information, however, is valuable in studying the structure of crystals. If, furthermore, we know the orientation of the oscillator in the molecules from the polarization spectra of solutions, we already know a great deal about the orientation of a definite molecular axis relative to the crystallographic axes. Such data may be a great help in determining the complete structure when new objects, crystalline or quasi-crystalline, for example, oriented and oriented biological structures, are being studied by x-ray structural analysis or other structural methods.

## §7. Search for Elliptically Polarized Luminescence of Molecular Crystals and Solutions

Perrin [156] and later Grisebach [157] carried out an experimental search for circular or elliptically polarized luminescence in solutions with circularly or elliptically polarizing exciting illumination. In no case did they succeed in observing such polarization. Possibly the reason for this was that Perrin studied only the plane molecules of dyes, in which the radiating oscillator is always linear.

As it seems to us, elliptically polarized luminescence may be expected in molecules having two spatially separated, coupled, mutually perpendicular radiating oscillators. Such molecules must, however, clearly be optically active in accordance with the well-known model of V. Kuhn [158]. Elliptical polarization of luminescence may be expected from molecules rotating the plane of polarization, in complete analogy with the appearance of circular dichroism in these. This latter effect was in fact observed in Cotton's experiment [159].

Optically active materials are also found among molecular crystals. Moreover, they may be divided into two groups. To the first group belong substances manifesting optical activity both in solution and in the crystalline state. Quite a lot of these are known. The second group includes crystals formed from optically inactive molecules. In this case the optical activity is a property of the crystal as a whole; it forms a feature of the crystal itself rather than the constituent molecules.

In this sense, the optical activity of such crystals is analogous to excitons. To such substances belongs, for example, benzyl, the rotatory power of which was experimentally studied by Chandrasekhar [160]. The theory of the optical activity of crystals of the second type was developed by Arganovich [161].

Qualitatively, we may explain the optical activity of crystals formed from inactive molecules by taking account of the interaction between the molecules. The linear oscillators of different molecules, oriented differently and mutually interlinked, correspond to Kuhn's two-oscillator model. The interaction forces in different materials may be very different. But in every case this is always associated with a departure from the oriented gas model.

The search for elliptically polarized luminescence should thus be conducted among substances simultaneously possessing luminescence and optical activity. Two cases of elliptically polarized luminescence are known in the literature. In sodium uranyl acetate crystals at liquid helium temperatures, Samoilov [162] observed circular dichroism in individual absorption lines, and also circular polarization of emission lines. Recently, these experiments were repeated and extended by Brodin and Dovgii [163].

Among molecular organic compounds, elliptically polarized luminescence was observed by Neunhoeffer and Ulrich [164] in a compound with very high optical activity, the sodium salt of 1,3,5-triphenyl-pyrazoline-sulfonic acid. This substance, on excitation by either natural or linearly polarized light, emits luminescence with quite a high degree of elliptic polarization, both in the crystalline state and in glycerine solution.

In searching for elliptical polarization, we used a well-known method for the analysis of polarized light [165]. Between the rotating Nicol analyzer and the luminescing object, we placed a $\lambda/4$ plate so that its principal directions were parallel to the directions of the maximum and minimum electric vector of the radiated light, the positions of which were determined by rotating the Nicol analyzer without the $\lambda/4$ plate. The distinction between purely linearly polarized and partly elliptically polarized light lies in that, in the first case, on introducing the plate, there should be no changes in the positions of the analyzer for which maximum and minimum intensities are reached. In the second case, however, the maximum and minimum intensity correspond to different positions of the analyzer with and without the $\lambda/4$ plate. A high-sensitivity photomultiplier detected the luminescence after passage through the Nicol analyzer. In order to eliminate any anisotropy of the photomultiplier cathode, this was rotated together with the Nicol analyzer. Careful control experiments were carried out in the apparatus by analyzing circularly and elliptically polarized light produced artificially by means of a λ/4 plate.

As a result of quite a long search, we selected for investigation the following optically active and luminescent substances: glycerine solutions of tryptophan (luminescence in the near-ultraviolet region with maximum at 330 to 340 mμ) and riboflavin (green luminescence with maximum at 550 mμ), and benzene crystals.

Working with large benzyl single crystals, it was possible to study polarization not only in toto but also over the whole luminescence spectrum.

The degree of polarization of the total luminescence from benzyl was 20% in a direction perpendicular to the optic axis and 5% in a direction parallel to the optic axis. The polarization changed somewhat over the luminescence spectrum. In the direction parallel to the optic axis, the polarization increased towards the short-wave part of the spectrum, reaching 20% at the end.

We shall give detailed consideration to the spectral distribution of the polarization of luminescence from crystals in Chapter V. At the moment we shall simply be interested in elliptical polarization.

Any rotation of the principal directions of the analyzer (corresponding to maximum and minimum intensity) on introducing the λ/4 plate could be detected to an accuracy of ±2%. Analysis showed that, to this accuracy, neither the total luminescence of the benzyl crystal, nor the luminescence in different parts of the spectrum (including the short-wave edge, in the direction of which the rotatory power increases) had any elliptic polarization, either at room temperature or the temperature of liquid notrogen. The polarization was partly linear. The same result was obtained (quite naturally) on analyzing the polarization of luminescence from a series of optically inactive molecular crystals (anthracene, stilbene, dibenzyl, phenanthrene, and tolan). Elliptical polarization was not found in the solutions of tryptophan and riboflavin either.

Thus the case of elliptic polarization of luminescence observed in [164] for an optically active molecular crystal and solution remains for the moment unique. One's attention is drawn to the exclusively high optical activity of the material studied in this paper. One feels that this may well be the reason why the sought effect has so far only been found in this substance.

CHAPTER IV

# MIGRATION OF EXCITATION ENERGY
# BETWEEN THE MOLECULES IN MOLECULAR CRYSTALS

DEPENDENCE OF THE POLARIZATION OF THE LUMINESCENCE

ON THAT OF THE EXCITING LIGHT

§ 1.   Migration of Energy Between Molecules, Resonance Transfer

The departure from the oriented gas model is connected with the interaction of molecules in the crystal. First of all, we must expect that this interaction will lead to a transfer of energy from excited to unexcited molecules. Without mentioning exciton processes, the existence of which in molecular crystals follows from theory and is confirmed by experiment, as indicated in Chapter I, exchange of energy between molecules in a crystal must already take place by means of resonance transfer, similar to that which takes place in solutions at high concentrations.

The phenomenon of resonance transfer (in older treatments called "inductive resonance") in solutions of fluorescent materials has been studied in detail by a number of investigators. Perrin [166] first proposed an explanation for certain experimental facts with the aid of a hypothesis of the migration of excitation energy. Vavilov [1] evolved a semiphenomenological theory describing all the concentration phenomena. The probability of energy transfer was calculated from experimental data from the same concentration effects. Further development of the theory in papers by Galanin [101], Förster [102], and Dexter [167] made it possible to link the probability of energy transfer with internal molecular properties, and thus with the mutual disposition of emission and absorption spectra. This connection was confirmed by experiment [103,168]. The distances at which the migration of energy takes place are small compared with the wavelength, but large enough for the wave functions of the molecules not to overlap. The migration of energy appears macroscopically in experimentally observable concentration depolarization and quenching, as well as in a reduction of the duration of the excited state with increasing concentration.

The most immediate and sensitive indicator of the migration of energy in solutions is concentration depolarization. Over a fairly wide concentration range, the polarization of fluorescence from a viscous solution remains unchanged, but, beginning from a concentration of the order $10^{-5}$ to $10^{-4}$ g/cm$^3$, it falls sharply. This phenomenon is naturally explained by the migrations of energy from excited molecules having a definite orientation and given excitation anisotropy to unexcited molecules disposed chaotically. Calculation [103] shows that the degree of polarization of the luminescence emitted by molecules after one act of excitation energy transfer is $P_1 = 1/42$, if the initial degree of polarization of the fluorescence of molecules excited directly by the light has the maximum value of $P_0 = 1/2$. The observed total polarization P of the fluorescence is composed of partial polarizations corresponding to successive acts in the process of excitation energy migration. In order to calculate this, we must know the probability of radiation after each transfer. This probability is calculated in Vavilov's theory, starting from the assumption that a definite probability of quenching is associated with the act of transfer. Hence we may calculate the partial intensities, and obtain a somewhat cumbersome expression for the total polarization. For the case of relatively small concentrations, when quenching may be neglected, the formula simplifies and gives a linear relation between polarization and concentration, which is confirmed by experimental data obtained in the low concentration region [169]. Another polarization effect in which energy migration appears is the depolarization of luminescence from viscous and solid solutions during the damping process after excitation has stopped. This was found experimentally in the case of luminescence from uranium glasses by Sevchenko [170] and for dye solutions by Galanin [103].

Besides the polarization of luminescence, energy migration also affects yield and duration. In Vavilov's theory, formulas for the concentration quenching and change in $\tau$ are obtained. The constants coming into this theory are taken from experiment. Experiment shows that the degree of overlapping of the absorption and luminescence spectra is directly related to the transfer probability.

In the classical sense, the interaction of the molecules is similar to coupled oscillatory systems. The resonance character of the transfer lies in that the eigenfrequencies of the interacting systems are supposed equal. In quantum language this means that the energies of the corresponding transitions are equal. In the case of wide spectra a condition of transfer is that they should overlap as much as possible. As already mentioned, the theory of excitation energy transfer was evolved by Galanin, Forster, and Dexter.

Let molecules $M_1$ and $M_2$ interact in a medium with refractive index n at a distance R. The interaction Hamiltonian equals H. The state of the system with wave function $\psi_1$ corresponds to the excited state of $M_1$ and the unexcited state of $M_2$, while $\psi_2$ indicates the reverse. Then the probability of a transition from one state into the other (i.e., the probability of transfer) in the first approximation of quantum-mechanical perturbation theory equals

$$w = \frac{2\pi}{\hbar} \, \rho_E \left( \int \psi_1^* H \psi_2 \, dv \right)^2,$$

where $\rho_E$ is the density of states with given energy E, which does not change during the transition. The interaction is assumed to be dipole–dipole, the operator H for this case being known. The probability of transfer may be connected with the probability of radiation and absorption in the molecules between which the transfer takes place, since the probability of transitions is in both cases connected with the matrix elements of the dipole moments. We thus obtain the following expression for the transfer probability:

$$w = \frac{9c^4 \Phi^2}{8\pi R^6 n^4 \tau} \int \frac{\alpha(\omega) \, F(\omega)}{\omega^4} \, d\omega,$$

where $\Phi$ is a coefficient depending on the mutual orientation of the dipoles, $\tau$ is the mean duration of the excited state of the molecules, $\alpha(\omega)$ is the spectrum of the molecular absorption index, and $F(\omega)$ is the normalized luminescence spectrum ($\int F(\omega) d\omega = 1$).

This expression gives us an opportunity of evaluating the distances between the molecules for which the transfer probability becomes comparable with the probability of emission, i.e., the transfer takes place effectively. For typical luminescent complex molecules, for example dyes, these distances are in order of magnitude tens of Angstroms. It is clear that between molecules in crystals, where the distances are of the order of 10 A, energy transfer must take place by the resonance mechanism, if not by the exciton.

The main method of experimentally investigating the energy migration process in solutions is the study of concentration effects: the fall in polarization, yield, and duration of luminescence with increasing concentration. The most obvious of these, perhaps, must be concentration depolarization, which appears to be a practically essential consequence of any act of energy transfer. The experimental study of concentration depolarization is complicated by the presence of reabsorption and secondary luminescence, which also depolarizes the luminescence. It is necessary either to carry out experiments with very thin layers of solutions [103], or to introduce corrections, for example, by the method proposed by Vavilov [171]. Quite a number of investigations (145-147, 172-178) have also been devoted to the migration of energy from the basic substance to the impurity in mixed molecular crystals. Such crystals as naphthalene with anthracene impurity, anthracene with naphthacene impurity, and a number of others have the property that the absorption spectrum of the impurity substantially overlaps the emission spectrum of the base substance. Even insignificant concentrations of the impurity ($10^{-5}$ to $10^{-4}$ g/g) leads to the luminescence of the base substance (for example, the strong anthracene luminescence) falling sharply, and that of the impurity (for example, the green luminescence of naphthacene) simultaneously increasing. Measurement of the relative luminescent yield of the base substance and the impurity for various concentrations of the latter also constitutes a fundamental experimental method of studying energy migration in mixed crystals.

An analogous phenomenon of energy migration from the solvent to a dissolved fluorescent substance in liquid solutions on photoexcitation was observed in [101, 179-181]. The same phenomenon was discovered earlier [182] on excitation by fast particles. The method of investigation in these cases reduces in essence to a careful quantitative measurement of the absorption coefficients and spectra, the luminescent yields, and mean duration of the excited state for the molecules of solvent and solute.

Unfortunately, all methods based on concentration effects or on the difference in optical properties of the main substance and the impurity, fruitful as they are for solutions and mixed crystals, are inapplicable for the observation and study of energy migration in pure one-component molecular crystals. In such cases other methods are required.

## §2. Relation Between the Polarization of Luminescence from Molecular Crystals and the Polarization of the Exciting Light [86,183,184, 185]

As a method of observing and studying the migration of excitation energy in molecular crystals, we examined the relation between the polarization of the luminescence and that of the exciting light. We now give the gist of this method.

We shall measure the degree of polarization of the luminescence of a single crystal with respect to the normal to its surface

$$P = \frac{I_1 - I_2}{I_1 + I_2}, \tag{IV.1}$$

where $I_1$ and $I_2$ are the luminescence intensity components polarized along the principal axes of the polarization measuring apparatus (usually the vertical and horizontal axes).

We excite the luminescence by linearly polarized light in the same direction as that in which it is viewed. The azimuthal angle of the electric vector of the exciting light can vary from 0 to $2\pi$.

If the crystal lattice contains two forms of molecular oscillators with different orientations, then, generally speaking, they are always differently situated with respect to the electric vector of the exciting light. Moreover, they will be excited with different probabilities. For example, if the electric vector of the exciting light is directed perpendicularly to the oscillators of one type of orientation, then only the oscillators of the other type will be excited.

The degrees of polarization (IV.1) of the light emitted by oscillators of different types will differ. Hence the degree of polarization of the total luminescence from the crystal must essentially depend on the position of the electric vector of the exciting light relative to the oscillators, i.e., relative to the crystal axes.

We must further stipulate that the orientation of the crystal itself relative to the axes along which the mutually perpendicular components $I_1$ and $I_2$, determining the degree of polarization of the luminescence, are polarized in the measuring system must be taken into consideration. This orientation is very important. Let the crystal, for example, be so oriented that the direction $I_1$ coincides with the bisector of the angle between the oscillators. It is not hard to understand that, in this case, the abovementioned dependence of the degree of polarization of the luminescence on the polarization of the exciting light should not exist. Actually, the oscillators with different orientations are excited with different probabilities, but the ratio of the projections on the $I_1$ and $I_2$ axes will be the same for each oscillator. Hence the degree of polarization of all the luminescence will also always be the same. This is the most unfavorable orientation for observing the sought relation between the polarization of luminescence and that of the excitation. Calculation shows that the most favorable position is that for which one of the oscillators coincides with one of the directions $I_1$ or $I_2$. Moreover, we should

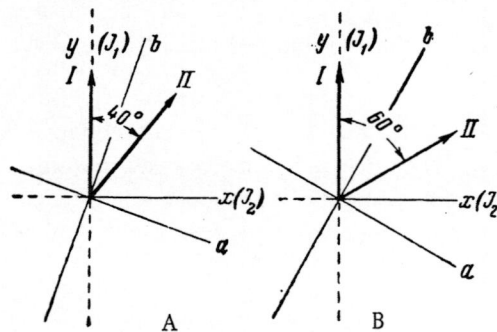

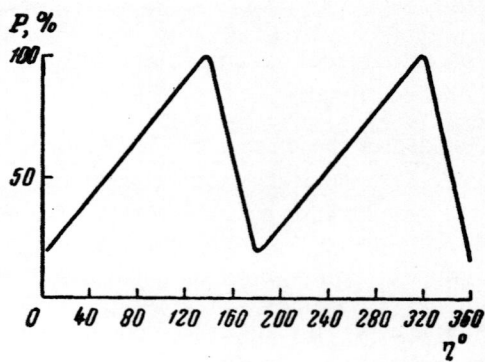

Fig. 41. Schematic disposition of the projections of the oscillators on the ab-plane for the orientation for which the relation between the polarization of the luminescence and that of the excitation is calculated.

Fig. 42. Calculated variation of the polarization of luminescence with the position of the electric vector in the exciting light, for the case depicted in Fig. 41 A.

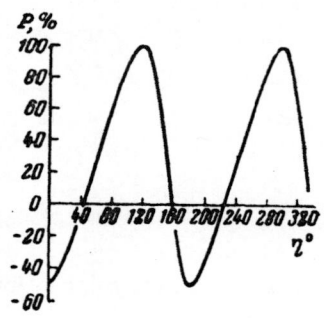

Fig. 43. Same as in Fig. 42, for the case depicted in Fig. 41B.

expect the sharpest variation in this case. For example, the polarization reaches 100% when one of the oscillators is vertical and the electric vector of the exciting light is perpendicular to the second oscillator.

By way of example, Fig. 41 shows schematically the disposition of the projections of the molecular oscillators on the plane perpendicular to which the excitation takes place and the luminescence is observed, for the two cases in which the angle between the projections is 40 and 60° and the orientation is the most favorable for studying the polarization relation in question. The first case corresponds to the ab-plane of anthracene, and the second to certain derivatives of anthracene in which the maximum of azimuthal variation of polarization for natural excitation is less than in anthracene.

Figures 42 and 43 show calculated graphs of the relation between the degree of polarization of the luminescence and the azimuthal angle $\eta$, reckoned from the horizontal, which determines the position of the electric vector in the exciting light. It is seen from the figures that the angular dependence has to be very sharp. Furthermore, the change in polarization is the more, the greater the angle between the oscillators.

We made measurements of this relationship for molecular crystals of various substances. The details regarding these measurements and the results obtained will appear later. At the moment we simply note the chief result: in every case we found a complete absence of any dependence of the polarization of the luminescence on that of the exciting light.

At first these measurements were performed with single crystals of microscopic dimensions (order of several microns), which can be obtained for many organic substances. It is much more difficult to obtain good single crystals of large dimensions. These must be prepared by sublimation, or by growing in a thermostatic bath or sealed test-tube. These methods are quite difficult, and by no means all substances succumb to crystal growing in this way. Operating with microcrystals thus substantially broadens the field of materials. It is true that we cannot by any means carry out all the required measurements with these—e.g., vary the wavelength of the exciting light, lower the temperature, etc. Measurements of the variation of the polarization of the total luminescence with that of the exciting light can, however, be made with microcrystals.

Comparing measurements of the polarization of luminescence made on large single crystals and microcrystals of the same material (in cases where both kinds were available) showed that, as a rule, the degree of polarization in the large crystals was somewhat larger than that measured in microcrystals. This may be brought

TABLE 1

| Substance | $P_{max}$, % |
|---|---|
| Anthracene. . . . . . . . . . . . . . . . . . . . . . . . . . | 39 |
| 9,10-dibromanthracene. . . . . . . . . . . . . . . . . . | 29 |
| 9,10-diphenyldiaminoanthracene. . . . . . . . . . . | 20 |
| Ortho-9,10-ditolyldiaminoanthracene. . . . . . . . | 12 |
| Meta-9,10-ditolyldiaminoanthracene . . . . . . . . | 9 |
| Para-9,10-ditolyldiaminoanthracene . . . . . . . . | (Almost unpolarized) |
| Ortho-9,10-dichlorphenyldiaminoanthracene . . . . | 31 |
| Meta-9,10-dichlorphenyldiaminoanthracene . . . . | 10 |
| Para-9,10-dichlorphenyldiaminoanthracene. . . . . | 28 |
| 9,10-diorthomethoxyphenyldiaminoanthracene . . . | 35 |
| 3-aminophthalimide. . . . . . . . . . . . . . . . . . . | 15 |
| 3-dimethylamino-6-aminophthalimide. . . . . . . . | 20 |
| Salicylic acid. . . . . . . . . . . . . . . . . . . . . . . | 37 |
| Carbazole . . . . . . . . . . . . . . . . . . . . . . . . . | 15 |
| Uranyl sulfate. . . . . . . . . . . . . . . . . . . . . . . | 20 |

about by the considerable aperture of the microscope, which records not only the total polarization in a direction perpendicular to the crystal surface, but also that in other directions making a considerable angle with the former. A certain amount of depolarization may also come from the edges of the crystals, which cannot be masked out under the microscope. In working with large crystals the edges were always masked out.

Measurements were made on a micropolarimetric system, the main components of which were a luminescent microscope and Cornu polarimeter.* This method was used to measure the polarization of luminescence in a large number of substances: anthracene, several of its derivatives, phthalimides, carbazole, and others. Among inorganic crystals we studied uranyl sulfate. Measurement of the azimuthal variation of polarization for natural excitation gives a certain value for the maximum degree of polarization. This value corresponds to the orientation for which the polarimeter axis $I_1$ coincides with the bisector of the angle between the oscillators (in the case of anthracene, the b-axis). As discussed above, it is just for this orientation that there is no dependence of the polarization of the luminescence on that of the exciting light. Measurements of this dependence were made for crystal orientations differing by 20 to 30° from the orientation corresponding to the maximum in the azimuthal variation. Such orientation should give a sharp variation, similar to that shown in Figs. 42 and 43.

Table 1 shows the maximum values of the polarization of luminescence in the measured crystals under natural excitation.

For the investigations in the micropolarimetric system, a number of control experiments were made, showing that the polarization measured experimentally was not seriously distorted by crystalline effects such as double refraction, reflection from faces, and so on. The calculations for anthracene made in the preceding chapter show that the depolarizing effect of the double refraction of the luminescence are very slight for measurements along the normal to the crystal surface. It was desirable to demonstrate this experimentally, as well.

The first experiment of this kind was to make measurements on crystals of the same substance, having different shapes and sizes. The effects of double refraction, reflection from the back surface, and so on, must clearly depend substantially on the latter. Experiment shows, however, that the maxima of the azimuthal curves for various crystals approximately coincide in magnitude, if reasonably homogeneous, undamaged, and strictly single crystals are used. The scatter of the maximum values about the mean is some 5 to 10%, which very slightly exceeds the mean relative experimental error. On the other hand, inhomogeneous crystals containing defects give low degrees of polarization. This also takes place in those cases in which an apparently homogeneous crystal in fact consists of two or more superimposed one on the other. In all cases in which a lower polar-

---

*The apparatus will be described in detail in Chapter VII.

ization was obtained than for other crystals of the same substance, higher magnification showed that the crystal was not single. At the same time, the polarization is substantially different for different materials, as seen in Table 1. In some cases it reaches the order of 40%; in others is it very small—around 3%.

The second group of control experiments comprised measurements with immersion. If the crystal is placed in a medium with a refractive index much nearer to that of the crystal than that of air, then clearly the refraction of the fluorescent light on emergence from the crystal is much less. Thus, if the measured polarization is the result of refraction and other purely crystalline effects, then the use of immersion must reduce it. On the other hand, if the polarization is caused by radiating molecules, and if refraction at the surface, double refraction, and so forth, introduce depolarization, then immersion will raise the degree of polarization.

As immersion medium we used glycerine ($n_D$ = 1.473) and a special nonfluorescing immersion oil ($n_D$ = 1.515). We made measurements on the same crystal with and without immersion. On immersion the degree of polarization never fell, but, on the contrary, somewhat increased (by about 10%).

These experiments showed that the crystal optical effects played no significant part in the conditions of our experiments.

Thus our measurements on the variation of the polarization of luminescence with that of the exciting light were carried out with the most favorable orientations of the crystals.

Figure 44 gives, by way of example, for several substances, the variation of the degree of polarization of the luminescence with the azimuthal angle $\eta$ of the Nicol through which the exciting light passes before striking the crystal. In all cases the polarization of the luminescence proved to be completely independent of the polarization of the excitation.

This result, at first glance paradoxical, may be explained if we assume that in homogeneous fluorescing crystals excitation migrates energy from the excited to the unexcited molecules. During the course of the excited state, the energy is redistributed among the molecules, so that, as a result, a definite number of molecules of every orientation is excited, although the initial number excited depends on the polarization of the excitation. Thus, radiation always takes place with a certain stationary distribution of excitation energy among the molecules. The polarization of the luminescence from the crystal thus retains no trace of the original anisotropy created at the moment of excitation.

The fact that the polarization of the luminescence is independent of that of the exciting light thus offers a method of observing and studying the migration of energy in one-component, impurity-free crystals. Since concentration methods are inapplicable in this case, the polarization method discussed takes on a special value.

At the same time, it is clear that without the aid of additional considerations and modes of investigation this method only offers the possibility of establishing the existence of energy migration, but so far does not help us decide whether this has an exciton or resonance nature.

With those substances of which we had available fairly large crystals (anthracene, 9,10-dibromanthracene, stilbene, etc.), we were able to carry out such investigations for various wavelengths of exciting light.

First of all, we measured the "polarization spectra" (if, in analogy with solutions, we so name the relation between the degree of polarization of the luminescence and the wavelength of the natural exciting light). Clearly, in molecular crystals, we should expect the polarization of the luminescence to be independent of the exciting wavelength. Actually, for isotropic excitation, the polarization of the fluorescence from a single crystal is completely determined by the aggregate orientation of the molecular radiation oscillators, and should not depend on the orientation of the absorbing oscillators, i.e., on the wavelength of the exciting light. The state of affairs is here analogous to the "spontaneous polarization" in

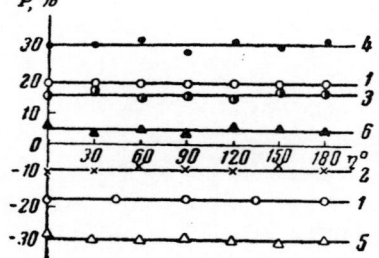

Fig. 44. Variation of the polarization of fluorescence from crystals with the polarization of the exciting light. 1) 9,10-diphenyldiaminoanthracene; 2) 9,10-dimetatolyldiaminoanthracene; 3) 9,10-diparachlorphenyldiaminoanthracene; 4) 9,10-diorthomethoxyphenyldiaminoanthracene; 5) anthracene; 6) 3-aminophthalimide.

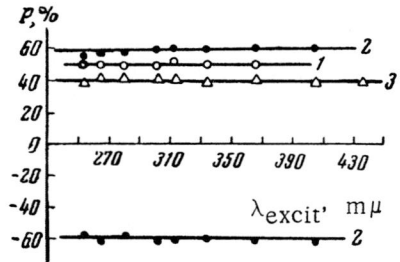

Fig. 45. "Polarization spectra" of crystals. 1) Anthracene; 2) 9,10-dibromanthracene; 3) 3-dimethylaminophthalimide.

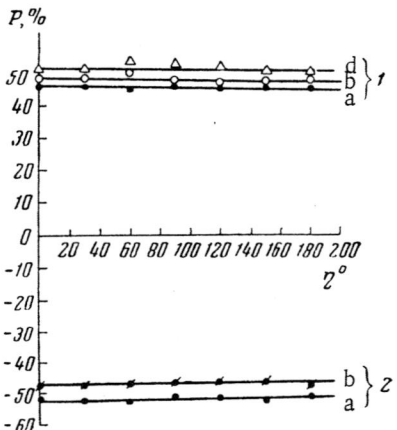

Fig. 46. Independence of $P(\eta)$ for various wavelengths of the exciting light. 1) Anthracene; 2) 9,10-dibromanthracene. Value of $\lambda_{excit}$: a) 365 m$\mu$; b) 313 m$\mu$; c) 280 m$\mu$; d) 265 m$\mu$.

partly oriented films colored by fluorescent substances with isotropic excitation (see Chapter II).

Figure 45 shows experimental "polarization spectra" for single crystals of several substances. The polarization of the fluorescence does not in fact depend on the wavelength of the exciting light.

By far less trivial results may be obtained from studying the variation of the polarization of luminescence with the polarization of the exciting light for various wavelengths of the latter. Figure 46 shows the results of measuring function $P(\eta)$ for anthracene and 9,10-dibromanthracene for various excitation wavelengths, corresponding to different absorbing oscillators, the long-wave oscillator directed along the transverse axis of the molecule, and the short-wave directed along the longitudinal axis. The different degrees of polarization on the graphs correspond to different orientations of the same crystal. It is seen that in no case does the polarization of the fluorescence depend on that of the excitation. From this we conclude that migration of energy between molecules always takes place, independently of which of the absorption oscillators is excited in the first place and of how it may be oriented in the molecule. It is evident that, as a result of the absence of any dependence of the polarization of the luminescence on that of the excitation, the polarization of the luminescence from a crystal must also be independent of the direction of the exciting light. This is in fact confirmed by experiment. We observed it qualitatively in a number of crystals. This was also mentioned in [61].

## §3. Effect of Double Refraction of the Exciting Light and Other Crystal Optical Factors, Their Calculation and Elimination [185,186,187]

The study of the polarization of luminescence in molecular crystals is always complicated by crystal optical effects, and above all by double refraction. In the preceding chapter the effect of double refraction of the luminescent light was considered. Now we shall turn serious attention to the double refraction of the exciting light and its relation to possible dependence of the polarization of luminescence on that of the excitation.

In fact, we must now consider whether we cannot explain the experimentally observed absence of any dependence of the polarization of luminescence on the polarization of the exciting light not so much by the migration of excitation energy as by the splitting of the exciting light in the crystal into two beams with electric vectors mutually perpendicular to one another. Will there not, as a result of this, be a certain degree of polarization of the luminescence, not depending on the state of polarization of the exciting light incident on the crystal?

The effect of double refraction of the exciting light may be explained in the following way. The components into which the exciting light is split are propagated with different velocities in the crystal. A certain path difference develops between them. The deviation of the beams is negligible, and they almost completely overlap. On interaction with the molecules undergoing excitation, the aggregate electric vector will therefore have a different direction at various points in its path. This result is identical with a rotation of the electric vector in the exciting light. Hence the electric vector will successively coincide with oscillators of different orientations (or, more precisely, with the projections of the oscillators on a plane perpendicular to the direction of the exciting beam), as a result of which they will all be uniformly excited, independently of the original position of the exciting electric vector. This section is devoted to an analysis of these circumstances.

First of all, the thickness of the crystals is very important. If we work with very thin single crystals, then the path difference mounting up as the exciting light passes through the crystals will be very small, and the rotation of the electric vector will be correspondingly slight. A sufficient condition is $d \cdot \Delta n \ll \lambda$, where d is the thickness of the crystal, $\lambda$ the wavelength, and $\Delta n$ the difference in refractive index for the components of the split beam. For example, in the case of an anthracene crystal having its surface coinciding with the ab-plane, the light incident perpendicular to this surface has a refractive index (for $\lambda = 436$ m$\mu$) of $n_{\parallel} = 1.97$ for the component polarized parallel to the b-axis, and $n_{\perp} = 1.81$ for the second component [137,138]. Thus, $\Delta n = 0.16$. For the thickness we obtain $0.16d \ll 0.44$ $\mu$, $d \ll 3$ $\mu$. This condition may be regarded as sufficiently well fulfilled for thicknesses of the order of 0.3 $\mu$. When we were working with microscopic crystals in the micropolarimeter apparatus, this condition was always more than fulfilled, as the microlaminas of the crystals had linear dimensions of the order of several microns, and thicknesses less than 0.1 $\mu$. On growing single crystal films by sublimation, however, the thickness of the crystals in many cases may reach several microns. It is true that a number of substances, for example anthracene containing naphthacene, form extremely thin films on sublimation. In working with large, thick single crystals, the laminas of which are cut from large blocks and have thicknesses of several tens of millimeters or more, we must not fail to make allowance for the double refraction of the exciting light.

It is important so to set up the experiments that the effect of double refraction of the exciting light is eliminated or allowed for. We shall first describe the results of such experiments, and then return to crystals of small thickness. The essence of one such experiment lies in the following. The splitting of the exciting light into two components can lead to equally probable excitation of oscillators in two different orientations only when these are disposed symmetrically with respect to the axes along which the double refraction components are polarized. This symmetry occurs, for example, in anthracene [137]. However, there are some crystals in which there is no such symmetry. After considering a number of crystals which we had available, we selected stilbene for these experiments. The orientation of the stilbene molecules in the lattice is known from x-ray structural data [14].

Calculations showed that, for present purposes, the most suitable cut was along the bc'-plane. The scheme of the projections of oscillators on this plane is given in Fig. 47. The directions of the electric vectors of the splitting components in biaxial crystals, as already mentioned, is determined by the well-known theorem of Fresnel [139]. The vectors lie in the bisector planes between the planes containing the direction of the beam and one of the optic axes. If the beam lies in the plane of the optic axes, then the two latter planes merge into one. One of the bisector planes merges with this plane, and the other is normal to it. In the stilbene crystal, the optic axes lie in the ab-plane. The a-axis is the bisector of the acute angle between them, and the b-axis is the bisector of the obtuse angle [149]. When the exciting light falls perpendicular to the selected bc'-plane, i.e., along the a-axis, the split beams will be polarized along c' and b. As seen from Fig. 47, the projections of the oscillators of the molecules are disposed asymmetrically with respect to these directions. In order to study the dependence of the polarization of the luminescence on that of the exciting light, we chose the orientation indicated in Fig. 47, for which the projection of one of the oscillators is vertical ($I_1$ axis vertical, $I_2$ axis horizontal). If now we excite the luminescence with light polarized along c' and along b, then clearly we must in these respective cases obtain different polarizations of the luminescence. Calculation shows that in the first case the degree of polarization is 88% and in the second 73%. The difference is 15%, and may easily be observed experimentally.

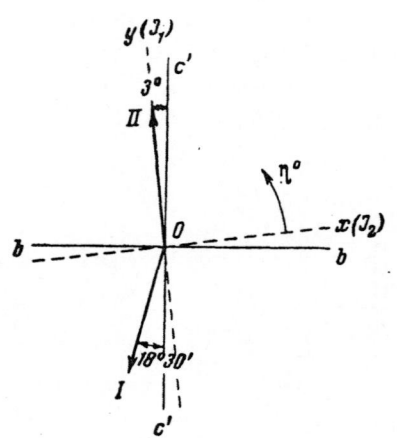

Fig. 47. Scheme of the projection of oscillators in a stilbene crystal on the bc'-plane. I and II: different projections of the oscillators.

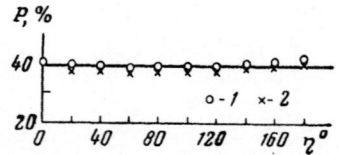

Fig. 48. Variation of the degree of polarization of the luminescence from a single crystal of stilbene with the directions ($\eta°$) of the electric vector in the exciting light. Cut through the bc'-plane (orientation of crystal shown in Fig. 47): 1) $\lambda_{lum} = 395$ m$\mu$; 2) $\lambda_{lum} = 424$ m$\mu$.

The measurements were carried out in a spectropolarimetric photoelectric system*; this was convenient in the present case, as it allowed the exciting light to be removed efficiently. This is extremely important in such experiments, as the least trace of exciting light may seriously distort the result. With stilbene, this is especially complicated by the fact that the spectrum lies in the violet and near ultraviolet region. Excitation was carried out by light from a PRK-4 lamp with a wavelength 365 mμ. In the luminescent spectrum, the polarization was measured for wavelengths 395 and 424 mμ. Control experiments were made without the crystal, using diffusing quartz plates. The results of the measurements appear in Fig. 48. It is seen that the polarization of the luminescence is completely independent of the position of the electric vector in the exciting light. This result can only be explained by the migration of energy.

The principle of the other group of experiments lay in directing the exciting light along the optic axis of the crystal. In these experiments, the objects of investigation were uniaxial single crystals of benzyl and biaxial single crystals of stilbene.

In the elementary cell of the benzyl crystal, which belongs to the hexagonal system, there are three molecules, helically disposed around a higher order axis coinciding with the optic axis [160,188, 189]. This is the growth axis of the single crystals. Thus, in laminas cut perpendicular to the growth axis, the optic axis is normal to the plane of the crystal, and the projections of the molecular oscillators on the plane of the crystal surface are disposed symmetrically around this axis, making an angle of 120° with each other (Fig. 49). It is not hard to calculate the variation of the degree of polarization of the luminescence in the direction of the optic axis with the angle $\eta$ determining the position of the electric vector of the exciting light in a plane normal to the optic axis. The calculated variation is shown in Fig. 50. Since for propagation along the optic axis there is no double refraction of the exciting light, this variation should be observable experimentally, if there is no migration of excitation between the molecules.

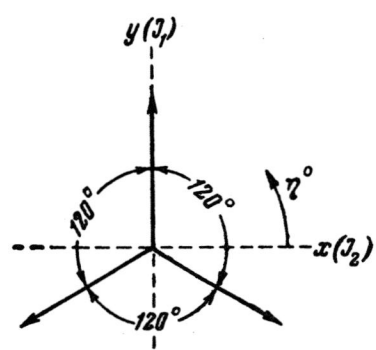

Fig. 49. Scheme of the projection of the oscillators in the benzyl crystal on a plane perpendicular to the optic axis.

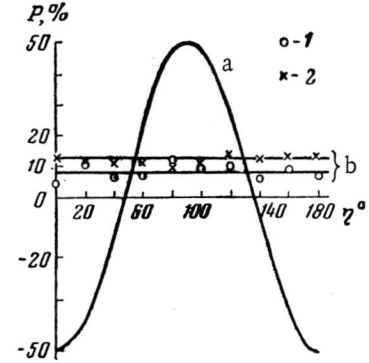

Fig. 50. Polarization of luminescence as a function of the polarization of the exciting light for benzyl (cut through plane perpendicular to the optic axis). a) Calculation; b) experiment. 1) $\lambda_{excit} = 436$ mμ; 2) 365 mμ.

It is fairly easy to separate the bright green luminescence of benzyl from the exciting light. The measurements were carried out in a visual apparatus with a Cornu polarimeter. The results for exciting light of wavelengths 365 and 436 mμ are also shown in Fig. 50. The excitation by the blue light is of interest in connection with the following optical features of the benzyl crystal [160]. For wavelengths of the order of 590 mμ the crystal is uniaxial positive. On passing to the short-wave end of the spectrum the double refraction diminishes, and at 420 mμ the crystal is optically isotropic. For lower wavelengths the double refraction changes sign. The crystal becomes uniaxial negative. Thus, on effecting the excitation with the light of wavelength $\lambda = 436$ mμ, we are working with a practically isotropic crystal. As seen from Fig. 50, in this case also, the polarization of the luminescence is completely independent of that of the excitation, and this can only be explained as a proof of the presence of energy migration in the crystal.

It is true that we must still make allowance for the fact that the benzyl crystal rotates the plane of polarization. The magnitude of this effect, however, is insufficient seriously to distort the calculated relation shown in Fig. 50. From data given in the paper mentioned [160], the rotary power for the blue region is around 45°/mm. The crystals used in our measurements had thicknesses of around 0.1 mm; in the experiments

_____

*This system is described in detail in Chapters V and VII.

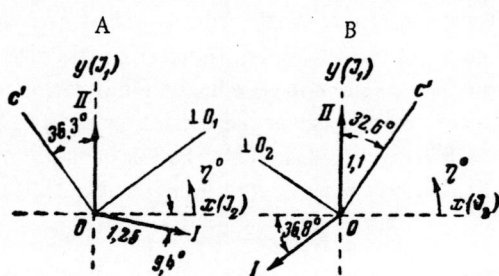

Fig. 51. Schemes of the projections of the oscillators in the stilbene crystal on planes perpendicular to the optic axes. A) $\perp O_1$; B) $\perp O_2$; I and II: projections of different oscillators.

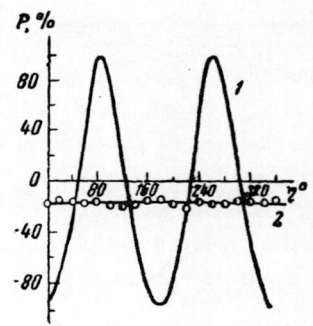

Fig. 52. Polarization of the luminescence as a function of that of the exciting light for stilbene. Crystal cut along a plane perpendicular to optic axis $O_1$; orientation shown in Fig. 51 A. 1) Calculated; 2) experimental.

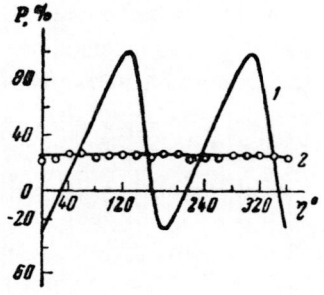

Fig. 53. Same as in Fig. 52, but for a cut perpendicular to the second optic axis $O_2$; orientation shown in Fig. 51 B.

in the micropolarimetric system they had thicknesses of the order of a few microns. The corresponding rotations lie between 5° and a fraction of a degree, and may be neglected.

A subject for similar experiments comprised biaxial stilbene crystals, belonging to the monoclinic system and having no optical activity. The position of the optic axes was determined in spherical single crystals from the conoscopical picture [149] for the violet end of the spectrum. Laminas were then cut from the sphere perpendicular to one or other of the optic axes. Excitation along the optic axes was effected by a parallel light beam with $\lambda = 365\ m\mu$. The polarization of the luminescence was measured for wavelength 414 m$\mu$. The experiments were carried out in the spectropolarimetric apparatus, special attention being paid to separating out the exciting light, since the excitation was initiated in exactly the same direction as that in which the luminescence was viewed. By means of monochromators and filters, every trace of the exciting light was successfully removed.

Since the position of the optic axes and the molecular axes relative to the crystallographic axes are known, it is not hard to calculate the projections of the oscillators and the angles determining their positions in planes perpendicular to the optic axes. The results are shown in Fig. 51, where the oscillator schemes for both cuts of the crystal are presented in the same most favorable orientations (one of the oscillators vertical) for which the calculations and experimental measurements of the variation of polarization were carried out. With natural excitation, these two cuts give very different polarizations of luminescence (−17 and +42% calculated, −18 and +30% experimental), so that they are easy to distinguish.

Figures 52 and 53 show the calculated variation of the polarization of the luminescence with the position of the electric vector in the exciting light. We see that the curves are extremely sharp and distinctive. Experimental results, however (presented in the same figures), show that the polarization of the luminescence is completely independent of that of the exciting light (just as in all the previous cases), and thus prove the existence of energy migration in the crystal.

In this experiment there is one complicating factor: we determined the position of the optic axes for violet light but excited the luminescence with near-ultraviolet. In view of the dispersion of the optic axes, the required axis for $\lambda = 365\ m\mu$ may deviate from that so found by several degrees. Determination of the axes for ultraviolet light by the ordinary visual means is impossible. We cannot excite with the mercury line $\lambda = 405\ m\mu$, as this wavelength lies near the long-wave end of the luminescent spectrum and almost all the luminescence appears in the anti-Stokes region.

In order to check that this inexactitude in determining the positions of the required optic axes did not distort the results, we carried out measurements after rotating the crystal through several degrees, from 2 to 10°, thus admitting the exciting light in directions close to the optic axes for violet light. Around 15 measurements were made for each cut, in a cone with an angular aperture up to 20°, completely guaranteeing that the optic axis should fall in the region examined. In all cases the polarization of the luminescence proved totally independent of the polarization of the exciting light.

At first glance it may seem that even a slight error in determining the positions of the optic axes might have serious consequences, as double refraction is only absent when the rays fall exactly along the optic axis.

It is easy to realize, however, that a small error of 1 to 2°, inescapable in experiment, is unimportant, since double refraction near the optic axis corresponds to a negligibly small path difference, and thus to a negligibly slight rotation of the electric vector of the exciting light.

Of course, we must also bear in mind the wavelength of the light in undergoing double refraction. In fact, if the excitation takes place in a strong absorption region, i.e., in a region where the dispersion curve rises sharply, possibly at very different rates for the separate components of the splitting, then the difference between their refractive indices may also be significant. An estimate of this must be made in each specific case.

For stilbene there are dispersion data in [46]. For the wavelength of $\lambda = 365$ m$\mu$ which we used for excitation, the refractive index for the component polarized $\parallel$ b is $n_{\parallel b} = 2.36$; for that polarized $\perp$ b it is $n_{\perp b} = 1.92$. Thus $\Delta n$ is 0.44. It is clear that for slight deviations from the optic axis the double refraction will be negligible.

For anthracene we already presented data calculated from [138]. According to these, for $\lambda = 436$ m$\mu$, $\Delta n = 0.16$. For $\lambda = 365$ m$\mu$ the values of the refractive indices may be taken from later results [34,190]. From the first of these, $\Delta n = 0.12$, from the second $\Delta n = 0.2$. Thus the difference in the refractive indices here is also slight. With regard to benzyl, on exciting with $\lambda = 436$ m$\mu$ we are working practically with an isotropic crystal.

Finally, in order to make allowance for all possible crystal optical effects, we must pay attention to the possible influence of dichroism. Even for the propagation of light along or close to the optic axis, when the path difference between the components of splitting is very small or entirely absent, there may be a rotation of the total electric vector if the intensity ratio of the components changes on account of dichroism.

In benzyl, dichroism in the blue region of the spectrum for cuts perpendicular to the optic axis is slight, which is quite natural in view of the high symmetry in the disposition of the absorbing molecules. In stilbene, for 365 m$\mu$, the dichroism for the ac' cut is very slight, but considerable for the bc' cut [191]. In cuts perpendicular to the optic axes we have an intermediate case.

In order to exclude the possible effects of dichroism, we must compare the polarization of the luminescence in two cases: i.e., for excitation by light having its electric vector coinciding with the direction of polarization of one or another of the splitting components, as determined by Fresnel's theorem (i.e., along and perpendicular to the c'-axis, respectively). In other words, we wish to use the so-called uniradial azimuths, for which only one component enters the crystal, as a result of which dichroism ceases to be troublesome.

The polarization of luminescence calculated for these two cases may be taken from the theoretical curves of Figs. 52 and 53. These data are given in Table 2.

The differences are considerable and could easily be observed in an experiment. But, as indicated above, they are completely absent.

The results of all these experiments indicate that the absence of any dependence of the polarization of luminescence from molecular crystals on the polarization of the excitation constitutes a proof of the migration of excitation energy in the crystals.

Let us now turn to the other possible means of avoiding the effects of double refraction of the exciting light, i.e., to the use of thin single crystals, which we mentioned earlier.

TABLE 2

| Direction of exciting light | Position of electric vector of exciting light | Degree of polarization of luminescence, % |
|---|---|---|
| Along optic axis $O_1$ | ∥ c' | −11 |
| | ⊥ c' | +18 |
| Along optic axis $O_2$ | ∥ c' | −33 |
| | ⊥ c' | +58 |

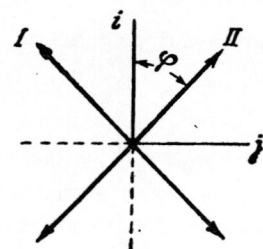

Fig. 54. Relating to the calculation of the permissible thickness of a crystal.

We indicated before that the effect of double refraction of the exciting light, which complicates our polarization picture, may be explained as a result of the rotation of the total electric vector owing to the different velocities of propagation of the splitting components in the crystal. Under this rotation, the electric vector of the exciting light may successively excite molecular oscillators of differing orientations. The selectivity of the original direction of the electric vector in the exciting light will thus be lowered.

It is evident that by working with very thin crystals we may eliminate this complication. The thickness required for this is determined from the condition that the angle of rotation should be small compared with the angle between the oscillator.

Since the components of splitting are coherent, strictly speaking, they cannot be treated as independent of one another inside the crystal, and in calculating the probability of excitation for various molecular oscillators they must necessarily be considered together. For simplicity, let the oscillators be oriented along the bisectors of the polarization axes of the splitting components (Fig. 54). Let the exciting light wave $E e^{ikr+i\omega t}$ be polarized in the direction II. In the crystal this splits into components polarized along i and j, $\mathbf{i}\beta E e^{ik_1 r}$ and $\mathbf{j}\beta E e^{ik_2 r}$, where

$$k_1 = \frac{\omega n_1}{c}, \qquad k_2 = \frac{\omega n_2}{c},$$

$n_1$ and $n_2$ being the refractive indices of the two components in the crystal, and $\beta = \cos \varphi$.

$$\text{I. } E_{\mathrm{I}} = \beta^2 E \left( e^{ik_1 r} - e^{ik_2 r} \right) = \beta^2 E \left[ e^{i(k_1 - k_2)r} - 1 \right] e^{ik_2 r},$$

$$I_{\mathrm{I}} \sim |E_{\mathrm{I}}|^2 \sim E^2 \left| e^{i(k_1 - k_2)r} - 1 \right|^2 \sim E^2 (k_1 - k_2)^2 r^2;$$

$$\text{II. } E_{\mathrm{II}} = \beta^2 E \left[ e^{i(k_1 - k_2)r} + 1 \right] e^{ik_2 r},$$

$$I_{\mathrm{II}} \sim |E_{\mathrm{II}}|^2 \sim 4E^2.$$

The condition in which we are interested must clearly be formulated in the following simple form:

$$I_{\mathrm{II}} \gg I_{\mathrm{I}}. \tag{IV.2}$$

Substituting the values obtained above, we have

$$\frac{(n_1 - n_2)^2 \omega^2}{4c^2} l^2 \ll 1,$$

where $l$ is the thickness of the crystal. Hence

$$l \ll \frac{2c}{(n_1 - n_2)\,\omega}.$$

For anthracene, for example, from the data of [34], $n_1 - n_2 = 0.12$ for exciting light with $\lambda = 365$ m$\mu$. In order of magnitude this is a typical value for molecular crystals.

Putting in our condition $c = 3 \cdot 10^{10}$ cm/sec and $\omega = 6 \cdot 10^{15}$ sec$^{-1}$, we obtain the condition $l \ll 0.8\ \mu$.

This more rigorous estimate gives us the upper limit of thickness. We must further estimate what error would actually accrue from the thinnest crystals with which we could possibly conduct experiments.

By sublimation in porcelain crucibles we obtained a large number of thin laminar anthracene crystals. Among these were some very thin ones, thinner than one micron. The thickness of the crystals (order of 0.5 $\mu$) was measured on the interference microscope MII-5. The mean measuring error was 0.05° (relative error 10 %).

We managed to select for measurement homogeneous, high-quality anthracene single crystals 0.3 $\mu$ thick. Let us calculate what error, associated with the rotation of the electric vector in the exciting light, could accure from this thickness. The optical path difference of the components is $l \cdot \Delta n = 0.3 \cdot 0.12 = 0.036 \mu \approx 0.1 \lambda$. This corresponds to a rotation of the electric vector through approximately 18°. It is evident that the "blurred" position of the electric vector lowers the selectivity of the preferred excitation of one form of oscillator rather than another. We may estimate this reduction with the help of the calculated curve P($\eta$) given in Fig. 42. The deviations of the electric vector from its fixed position at the maximum and minimum of the curve must lead to a fall in the range of variation of the degree of polarization of the luminescence. From this curve we may calculate that for deviations of 18° the polarization may change at least by 35 to 80%. This is a very large change, and much smaller changes can be detected experimentally.

The relation P($\eta$) for thin crystals of anthracene was measured by means of the spectropolarimetric photoelectric apparatus, which made it possible to measure the polarization for any wavelength of luminescence. This was very important in the present case in order to remove traces of exciting light, which, however small, might fundamentally distort the results.

The exciting light passes through a polaroid capable of rotation around an axis coinciding with that of the exciting beam, and falls on the crystal. The degree of polarization P of the luminescence from the latter is measured as a function of azimuth angle $\eta$ of the polaroid through which the exciting light passes. The wavelength of the luminescence in the measurements was $\lambda_{lum} = 419$ m$\mu$. Control experiments showed the complete absence of traces of exciting light.

The excitation was effected with light of various wavelengths: 365, 313, 302, and 254 m$\mu$. The results are shown in Fig. 55. The various straight lines correspond to various wavelengths of exciting light and slightly different crystal orientations. All the orientations were close to the most effective (one of the oscillators vertical). The position of the b-axis was determined by means of preliminary measurement of the azimuthal variation of polarization for natural excitation.

It is seen from the figure that in all cases the polarization of the luminescence is independent of that of the exciting light. Since double refraction is here absent, this fact may be considered as a proof of energy migration. The existence of this independence for all wavelengths of exciting light indicates that migration always takes place, whichever absorption oscillator is excited.

Thus we may conclude that the absence of any dependence of P on $\eta$, after making allowance for extraneous crystal optical factors, proves the existence of energy migration between the molecules in pure one-component molecular crystals.

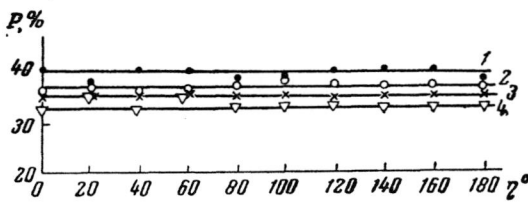

Fig. 55. Degree of polarization of luminescence from a pure anthracene single crystal as a function of azimuthal angle $\eta$ of the electric vector in the exciting light for various wavelengths of the latter. 1) $\lambda_{excit} = 302$ m$\mu$; 2) $\lambda_{excit} = 365$ m$\mu$; 3) $\lambda_{excit} = 313$ m$\mu$; 4) $\lambda_{excit} = 254$ m$\mu$.

The only investigation in which the migration of energy in a one-component crystal has been observed by another method is that of Simpson [192]. This method is as follows. A pure anthracene single crystal was directly excited by ultraviolet light; on the opposite side was placed a detector layer (anthracene containing a trace of naphthacene). The luminescence of the naphthacene served as an indication of a transfer of energy from the first crystal to the detector. On the experimental side, these experiments were quite difficult, requiring many quantitative control experiments.)

The polarization method of observing migration is simple and easily carried out for any crystals. In one-component crystals, however, it essentially only reveals the migration of energy between molecules of different orientations, and gives no information on the nature of the migration and its parameters (diffusion length, etc.). However, by varying this method and making use of its delicacy (its "molecular level") valuable information may be obtained.

In the following sections we shall show some of the possibilities of the polarization method.

### §4. Migration of Energy Between Impurity Molecules in Molecular Crystals [184,187]

It is of considerable interest to apply the above method of observing energy migration in crystals (from the constancy of the polarization of luminescence with respect to the polarization of the excitation) to mixed crystals (such as anthracene containing naphthacene), and by this means attempt to observe energy migration between impurity molecules at distances depending on the impurity concentration, which for small concentrations many times exceed the lattice constant.

As already discussed, a large number of investigations have proved the existence in such crystals of energy transfer from the base material to the impurity. But so far, no method has been devised for detecting energy migration between molecules of the impurity itself.

It is evident that in principle our method allows this to be done. In using it, we may consider the impurity as an interstitial lattice with a many-times magnified lattice constant. This lattice, of course, will have only orientational and not translational symmetry. Measurements of polarization over the luminescent spectrum of mixed anthracene-naphthacene crystals* have shown that the orientation of the interstitial naphthacene molecules differs little from that of the anthracene molecule in the main lattice.

In order to use our method, we must excite only the impurity molecules without exciting the base substance. In the case of an anthracene-naphthacene mixture, this can be done. The mercury line 436 m$\mu$ is strongly absorbed by naphthacene and lies far beyond the limits of the anthracene absorption spectrum.

The first of these experiments which we carried out appeared in [184]. Anthracene crystals containing naphthacene (concentration of naphthacene $10^{-4}$ g/g) were obtained by sublimation. Measurements were made in a visual apparatus with a Cornu polarimenter. In the polarimeter was an OG-4 filter, passing only the naphthacene fluorescence and completely cutting off that of anthracene. Excitation was effected both in the absorption band of anthracene (365 m$\mu$) and in that of the naphthacene itself (436 m$\mu$). The corresponding mercury lines were separated out by filters. In the first case, the anthracene molecules were chiefly excited; energy passed from them to the naphthacene molecules and was emitted by the latter. Clearly, here, the degree of polarization P of the fluorescence of naphthacene should not depend on the azimuth angle $\eta$ determining the position of the electric vector in the exciting light; this was in fact observed experimentally (Fig. 56 a). This once more confirms the existence of migration from the anthracene to the naphthacene, as shown earlier by various authors using other methods.

Figure 56 b also shows the total variation of the polarization of the polarization of naphthacene luminescence with the position of the electric vector of the exciting light for direct and selective excitation of the naphthacene itself ($\lambda_{excit} = 436$ m$\mu$). This indicates the existence of energy migration between the naphthacene molecules at distances several times exceeding the lattice constant. This is essentially a new result.

However, before asserting this with full certainty, we must again consider, for this particular case, the possible effects of the complicating factors mentioned in the foregoing section.

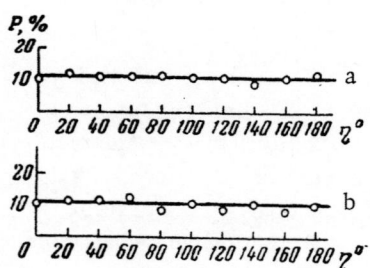

Fig. 56. Polarization P of the naphthacene luminescence in anthracene as a function of the position of the electric vector in the exciting light ($\eta$). a) Anthracene absorbing; b) naphthacene absorbing.

_____

*We shall treat these measurements in more detail in Chapter V.

Evaporated anthracene-naphthacene films are always obtained very thin, though in the investigation cited their thickness was not strictly regulated. We were not able to apply the excitation along the optic axis in this case. For this, crystal laminas cut perpendicular to the optic axis were required, and these in their turn called for initially rather large blocks of single crystals, in which the positions of the optic axes could be determined by one of the usual crystal optical methods. There were no such crystals of naphthacene-containing anthiacene at our disposal. For this reason, in a later treatment [187], we made detailed and careful measurements on thin crystals of anthracene containing naphthacene. For the measurements we used crystals of very small thickness (order of 0.5 $\mu$) obtained by sublimation. From a large number of evaporated crystals we chose the thinnest, from 0.3 to 0.6 $\mu$. The thickness was measured in the interference microscope MII-5. The mean measuring error was 0.05 $\mu$. The refractive indices of the splitting components of the exciting light must in this case be taken for wavelength 436 m$\mu$ ($n_{\parallel b} = 1.97$ and $n_{\perp b} = 1.81$; $\Delta n = 0.16$). An estimate of the thickness which may be considered small for this wavelength was made on p. 74. The following result was obtained: $l \ll 3\,\mu$. May we regard the crystals used in our experiments as thin enough? Let us make an estimate for the maximum thickness of 0.6 $\mu$.

As the degree of polarization for natural excitation in naphthacene is somewhat smaller than in anthracene, the angle between the molecular oscillators of naphthacene is somewhat greater than in a pure anthracene crystal. We must use the curve of Fig. 43, which shows the variation of the degree of polarization of the luminescence with the azimuthal angle of the electric vector of the exciting light, calculated on the assumption that there is no energy migration, for the disposition of oscillators shown schematically in Fig. 41B. As may be seen from this figure, the expected changes in polarization are very large: from -50 to +100%. But in this calculation we have still not taken account of double refraction, which "blurs" the anisotropy of the excitation.

For a thickness of 0.6 $\mu$ the path difference is $l \cdot \Delta n = 0.096\,\mu$, and this path difference corresponds to an angle of rotation of the electric vector through approximately 30° (rotation through 45° would correspond to a path difference of $\lambda/4$, but our path difference is less than this). It follows from Fig. 43 that the deviation of the electric vector from its fixed position at the maximum and minimum of the curve may lead to a fall in the range of variation of the degree of polarization of the luminescence. Verifying calculations show that the polarization will vary approximately from −30 to +85%. This is a large change, but much smaller ones can be noticed in experiment. This is how the matter stands even for crystals with the largest thickness used.

The concentration of naphthacene was determined directly in the thin crystals selected for investigation by a spectral method. By planimetering the anthracene and naphthacene spectra, the relative integral intensity of their luminescence was determined, and with the help of published data [146,193], the concentration was deduced.

We measured three single-crystal samples of the best quality selected from a large number of prepared crystals. The concentration of naphthacene in these was $1 \cdot 10^{-3}$, $2 \cdot 10^{-4}$, and $6 \cdot 10^{-5}$ g/g.

The excitation was produced by the light of the mercury line of wavelength 436 m$\mu$, separated out by a monochromator, a dense filter for the line (from a collection of mercury line filters), and by a liquid filter, a concentrated solution of $NaNO_2$. This wavelength, as mentioned earlier, is only absorbed by the naphthacene and lies beyond the limits of the anthracene absorption spectrum. Naturally, for such small thicknesses and concentrations, the intensity of the directly excited naphthacene luminescence is extremely weak.

The most important experimental problem was the reliable separation of the exciting light. The measuring procedure primarily consists of passing the exciting light through a polaroid capable of rotation around an axis coinciding with that of the exciting beam and then onto the crystal. The degree of polarization of the luminescence from the crystal is measured as a function of the azimuthal angle $\eta$ of the polaroid through which the exciting light passes. The polarization was measured by means of a spectropolarimetric photoelectric system, which enabled measurements to be made for any wavelength of the luminescence. This is extremely important, since the least trace of exciting light with variable linear polarization, as we can easily understand, can fundamentally distort the results. It is convenient to measure at $\lambda_{lum} = 495$ and 530 m$\mu$, corresponding to the maxima in the naphthalene spectrum. In both cases, however, it is difficult to avoid the background from the mercury lamp, since in the first case the 492 m$\mu$ line and in the second case the 546 m$\mu$ are situated quite close.

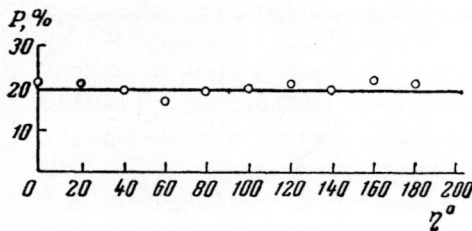

Fig. 57. Polarization of the naphthacene fluorescence as a function of azimuthal angle $\eta$ of the electric vector in the exciting light. $\lambda_{excit} = 436\,m\mu$; $\lambda_{lum} = 532\,m\mu$.

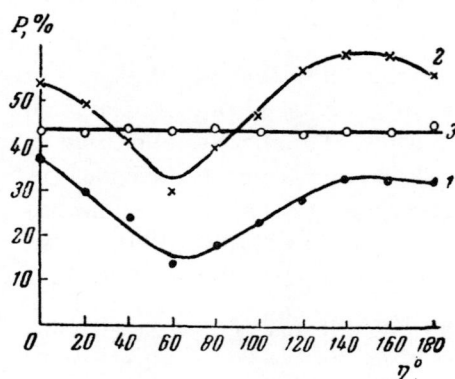

Fig. 58. Results of experiments on measuring the polarization of naphthacene luminescence as a function of the azimuthal angle $\eta$ of the electric vector in the exciting light. 1) One of the oscillators vertical (Fig. 41B); 2) b-axis vertical; in both cases $\lambda_{excit} = 436\,m\mu$, $\lambda_{lum} = 530\,m\mu$; 3) $\lambda_{excit} = 365\,m\mu$; $\lambda_{lum} = 495\,m\mu$; orientation corresponds to Fig. 41 B.

From this point of view, it is best of all to measure for $\lambda_{lum} = 508\,m\mu$, where the background is least. But here, unfortunately, there is a minimum in the luminescence spectrum, and the intensity of luminescence is several times weaker than at the maxima.

The choice of a definite wavelength in the luminescence spectrum for measurement depends on the ratio of the luminescent intensity to that of traces of exciting light. If the latter is in all a few percent, then we can work in the amplification range which will be sensitive only to the wanted signal. We can convince ourselves that no traces of exciting light are present by means of control experiments in which the crystal is replaced by a mat diffusing plate of fused quartz. For the largest of the concentrations used ($10^{-3}$ g/g), it proved possible, in view of the still-bright luminescence, to remove all traces of excitation and measure the function $P_{lum} = f(\eta)$. The results are shown in Fig. 57 (orientation of the crystal corresponds to Fig. 41B). As seen from the figure, the degree of polarization of the luminescence does not depend on the position of the electric vector in the exciting light. From this we conclude that excitation energy transfer takes place between naphthacene molecules separated from one another by a distance of 10d (d = anthracene lattice constant, mean value 8 A [11]).

Unfortunately, with crystals having a lower concentration of naphthacene, it proved impossible to separate the weak luminescence from the traces of exciting light. Figure 58 (curve 1) shows the result of an analogous experiment for a crystal with a naphthacene concentration of $2 \cdot 10^{-4}$ g/g. As we can see, there is a variation of $P_{lum}$ with $\eta$. But the question arises whether this may not be due simply to traces of exciting light. An answer to this can be given with the aid of the following control experiment.

As we have already indicated, in these experiments great significance attaches to the orientation of the crystal relative to the directions in which the mutually perpendicular components $I_1$ and $I_2$, which determine the degree of polarization of the luminescence, are polarized in the measuring apparatus. For example, for the orientation in which the direction $I_1$ coincides with the bisector of the angle between the oscillators, there should be no variation of $P_{lum}(\eta)$, even in the absence of energy migration. The sharpest variation of $P_{lum}(\eta)$ corresponds to the orientation in Fig. 41 B.

In order to determine what causes the variation of P observed in the last experiment for the orientation in which one of the oscillators is vertical, we must carry out a control experiment with the other orientation, in which the bisector of the angle between the oscillators coinciding with the crystallographic b-axis is vertical. These results are also shown in Fig. 58 (curve 2). There is the same kind of $P(\eta)$ variation as in curve 1. It follows that this variation is of no interest, being caused by nothing else but traces of exciting light. For excitation by light with wavelength $\lambda = 365\,m\mu$ (absorbable by anthracene), in the same crystal, the degree of polarization of the naphthacene fluorescence appears independent of $\eta$ (Fig. 58, straight line 3). This result causes no surprise; energy transfer from anthracene to naphthacene is already proved by the very ratio of the luminescent yields of these two substances, and needs no additional proof. But at the moment, this result interests us as to the methodology. It plainly shows that for sufficient brightness of the luminescence this method is feasible. For excitation of the naphthacene itself, even in such small concentrations, it is not suitable.

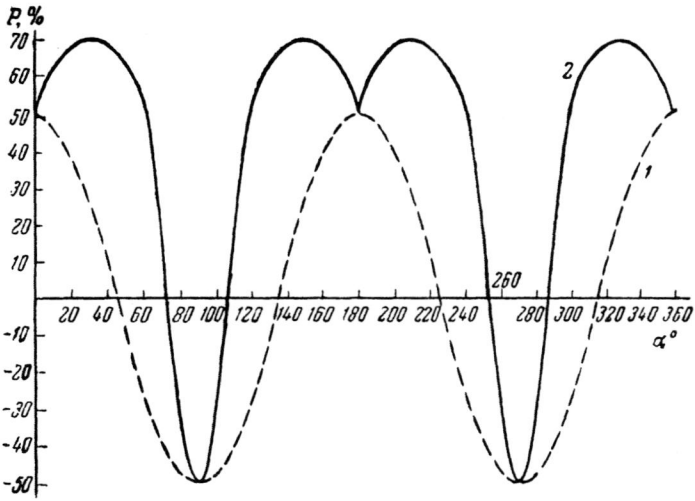

Fig. 59. Azimuthal variation of the degree of polarization of naphthacene luminescence in an anthracene-naphthacene crystal with surface ab for (1) natural and (2) vertically polarized excitation, calculated on the assumption that there is no migration; $\alpha$ is the azimuthal angle of the crystal.

In view of this, we developed a somewhat altered measuring method; this was based on the following considerations. The most troublesome aspect for us is not that the exciting light becomes mixed with the luminescence, but that its polarization changes. Let us consider what will happen if we fix the polaroid immovably in the exciting beam (say, with the electric vector vertical) and rotate the crystal itself around a horizontal axis. As far as the crystal-polaroid system is concerned, it all the same whether the crystal rotates with respect to the polaroid or vice versa. But for the measuring system it is not quite the same thing whether the crystal rotates or remains stationary. During the rotation of the crystal, two forms of variation of polarization may superimpose: the normal azimuthal variation, and another variation depending on the polarization of the excitation. The question arises how to separate these.

Separation may be achieved by comparing the experimental azimuthal variation with the calculated variation for natural and polarized excitation. If there is energy migration between the molecules, then these two cases of excitation will be identical in their results, and the azimuthal variations must coincide. If, however, there is no energy transfer between the molecules, then the probability of the excitation of differently oriented oscillators will depend on the azimuth angle of the crystal relative to the electric vector of the exciting light, which is immovable, and the azimuthal variation for linearly polarized and natural light must be different.

Figure 59 shows the relation calculated for this case between the degree of polarization of the luminescence and the azimuthal angle of the crystal for natural and vertically polarized excitation. Between these there is a sharp qualitative distinction, making it possible to distinguish this case from that in which energy transfer occurs between the molecules, for which there is no difference between the azimuthal variations for natural and polarized light.

The method described was used for measurements on crystals with naphthacene concentrations $2 \cdot 10^{-4}$ to $6 \cdot 10^{-5}$ g/g. The results appear in Figs. 60 and 61. It is seen from the figures that the form of the azi-

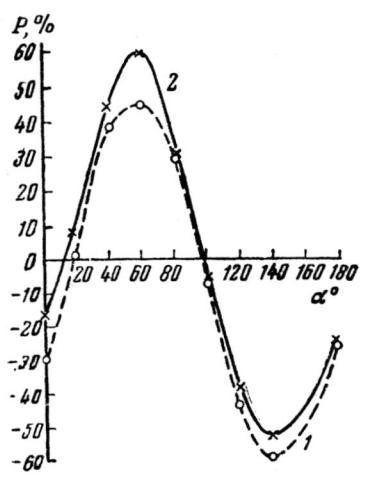

Fig. 60. Azimuthal variation of the degree of polarization of naphthacene in an anthracene crystal containing naphthacene concentration $C = 2 \cdot 10^{-4}$ g/g. 1) Natural excitation; 2) vertically polarized excitation; $\lambda_{excit} = 436$ m$\mu$; $\lambda_{lum} = 529$ m$\mu$.

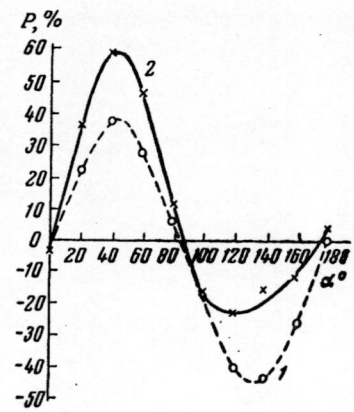

Fig. 61. Same as in Fig. 60 for a crystal with concentration $6 \cdot 10^{-5}$ g/g. $\lambda_{excit} = 436$ mμ; $\lambda_{lum} = 495$ mμ.

muthal variations are qualitatively entirely the same for natural and vertically polarized excitation, respectively. The fact that for the polarized excitation the curves are shifted, without distortion, along the ordinate axis in the direction of more positive polarization is due to the presence of traces of exciting light with constant polarization.

Thus these results show that, even with the concentrations in question, there is energy transfer between the naphthacene molecules present in anthracene lattice. Table 3 shows the calculated mean distances between the naphthacene molecules corresponding to the concentrations employed (the quantity d is the mean distance between the molecules in the anthracene lattice, i.e., the mean lattice constant from x-ray structural data [11], equal to 8 A).

As seen from Table 3, energy migration between naphthacene molecules takes place at quite considerable distances. The existence of free excitons in the naphthacene "sublattice" is impossible, if only because of the great mean distances between the molecules.

Only for very large impurity concentrations (order of percent or tens of percents), when the quantity of impurity becomes comparable with that of the base substance, have the so-called "impurity excitons" been observed (in certain recent investigations in Kiev [194,195]). As impurity, a deuterium-containing version of the base material was used, so that the properties (in particular energy levels) of base and impurity were very much the same. In the case of our substances and the concentrations which we used, the existence of free excitons in the naphthacene sublattice is completely excluded.

Might not the migration perhaps be caused by a resonance, inductive mechanism, as in liquid solutions?

The theory of the resonance transfer of energy [167,196,197]* gives, for the case under discussion, the following expression for the transfer probability per unit time for an excited molecule to an unexcited one situated at a distance R:

$$w\,(R) = \frac{9}{10\pi}\left(\frac{\bar{\lambda}}{2\pi n}\right)^4 \frac{\bar{\alpha}}{\tau_e}\frac{1}{R^6}\,,$$

where $\bar{\lambda}$ is the mean wavelength of the overlap region of the absorption $\alpha(\lambda)$ and emission $E(\lambda)$ spectra; n is the refractive index in this spectral region, $\bar{\alpha} = \dfrac{\int \alpha\,(\lambda)\,E\,(\lambda)\,d\lambda}{\int E\,(\lambda)\,d\lambda}$, and $\tau_e$ is the natural lifetime of the excited molecule. Hence we may calculate the distance $\bar{R}$ at which, for the excited molecule, there will be equal probabilities of emission and transfer. This distance will also determine the order of the spacing at which resonance transfer takes place.

TABLE 3

| Concentration of naphthacene in anthracene, g/g | Mean distribution between naphthacene molecules | |
| --- | --- | --- |
| | In multiples of lattice constant d | In A |
| $10^{-3}$ | 10 | 80 |
| $2 \cdot 10^{-4}$ | 17 | 135 |
| $6 \cdot 10^{-5}$ | 26 | 208 |

*See §1 of this chapter.

The absorption and emission spectra of naphthacene in anthracene overlap comparatively little. Data on these spectra were taken from [62] and [147]. The values of $\bar{R}$ calculated from these independent data were quite close (8 and 10 A). These lengths are considerably less than the actual distances at which migration takes place, according to our experimental results. Evidently the medium separating the naphthacene molecules, i.e., the anthracene lattice, takes part in the process of energy transfer between naphthacene molecules by means of "virtual" excitons [198].

We may also give another explanation for the experimental results, on the assumption that the naphthacene molecules occur in the anthracene lattice not singly but in complete cells, i.e., pairs of molecules with different orientations, or even more numerous groups, forming microcrystalline inclusions of naphthacene in the anthracene, inside which migration will take place as in ordinary one-component crystals. Against this assumption there are several objections.

First, it is known that crystalline naphthacene has a very small luminescent yield, practically none at all.

Secondly, interstitial pairs would imply that the insertion of a single molecule would lead to such local lattice distortion that the arrival of a second molecule alongside would be facilitated. We should accordingly expect a high solubility of naphthacene in anthracene, and the possibility of creating solutions of high concentration. In fact, however, naphthacene is only easily accepted in small concentrations (less than $10^{-2}$ g/g). It is also significant that in growing single crystals by sublimation, the concentration of naphthacene in the end-product (in the sublimed flakes), as determined by a spectral method, is always less than that in the original material from which sublimation took place.

These objections, however, are not entirely convincing. Thus the interpretation of the experiments described reduces to a dilemma. Either they must be explained by paired interstitial impurity molecules, or else they prove the existence of energy migration between impurity molecules at the distances indicated, where in the latter case the migration cannot be explained either by the resonance or exciton mechanism, and has some peculiar nature.

In concluding this section, we note that thin (order of tenth parts of a micron) crystals are also very valuable for these experiments in that they allow practically no reabsorption and secondary luminescence, which in principle may also complicate the results.

The effect of reabsorption is especially fully eliminated in the case of naphthacene in anthracene, where, in view of the negligible thickness, the low concentration, and the weak overlapping of the spectra, the influence of secondary luminescence is completely excluded. Moreover, in our other experiments also, this factor was taken into account and measures taken to insure that it should not noticeably affect the results. For example, in the work with the stilbene crystals cut perpendicularly to the optic axes, the crystal thickness was considerable, of the order of 1 mm. Reabsorption and secondary fluorescence cannot, in general, be neglected in this case. Their effect on the results must be estimated. As mentioned before, in this case the polarization was measured not for the total luminescence but for the spectrally resolved luminescence at $\lambda_{lum} = 414$ m$\mu$.

We made measurements of the stilbene absorption spectrum in [191].* The absorption at 414 m$\mu$ was very small. Depolarization of the luminescence as a result of the reabsorption dichroism is hence also small. However, reabsorption may lead to the same consequences as energy migration, i.e., to depletion of the initial anisotropy created by the polarization of the exciting light. The polarization of the secondary luminescence will not depend on that of the excitation. If the proportion of secondary luminescence in the total flux is great, this will considerably level out the expected variation of $P(\eta)$.

However, simple calculation with the aid of data on the absorption and luminescence spectra of stilbene shows that, in the 414 m$\mu$ region, secondary luminescence forms less than 10% of the total flux. Thus its inclusion cannot substantially reduce, far less completely level out, the expected sharp variation of the polarization of luminescence with that of the exciting light (see Figs. 52 and 53). Hence reabsorption certainly cannot lead

---

*More will be said of these data in Chapter V.

to the observed independence of $P(\eta)$ even in the case of thick crystals. Such independence can only be explained by means of energy migration.

## §5. Energy Migration in Molecular Crystals at Low Temperatures

The probability of excitation energy transfer in a crystal from one molecule to another by means, for example, of a localized exciton is determined by the exchange integral of the wave functions of the vibrational states corresponding to the localization of excitation in the two molecules [9]. In the case in which this integral vanishes, the excitation cannot pass over together with local deformation, and remains "frozen" in the one molecule. We shall approach this condition as the temperature falls. It is thus interesting to apply our polarization method to the study of energy migration in single component molecular crystals at low temperature.

On the other hand, it was shown in experimental studies of Lyons [199] and Faidish [176] that, with falling temperature, the efficiency of energy transfer in mixed crystals from the base substance to the impurity (from anthracene to naphthacene) diminishes. This diminution occurs even at liquid-nitrogen temperatures, but at liquid-helium temperature it is extremely significant. The transfer efficiency was studied in these investigations by the usual method for mixed crystals, i.e., from the ratio of the anthracene and naphthacene luminescent yields.

In view of this, it is interesting to apply our method also to mixed crystals at low temperatures, in particular to determine whether energy migration occurs between the impurity molecules themselves under these conditions.

We measured the variation of the degree of polarization of the luminescence with that of the excitation by the ordinary method for pure anthracene and anthracene containing naphthacene ($C = 10^{-3}$ g/g) in thin crystals, and also for a stilbene crystal cut perpendicular to the optic axis and excited along the optic axis at the temperature of liquid nitrogen. The crystals were placed directly in the liquid nitrogen, in a dewar with a transparent zone, through which the excitation was effected and luminescence measured. For stilbene and anthracene, the excitation was by means of light with $\lambda = 365$ m$\mu$.

In the mixed anthracene-naphthacene crystal, excitation was effected both in the absorption region of the base substance (anthracene) ($\lambda = 365$ m$\mu$), the polarization of the anthracene and naphthacene luminescence being measured, and also in the absorption region of the naphthacene impurity ($\lambda = 436$ m$\mu$), the polarization of

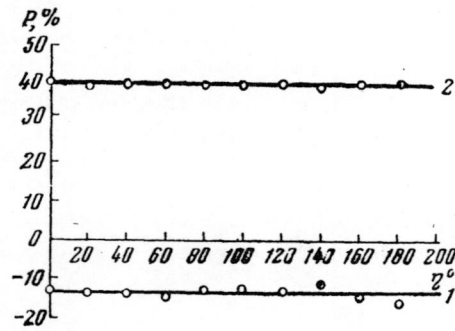

Fig. 62. Polarization of luminescence as a function of polarization of exciting light at −196°C. 1) Stilbene single crystal, $\lambda_{excit} = 365$ m$\mu$, $\lambda_{lum} = 403$ m$\mu$; orientation of crystal such that one of the oscillators is vertical; 2) anthracene single crystal 0.3 $\mu$ thick, $\lambda_{excit} = 365$ m$\mu$, $\lambda_{lum} = 419$ m$\mu$; orientation corresponds to Fig. 41 A.

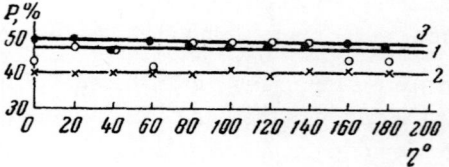

Fig. 63. Polarization of luminescence as a function of polarization of exciting light in an anthracene crystal containing naphthacene at −196°C ($C = 10^{-3}$ g/g). 1) $\lambda_{excit} = 365$ m$\mu$, $\lambda_{lum} = 419$ m$\mu$ (anthracene is excited and luminesces); 2) $\lambda_{excit} = 365$ m$\mu$, $\lambda_{lum} = 508$ m$\mu$ (anthracene excited, naphthacene luminesces); 3) $\lambda_{excit} = 436$ m$\mu$, $\lambda_{lum} = 495$ m$\mu$ (naphthacene is excited and luminesces).

the naphthacene luminescence only being measured. The results appear in Figs. 62 and 63. In all cases the polarization of the luminescence remained independent of the polarization of the excitation.

Thus there is quite efficient energy migration at a temperature of −196°C between the molecules of the main lattice, from these to the impurity, and also between the impurity molecules themselves at a distance of the order of 10 d.

## §6. On the Polarization of Luminescence from ZnS Single Crystals [200]

The methods of polarized luminescence may be applied to the study not only of molecular crystals but also of crystals of other types: for example, ionic.

We investigated the polarization of the luminescence from zinc sulfide single crystals activated by copper, which were prepared and kindly presented to us for investigation by E. I. Panasyuk [201]. The degree of polarization of the single crystals was measured by means of a micropolarimetric system.* Besides these, we had at our disposal several crystals about 12 mm in size which could be studied without a microscope. The single crystals were not very perfect, containing many internal cracks, and could not be examined crystal optically, the interference picture in a converging beam passing through the crystal placed between Nicols being quite obscure. However, the crystals emitted polarized luminescence. It is reasonable to suppose that the cracks were not accompanied by displacement of parts of the crystal and serious deformation.

The existence of polarized luminescence under natural excitation suggests that the centers of luminescence under natural excitation suggests that the centers of luminescence are anisotropic (whatever their nature may be), and that their orientation in the crystal is also at least partly anisotropic. Zinc sulfide crystals may belong to two systems, cubic and hexagonal; they also give mixed forms. In the cubic crystals (sphalerite), as a rule, there cannot be any polarization of the luminescence under natural excitation. There can only be the so-called "latent anisotropy" [202]. Hence polarization under isotropic excitation can in general only come from the hexagonal crystals (wurtzite).

The wurtzite structure of zinc sulfide comprises hexagonal close packed sulfur atoms, half of the tetrahedral spaces of which are occupied by zinc atoms. The interatomic distances (lattice constants) are the same for the a- and b-axes (3.81 A), but very different for the higher denomination c-axis (6.23 A) [203]. With such a difference in the lattice constants, we may unequivocally assert that the higher denomination axis will be the growth axis for the single crystals. Moreover, the exfoliation of the single crystal laminas will occur perpendicularly to the growth axis. This can be understood, since the greater interatomic distance corresponds to the weaker bond strength. Thus we may take it that the surface of the single-crystal laminas being studied is perpendicular to the c-axis.

On the other hand, it is known that crystals of the hexagonal system are optically uniaxial, the optic axis coinciding with the higher denomination c-axis (see, for example, [204]).

Thus there are good grounds for believing that by exciting the crystal in a direction perpendicular to the surface (coinciding with the direction of viewing the luminescence), we shall admit light to the crystal along the optic axis. Hence we can exclude double refraction of the exciting light and all the accompanying difficulties.

Measurement of the azimuthal variation of polarization on exciting by natural light enables us to find the orientation of the crystal for which the degree of polarization is maximum. This maximum value was of the same order for various crystal samples, approximately equal to 20%.

Both the azimuthal variation and the maximum degree of polarization were the same when the luminescence was excited by light of various wavelengths: 436, 405, 365, and 313 mμ.

There is one published paper [205] in which data are also given for the polarization of luminescence from activated ZnS crystals. In order of magnitude, the degree of polarization agrees with that which we measured,

*A description of this apparatus is given in Chapter VII.

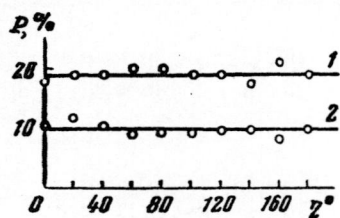

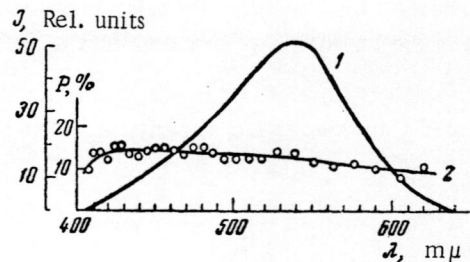

Fig. 64. Degree of polarization of the luminescence P from a ZnS single crystal as a function of the position ($\eta$) of the electric vector of the exciting light ($\lambda_{excit} = 365\,m\mu$). Straight lines 1 and 2 relate to two different orientations of the sample.

Fig. 65. Degree of polarization of the luminescence P from a ZnS single crystal as a function of luminescence wavelength. 1) Luminescence spectrum; 2) Degree of polarization.

but the author pointed out that his measurements were made with a lamina having the optic axis lying in its own plane. In viewing along the axis, no polarization was observed. There is yet another unexplained circumstance in this paper. It is clear that the absolute values of the positive and negative extrema of the azimuthal variation should coincide. This was fulfilled in our own experiments, but in the paper cited the data failed to agree with this. The author measured the azimuthal variation for polarized excitation, the electric vector of the exciting light being fixed and the crystal rotating. This gave the following results: for polarized excitation, the negative extremum was −3% and the positive +20%; for natural light, 12%. With the measuring arrangements used by the author, the cause of this is uncertain; it may be the result of crystal imperfection or variation of the polarization of the luminescence with that of the exciting light, or may be caused by an admixture of polarized light with the luminescence being measured.

We particularly investigated whether the polarization of the luminescence from ZnS single crystals depended on the position of the electric vector in the exciting light. In our measurements, the crystal was oriented similarly all the time, but rotating around an axis coinciding with the direction of excitation and viewing, was the Nicol through which the exciting light passed. The results of these measurements are shown in Fig. 64. These show that the polarization of the luminescence is independent of that of the exciting light. Furthermore, the degree of polarization is the same as that excited by natural light. At first glance this contradicts the Antonov-Romanovskii [206] proposition that, on excitation by polarized light, centers in a definite orientation are ionized (orientation determined by the position of the electric vector in the exciting light), and that electrons can recombine only with these centers, so that as a result the polarization of the luminescence must depend on the polarization of the excitation. But when we consider that, besides the migration of electrons, migration of holes takes place in a crystal phosphor (i.e., migration of states of ionization from center to center), the contradiction vanishes.

The fact that the polarization of luminescence from ZnS single crystals is independent of the polarization of the exciting light may thus be interpreted as the result of excitation energy migration about the crystal. But this result can give no indication as to the mechanism of this migration (for example, exciton or electron-hole).

Measurement of the degree of polarization over the luminescent spectrum of ZnS single crystals on the spectropolarimeter showed that, within the limits of experimental error, the polarization was practically the same over the whole spectrum (Fig. 65).

Qualitative observations in the Cornu polarimeter showed that the afterglow after the cessation of excitation had the same degree of polarization as the luminescence during excitation.

The results set out in this section show that polarization methods may prove useful even in studying inorganic crystals.

CHAPTER V

# THE PART OF FREE EXCITONS IN THE LUMINESCENCE
# OF MOLECULAR CRYSTALS

STUDY OF THE RELATION BETWEEN THE DEGREE OF POLARIZATION

AND THE WAVELENGTH OF THE LUMINESCENCE [185,191,207,208]

## §1.  Luminescence of Molecular Crystals and Free Excitons

In Chapter I we set out in detail the theory of free and localized excitons in molecular crystals.

In the case of free excitons, one transition in the molecule corresponds to two allowed transitions in the crystal, owing to the Davydov splitting. This splitting is caused by resonance interaction between the molecules, and by the fact that the molecules have different orientations in the elementary cell. The two terms formed in the crystal have different polarizations, but both are polarized along definite crystallographic axes. In the monoclinic system, one term is usually polarized so that the vibrations of the electric vector are parallel to the b-axis, and the other term perpendicular to the b-axis.

A free exciton is an excited state of the crystal as a whole, and for just that reason its polarization is determined by the symmetry of the crystal lattice. The polarization of transitions associated with localized excitons, however, is mainly determined by the orientation of the molecules in the lattice, i.e., it is connected with the oriented gas model. Hence arises the possibility of experimentally distinguishing between localized and free excitons from their polarization properties.

Generally speaking, there may be two relative dispositions of the terms for free and localized excitions: the former may be above the latter or the latter above the former. Certain considerations discussed by Rashba [20] regarding the terms of deforming excitations, and in equal measure valid for localized excitons, would suggest the first possibility. The decision must evidently rest with experiment.

A number of considerations, discussed in detail in Chapter I, lead to the conclusion that luminescence from states corresponding to free excitons should be restricted, especially at low temperatures.

As a result of the translational symmetry of the crystal, only those transitions are allowed for which the condition $k = Q$, where $k$ is the vector of the exciton and $Q$ the vector of the light, is fulfilled. For visible and ultraviolet light, the vector of the light wave may be neglected ($Q \sim 0$). Then the condition becomes $k = 0$.

There are theoretical indications that, as a rule, the bottom of the exciton band corresponds to a point $k \neq 0$, while the point $k = 0$ is situated some distance off. At low temperatures, excitons will be situated mainly at the bottom of the exciton band. States with $k = 0$ will be weakly filled. As a result of this, luminescence from free exciton states will be restricted. Even if transitions take place from points close to the bottom of the band and satisfying the selection rules, their intensity will be weak. With rising temperature we may expect the restriction on this luminescence to be lifted. The restriction may also be removed by de-excitation with simultaneous emission or absorption of a phonon, or at crystal lattice defects, the excess quasi momentum $\hbar(k-Q)$ being passed over to the defect or phonon.

From the theoretical considerations and reasons indicated, there arises a clear experimental problem: by using the polarization-spectral method of investigation, to try and observe whether free excitons take part in the luminescence of molecular crystals, and to study their role.

The study of polarization must extend over the whole luminescence spectrum in order to seek differences of polarization for different luminescent wavelengths. The sharp difference in the polarization of localized and free excitons would make us expect polarization in this case to be a reliable indicator, facilitating the observation of the presence or absence of radiation from the free exciton state, and to determine in what part of the spectrum this appears.

In Chapter I we discussed the work of Shpak and Sheka [32], who observed, in the luminescence of particularly pure naphthalene single crystals at low temperatures, narrow bands sharply polarized along the crystallographic a- and b-axes. These bands may be ascribed to radiation from free excitons. The method of these authors was imperfect in that their polarization observations were qualitative. By this method, one can only observe narrow and sharply polarized lines and bands in the spectrum, which is possible only at extremely low temperatures. On the other hand, it is just at these low temperatures that we may expect particular difficulty in the way of exciton luminescence. In studying complex spectra, one must use a method offering the possibility of quantitatively and fairly exactly measuring the degree of polarization corresponding to various wavelengths in the luminescence. Unfortunately there are hardly any such investigations in the literature. The need for such experiments was presented to us earlier in considering the polarization diagrams of crystals (see Chapter III, § 6).

The search for causes underlying the quantitative discrepancy between calculated and experimental polarization diagrams lead to the idea of measuring the degree of polarization over the luminescence spectrum, in order to establish whether there might not be substantial changes in certain spectral regions which might lead to the observed fall in the absolute magnitude of the degree of polarization of the total (not spectrally analyzed) luminescence. Such sharp changes may be expected if one assumes that radiation of free excitons appears in certain regions of the spectrum.

We can in fact easily understand that a fall in the calculated diagrams (more exactly, their approach to the axis of abscissas) may be expected without any accompanying distortion, if we assume that the total luminescence of the crystals contains not only luminescence with a space distribution corresponding to the oriented gas model but also luminescence polarized along one or other of the crystallographic axes. This second kind of luminescence will for certain crystal orientations be strongly polarized, but its polarization will not depend on the viewing direction.

Let us consider, for example, the positive stilbene diagram (crystal plane ab). In this case the a-axis is set almost vertically, and the b- and c'-axes lie in the horizontal plane. If there is any luminescence polarized along one of these axes b or c', then in measuring the polarization diagram this luminescence will give, for any direction, a negative degree of polarization (100%). It is evident that the addition in this way of polarized luminescence, which possesses polarization in accordance with the oriented gas model (luminescence of localized excitons), will lead to the experimental polarization diagram being lowered with respect to the calculated, but without substantial distortion. On rotating the crystal by 90° (that is, going over to the negative diagram of the same crystal), we obtain a polarization of +100% for the second kind of radiation for all directions of observation in the ac'-plane, if this radiation is polarized along the b-axis. As a result, the negative polarization diagram of the total luminescence will approach the axis of abscissas, in accordance with experiment.

By analyzing in this way the whole assembly of measured diagrams, we may establish the crystallographic axes along which the luminescence of the second type is polarized, and compare the result with that given by theoretical calculations using group theory, as set out in § 3 of Chapter I.

Analysis of all the stilbene polarization diagrams leads to the conclusion that in the luminescence there may be components polarized along the b-axis and also along the c'. This agrees with theoretical results of Lubchenko [13] (see § 3, Chapter I). These ideas may be confirmed or denied experimentally by studying the spectral variation of the polarization of luminescence with its wavelength. Thus our former experimental results have led us to the setting up of this new problem.

To this end, together with V. I. Gribkov, we constructed and assembled a spectropolarimetric photoelectric system based on the UM-2 monochromator and doubly refracting prism with a large splitting angle (up to 12°). The mutually perpendicular polarization components were detected by two photomultipliers, the photocurrents passed to a differential two-channel amplifier, and either each component separately or their difference measured directly by a sensitive meter. Thus the apparatus simultaneously measures the degree of polarization of the luminescence and its intensity.

A more detailed description of the apparatus, especially its optical part, and the method of using it are given in Chapter VII. The electronic part of the system was constructed and set up by V. I. Gribkov and is described in detail in his paper.

In the following sections we shall set out experimental results mainly obtained with this system. A number of the results already given in Chapter IV were obtained with the same system.

## §2. Variation of the Polarization of Fluorescence from Stilbene Single Crystals with the Wavelength of the Radiation

First of all, we shall present the results of experiments carried out at room temperature. The first investigations were conducted on a hemispherical single crystal (especially suitable for measuring polarization diagrams). The plane of the cut across the sphere was the ab. The orientation of the crystal was such that the a-axis was vertical and the b-axis horizontal. The crystal was excited from the plane side in the center of the sphere through a fine diaphragm by the mercury line 365 mμ. Measurement of the spectrum and spectral variation of polarization was made in a direction perpendicular to the ab-plane.

In the same sample we were able to measure polarization diagrams for various radiation wavelengths. In measuring the polarization diagrams the crystal was rotated around a vertical axis through various angles.

Figure 66 shows the luminescence spectrum and the spectral variation of polarization for this stilbene crystal. The spectral sensitivity of the apparatus was calibrated from a band-lamp with known color temperature. Calibration was troublesome in the very short wave part of the spectrum (around (360 mμ), so that the edge of the measured spectrum was only qualitatively adjusted to the spectral sensitivity.

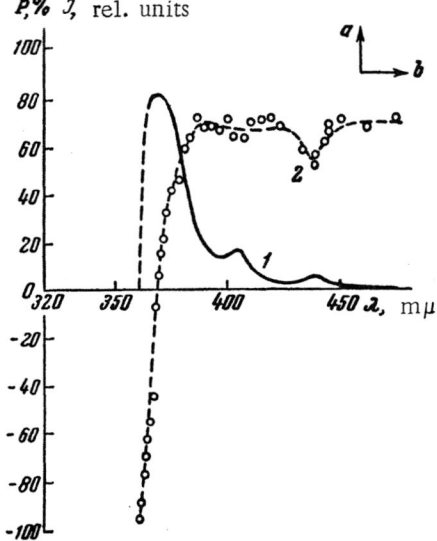

Fig. 66. 1) Luminescence spectrum; 2) spectral variation of polarization for a hemispherical stilbene single crystal. Orientations of axes are shown by arrows.

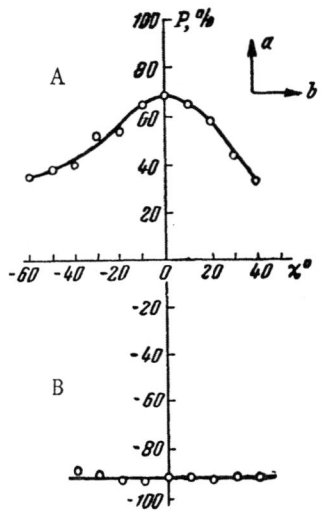

Fig. 67. Polarization diagrams of hemispherical stilbene single crystal for various radiation wavelengths. Cut through ab-plane. Orientation of axes shown by arrows. A) λ = 415 mμ; B) λ = 362 mμ.

It turned out that over a large part of the spectrum (in the long-wave region) the degree of polarization had a positive and approximately constant value (around +70%), but in the short-wave band, and especially at the short-wave edge, the polarization rapidly diminished, changed sign, and attained very large (almost limiting) negative values (up to −95%).

As already indicated, on the basis of an examination of the polarization diagrams for the total (not spectrally analyzed) luminescence, we considered the suggestion that the free excitons taking part in the luminescence of stilbene were polarized along the b- and c-axes. Thus, bearing in mind the orientation of the crystal in our experiment, we may conclude that the short-wave luminescence with polarization around −100% is caused by free excitons with polarization along the b-axis, and the long-wave luminescence by localized excitons. The polarization in the long-wave region in absolute magnitude tends towards the value calculated from the oriented gas model, though it is still quite a long way from this.

This conclusion may be checked by measuring the polarization diagrams in various parts of the spectrum. These diagrams are shown in Fig. 67 (for the short-wave edge $\lambda = 362 \, m\mu$ and the long-wave region $\lambda = 415 \, m\mu$). The results confirm our expectations. The short-wave radiation diagram shows independence of the observation angle, as in fact should be the case for luminescence polarized along the b-axis. The long-wave diagram has a form very well agreeing with that calculated from the oriented gas model, but is nevertheless still less than the calculated curve in the absolute magnitude of the polarization (see Fig. 34). Thus the polarization of the luminescence from localized excitons still differs from that given by the oriented gas model, though it retains the main features of this. This fact and the conclusions following from it were discussed in detail in §6 of Chapter III, which was devoted to the polarization diagrams of crystals.

Since the polarization of the long-wave luminescence does not coincide with calculated values, it is pointless to attempt a quantitative calculation of the relative intensity of the two forms of luminescence (from the states of the localized and free excitons) on the basis of the total and partial degrees of polarization. We can only state that the additional luminescence polarized along the b-axis is situated at the short-wave edge of the luminescence spectrum and forms an extremely small part of the whole luminescence, not more than a few percent.

These experiments accordingly lead to the following results:

1. They reveal the presence of luminescence from free excitons in the luminescence of molecular crystals.

2. They show that this luminescence is actually polarized along the optic axes. The identity of the axis agrees with that calculated by group theory methods.

3. This luminescence is situated at the short-wave edge of the spectrum, i.e., the level of the free excitons lies higher than that of the localized excitons, in agreement with theoretical considerations.

4. The intensity of this luminescence is very slight, which confirms the restrictions on exciton luminescence also predicted by theory.

These results are interesting and important. In view of this we made a detailed study of the phenomenon in question, taking account of all possible factors which might complicate it. First of all, we measured the spectral distribution of the polarization of the luminescence with excitation $\lambda = 365 \, m\mu$ for stilbene crystals cut in various crystallographical planes. These experiments were conducted on single-crystal, plane, parallel plates. The results are shown in Figs. 68-70. The polarization in the long-wave part of the spectrum was in all cases roughly constant and corresponded to the luminescence of localized excitons. The sharp changes of polarization at the short-wave edge of the spectrum corresponds, as clearly shown by comparing the results for various orientations of the crystals, to luminescence polarized along the b-axis and caused by free excitons.

At the same time, no luminescence was observed polarized along the c-axis. There is nothing very surprising in this, as the polarization in the long-wave part of the spectrum is still considerably less than would follow from the calculated polarization diagrams. Hence we could hardly expect that all the conclusions based on analysis of the polarization diagrams and the assumption regarding the two forms of luminescence should be completely justified.

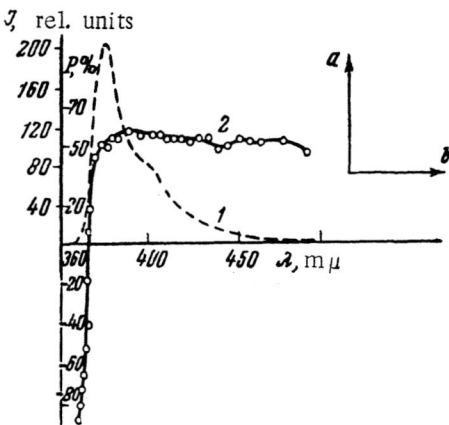

Fig. 68. 1) Luminescence spectrum; 2) spectral distribution of polarization for a stilbene single crystal. Cut ab. Excitation with $\lambda = 365$ m$\mu$. Arrows indicate orientation of crystallographic axes.

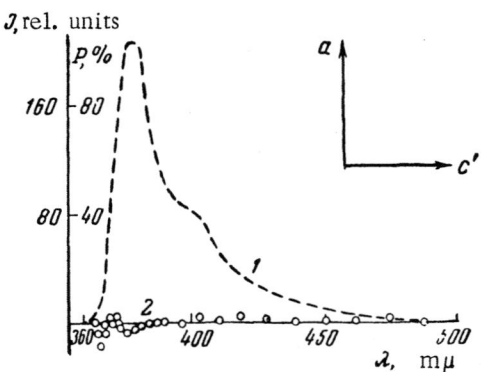

Fig. 69. Same as in Fig. 68. Cut ac'.

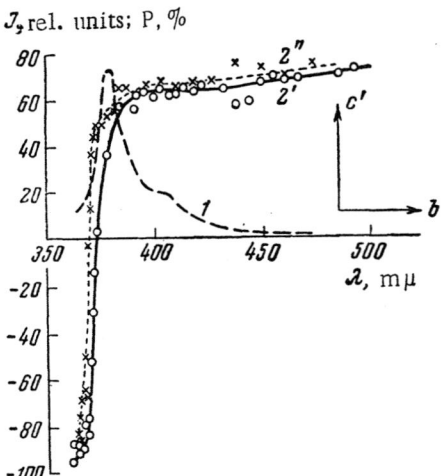

Fig. 70. Same as in Fig. 68. Cut bc'; 2') Natural excitation; 2") vertically polarized excitation.

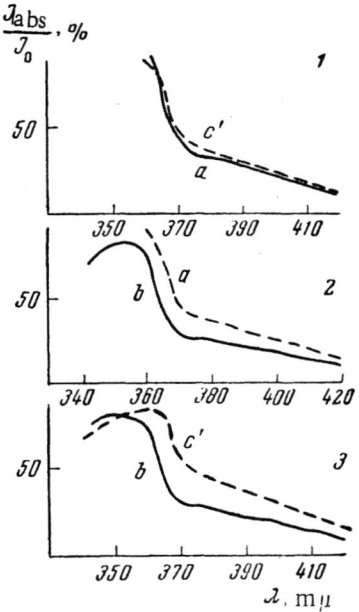

Fig. 71. Dichroism spectrum of stilbene crystals. 1) cut ac', d = 0.92 mm; 2) cut ab, d = 1.17 mm; 3) cut bc', d = 1.15 mm.

Nevertheless, luminescence polarized along the b-axis was observed in all the experiments listed; these were carried out on a large number of samples and were clearly reproducible. However, misgivings arise as to whether the observed effect may be caused (or distorted) by the dichroism of the crystals.

First of all, traces of exciting light might be introduced at the short-wave end of the luminescence spectrum. Direct exciting light could not fall into the slit of the monochromator, as checked by control experiments, but light scattered at the crystal surface might do so. The excitation was effected by natural light, but it could become partly polarized as a result of the dichroism of the crystal.

With the aid of the SF-4 spectrophotometer, we measured the dichroism spectra of the stilbene crystal plates under investigation. The results are shown in Fig. 71. Absorption in a region coinciding with the short-wave part of the luminescence spectrum was approximately the same for light polarized along the a- and c'-axes, but less for b. Hence dichroism must certainly be taken into account. In order to check how traces of exciting light might affect the situation, we made the following experiments.

First, we measured crystals of different thicknesses (thickness varying 10 times or more). It turned out the thickness had no effect on the relation between the polarization and the wavelength of the luminescence. Second, we effected the excitation with light linearly polarized in a direction perpendicular to the b-axis of the crystal (i.e., the polaroid extinguished the component of the exciting light passing preferentially through the crystal). The spectral distribution of the polarization was still unchanged (see Fig. 70, curve 2").

These experiments show that the effect cannot be caused by traces of exciting light. In order to be still more convinced of this, we conducted experiments with short-wave excitation (302 and 254 mμ); control experiments with nonfluorescent mat quartz plates convinced us of the complete absence of any traces of exciting light in the region of the luminescent spectrum. These measurements were quite difficult (small excitation intensity), but the sensitivity of our apparatus nevertheless allowed them to be made.

Certain data obtained, for example, for excitation wavelength 302 mμ and crystal plane bc' are shown in Fig. 72. Clearly the result is in general features the same as for excitation wavelength 365 mμ. In view of the vastly diminished intensity, it was here impossible to extend right up to the edge of the short-wave part of the spectrum as in the case of excitation by $\lambda = 365$ mμ. Thus we may take it that the sharp change in polarization near the edge of the spectrum is not linked with traces of excitation light. We must further take into consideration the reabsorption dichroism of the luminescence.

In order to introduce corrections for the change in the true polarization of luminescence as a result of reabsorption dichroism, we may use the well-known formulas for reabsorption corrections derived by Förster [102] and already applied to molecular single crystals [61]:

$$\rho = \frac{K_b}{K_{c'}} \rho',$$

where $\rho' = I_b/I_c$ is the measured polarization ratio [$P' = (1-\rho')/(1+\rho')$], $\rho$ is the ture polarization ratio [$P = (1-\rho)/(1+\rho)$], and P is the degree of polarization.

$$K_b = \frac{1 - e^{-\mu_{1b}d}}{e^{-\mu_{2b}d} - e^{-\mu_{1b}d}} \cdot \frac{\mu_{1b} - \mu_{2b}}{\mu_{1b}},$$

where d is the thickness of the crystal, $\mu_{1b}$ the absorption coefficient of the exciting light (302 mμ) polarized along the b-axis, and $\mu_{2b}$ is the absorption coefficient of the luminescent light at the given point of the spectrum, polarized with respect to b.

The expression for $K_{c'}$ is entirely analogous. These formulas relate to the case in which the luminescence of the crystal is excited from behind. From these formulas corrections were calculated in the reabsorption region. The absorption coefficient for 302 mμ for stilbene was taken from [46]. The spectral vibration of polarization at the edge of the luminescence spectrum, making allowance for the corrections in question, appears in Fig. 72 (curve 2). It is seen from this that the sharp change in polarization at the edge of the spectrum remains even after introducing all the corrections and removing all the complicating factors mentioned. As already discussed, this fact indicates that luminescence from free excitons is present in the stilbene crystal luminescence.

The question arises as to why only radiation polarized along the b-axis appears in the luminescence, and none polarized perpendicular to the b-axis. It is difficult at the

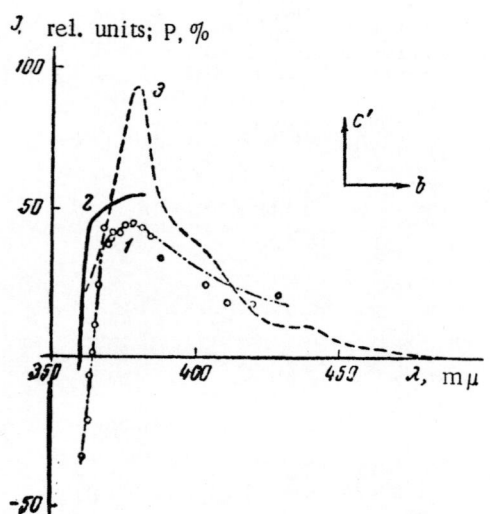

Fig. 72. Spectral distribution of the polarization of luminescence from a stilbene crystal. Cut bc', excitation with $\lambda = 302$ mμ. 1) Measured; 2) with corrections of dichroism; 3) luminescence spectrum.

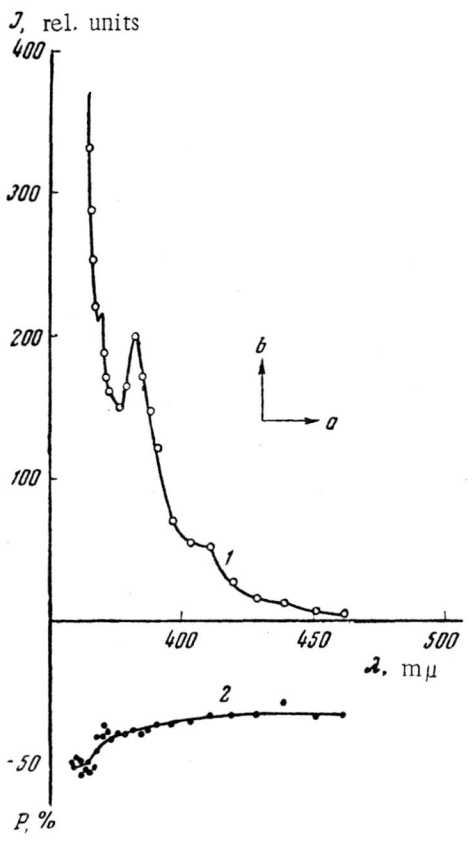

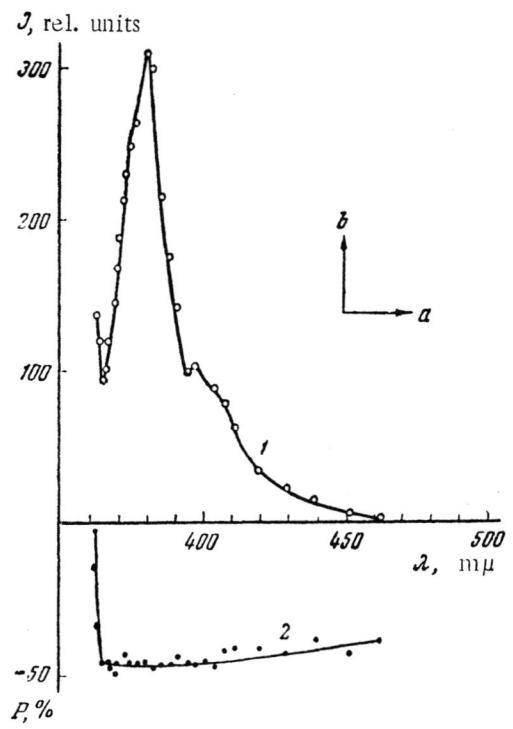

Fig. 73. 1) Luminescence spectrum and 2) spectral variation of polarization of a stilbene single crystal at liquid nitrogen temperature. Surface of the plate is the ab-plane; $\lambda_{excit} = 313$ m$\mu$. Orientation of the axes shown by arrows.

Fig. 74. Same as Fig. 73, for the same crystal at room temperature.

moment to answer this fully. It may be that the component polarized ⊥b is simply of low intensity. The intensities of the Davydov splitting components for stilbene have not been calculated. In general their ratio may vary.

It is also possible that here an essential part is played by the orientation of the molecular oscillators. As shown by Davydov (see § 3, Chapter I), the splitting components of the molecular term only both appear when the projections of the radiating molecular oscillators with different orientations to the corresponding crystallographical planes form angles different in sign and magnitude with the crystallographic axes.

In the case of stilbene this certainly happens for the ac' cut. The projections of the oscillators are disposed on different sides of the a-axis and make a considerable angle with each other (64°). For the other cuts this angle is small: thus for the ab cut it is only 9°, and the projections of the oscillators lie on the same side of the a-axis. Here we can expect only one splitting component, polarized along one of the axes. Evidently it is just this which offers the possibility of observing it experimentally, since the splitting is small, and at room temperature the separate transitions are spread out into a complex spectrum, so that the bands overlap, and, if they have linear polarization along mutually perpendicular axes, unpolarized luminescence will be observed in the given spectral region (cut ac'). If, however, only one of the splitting components appears in the luminescence, then linear polarization may be observed. In view of these and a number of other considerations, it seemed interesting to pursue these investigations at low temperatures as well.

Experiments were carried out at liquid nitrogen temperature. The crystals were placed directly in the liquid nitrogen in a dewar with a transparent zone. In these experiments the cooling involved special difficul-

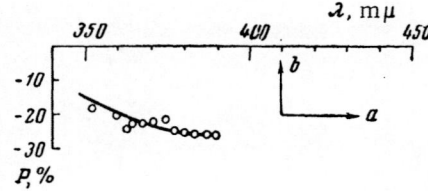

Fig. 75. Short-wave edge of the spectral distribution of polarization of the luminescence from a stilbene single crystal (ab-plane) at liquid nitrogen temperature according to spectrophotographic data, $\lambda_{excit} = 313$ m$\mu$.

ties, since the cracking of the single crystals had an effect on the magnitude of the polarization, though not harming the spectral investigations.

The method developed for cooling the crystals with due allowance for the circumstance is described in Chapter VII. Measurements were made in the spectropolarimetric system.

Figure 73 shows the spectral variation of the polarization of luminescence from a stilbene crystal (surface parallel to the ab-plane) at liquid nitrogen temperature for excitation by $\lambda = 313$ m$\mu$. At this temperature the following facts are observed (cf. Fig. 74): the luminescence band with $\lambda_{max} = 380$ m$\mu$ is weakened, a new strong luminescence band arises, and (this is especially interesting for us) there is a complete disappearance of the sharp change in polarization at the short-wave edge of the luminescence spectrum which is always observed at room temperature and has a tendency to change the sign of the polarization. For control, we repeated the former experiments with excitation $\lambda = 313$ m$\mu$ on the same sample at the same position in the dewar, but at room temperature after the evaporation of the nitrogen. The results are shown in Fig. 74, from which it may be seen that the former luminescence spectrum and sharp change of polarization at the short-wave edge of the spectrum are completely restored at room temperature.

In our apparatus it was impossible to measure the extreme short-wave edge of the luminescence spectrum; we could only measure the long-wave drop of the second band; the apparatus could not be calibrated any further (calibration was effected from the spectrum of a bandlamp with known color temperature). In measuring the polarization we were able to go as far as $\lambda = 359$ m$\mu$. In order to measure the extreme short-wave edge as well, we photographed the spectra of the same sample in a quartz dewar with optical windows at room temperature and liquid nitrogen temperature by means of a Fyuss quartz spectrograph. Excitation was carried out with lines 313 or 302 m$\mu$. The control experiments showed that the excitation light did not interfere in the region of the luminescence.

Figure 75 shows the spectral variation of the polarization at liquid notrogen temperature calculated from the density of the spectra photographed through an analyzer oriented in the one case vertically and in the other case horizontally. The density was measured in the microphotometer.

For calculating the degree of polarization (a relative quantity), graduation of the apparatus from a source with known spectral distribution was not needed. The precision of the spectrographical method, as we know, is not great. We repeatedly photographed the components of the spectrum and measured each spectrum on the microphotometer several times. The data presented are the mean results of several tens of independent measurements.

It is seen from Fig. 75 that right up to the very edge of the spectrum there is no sharp change in the polarization of the luminescence. The absolute value of the polarization in the long-wave region of the spectrum in this experiment appeared to be somewhat lower than in the previous apparatus. Evidently this is a chance circumstance. It may be explained, for example, by the not quite exact orientation of the crystal in the dewar.

Thus the sharp change of polarization at the edge of the spectrum, linked, as we supposed, with the luminescence of free excitons, vanishes at low temperatures.

It has already been indicated that there are several theoretical data confirming the difficulty of obtaining exciton luminescence at low temperatures.

The vanishing of exciton luminescence as temperature falls may also be explained by means of a concept developed by Rashba [19,20].

In Fig. 3 (Chapter I), taken from Rashba's work, appear potential curves of ground and excited states, the latter corresponding to deforming excitations and also to localized excitons (P) and free excitons ($P_0$). Transitions on excitation into a $P_0$ state (exciton absorption) will be accompanied by radiationless transitions into the

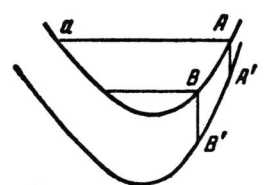

Fig. 76. Potential curves of ground and excited states.

P state. The reverse transitions are possible at high but not low temperatures. Thus lowering the temperature should actually lead to the vanishing of free exciton luminescence.

On the other hand, lowering the temperature changes the form of the luminescence spectrum; it weakens the band with maximum 380 m$\mu$, while the new, short-wave band increases. This circumstance may be unconnected with reverse transitions to $P_0$ and transitions with radiation from this state. The short-wave shift with falling temperature may be explained by considering only the potential curves of the ground and one excited (deforming excitation or localized exciton) state. Analogous behavior of the fluorescence spectrum with falling temperature is found in a large number of fluorescent substances, both in solution and in crystals. We proposed [86] the following explanation of this phenomenon.

The graphical scheme of the radiation process is usually represented by means of potential curves. This method is strictly valid only for diatomic molecules. For polyatomic molecules one must operate with multi-dimensional potential curves. For a qualitative description of the phenomenon, however, we may use a plane sketch, which in this case has the sense of a cross section through the potential surfaces by some coordinate plane corresponding to the most important configuration parameter.

Since the molecule possesses different electron energies in the ground and excited states, and the bonds between the atoms have an electron nature, the potential curve of the excited state, generally speaking, is always deformed in comparison with the potential curve of the ground state, and their curvature differs. Experimental data show that almost always this difference must be similar to that in Fig. 76, the ground state having the greater curvature, i.e., it rises more sharply and the curves converge. This, in particular, explains such a general phenomenon as the temperature quenching of luminescence. At room temperature, the molecules have an equilibrium distribution with respect to the vibrational levels of the excited state, schematically represented by level aA. At low temperatures the equilibrium distribution will be different, and corresponding to this the level bB lies lower. This will lead to an increase in the energy of the radiated quantum (BB' instead of AA'), i.e., to a shift of the fluorescent spectrum to the short-wave side.

The existence of a shift in the fluorescence spectrum of stilbene crystals to the short-wave side on lowering the temperature from room to −180°C was first observed by Lewis and his colleagues [209].

## §3. Variation of the Polarization of Fluorescence with the Wavelength of the Radiation for Anthracene Single Crystals and Certain Other Substances

The luminescence spectra of anthracene crystals have been studied by many authors [37,38,39,61,71,72, 74, etc.], but quantitative measurement of the spectral variation of the degree of polarization over the whole luminescence spectrum has not been carried out. It is true that in [38, 39, and 61] the mutaually perpendicular components of the spectrum were photographed through an analyzer, but in [61] the short-wave edge of the spectrum was not measured at all, and in [39] the result was obtained only qualitatively. In [38] the polarization ratio was measured at seven points of the luminescence spectrum (the shortest wavelength was $\lambda = 400$ m$\mu$, with P = 80%). It was noted in both [38] and [39] that the violet edge of the luminescence spectrum was strongly polarized, the dominant component being directed along the b-axis. An analogous result was noted in [72, 74], in which the photographic method was also used.

We carried out quantitative measurements of the variation of the degree of polarization with the radiation wavelength over the whole luminescence spectrum right up to the extreme short-wave edge. The measurements were made with the same spectropolarimetric apparatus on single-crystal plates obtained by sublimation in closed crucibles. The surface of the plates was parallel to the ab-plane. The measurements were effect at room and liquid notrogen temperatures.

It was considerably simpler to cool anthracene crystals than stilbene. Evidently on account of their small thickness they will not crack even on sudden cooling. Hence in this case cooling could be carried out without the precautions found most necessary in the case of stilbene.

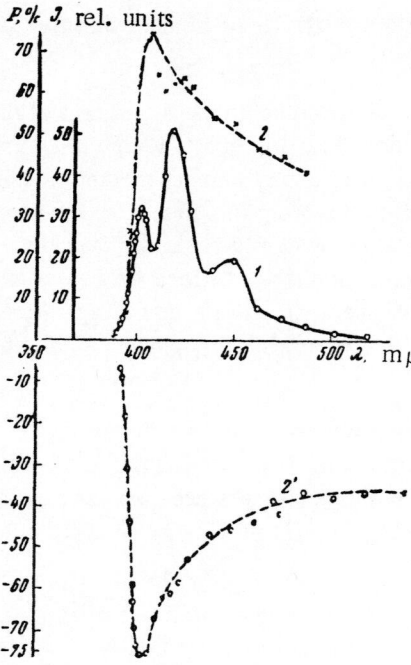

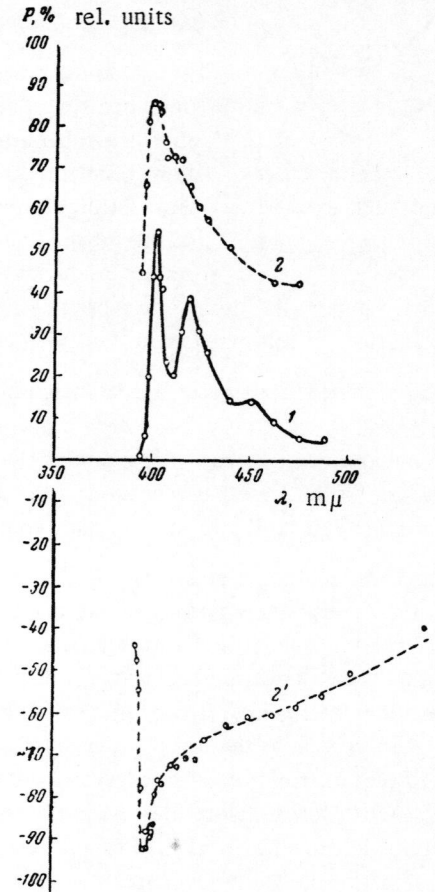

Fig. 77. 1) Luminescence spectrum and 2) spectral variation of polarization for anthracene at room temperature. Polarization spectra 2 and 2' relate to two orientations of the crystal differing by an angle of 90°; the first corresponds to maximum positive and the second to maximum negative degrees of polarization. Excitation from in front with light λ = 365 mμ.

Fig. 78. Same as Fig. 77, for liquid nitrogen temperature.

The results are presented in Figs. 77 and 78. The upper polarization spectrum corresponds to the orientation of the crystal for which the b-axis is vertical and the a-axis horizontal; the lower spectrum differs from this by 90°.

The variation of polarization with wavelength in anthracene is of a more complex character than in stilbene. In the region of the short-wave band, a sharp polarization maximum is noticeable. At room temperature this reaches 75%, but at low temperature it becomes still sharper and reaches 90%. Then, at the short-wave end of the band, the polarization falls almost to zero at room temperature, but not quite so strongly at low temperature. We can try to explain these results in the following way.

The long-wave bands in the anthracene spectrum are caused by localized excitons and the short-wave band by free excitons. Free excitons with transitions polarized along the b- and a-axes take part in the radiation. To these correspond different (though close) excited levels. The first is situated below the second, but both are above the localized exciton level. The bands polarized along a and b overlap considerably, so that the polarization at the very end of the spectrum does not reach large negative values, but only falls to zero. At high temperature the transitions from the level corresponding to the a-axis are more efficient (i.e., thermal transitions from b to a are efficient); this partly explains the lowering of the polarization diagram of the total luminescence in comparison with that calculated from the oriented gas model.

At low temperature, it appears that redistribution takes place between the intensities of these bands, and that polarized along b becomes the more effective (analogously to the behavior of the α and β bands in the persistent luminescence of complex molecules at high and low temperatures).

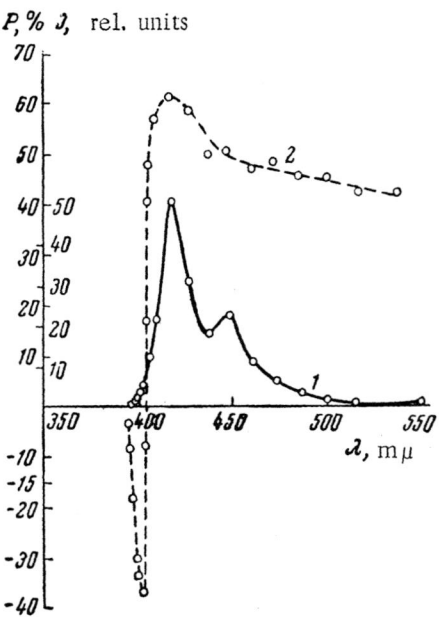

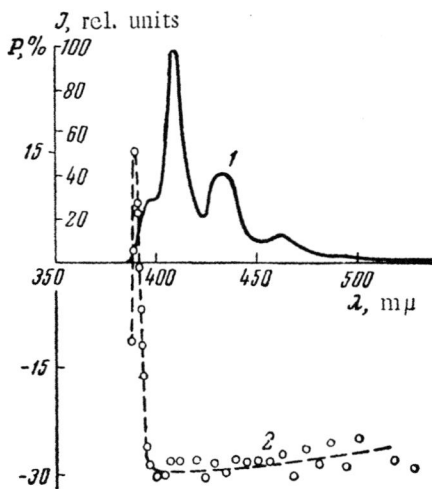

Fig. 79. Same as Fig. 77, for room temperature and excitation from behind.

Fig. 80. 1) Luminescence spectrum and 2) spectral variation of polarization of a phenanthrene crystal at room temperature. Orientation corresponds to maximum negative polarization. Excitation from behind with natural light, $\lambda = 365 \, m\mu$.

In the case of anthracene, in contrast to stilbene, the sharp polarization changes at the edge of the spectrum, as seen from the results shown, do not vanish at low temperature. This may be explained by the fact that here, as may be easily calculated from the position of the bands in the spectrum, the lowering of the level of localized excitons relative to those of free excitons is roughly a factor of two less than in stilbene, or less than the height of the potential barrier separating the minima of the potential curves for the excited states corresponding to free and localized excitons ($P_0$ and P in Fig. 3, Chapter I).

We must consider the possible effect of reabsorption on the experimental results obtained. At first glance it seems that reabsorption in the anti-Stokes region cannot influence the polarization of the luminescence, since in the region of the observed sharp polarization changes the luminescent yield is close to unity [52] and begins to fall only from 420 $m\mu$ to the long-wave side.

Thus the reabsorption energy passes almost entirely into secondary luminescence having the same polarization as the primary. The polarization of the luminescence may, however, change on account of the reabsorption dichroism. The dichroism of anthracene has been studied by various authors [39,190], their data being in good agreement. The dichroism is large in a narrow spectral range, 400 to 398 $m\mu$ (absorption along the b-axis predominates), and falls to zero at 396 $m\mu$. As a result of this, when the b-axis is vertical the luminescence may be depolarized; the positive component falls away in the primary luminescence more effectively than it appears in the secondary. Simple calculation shows that for a single complete reabsorption and subsequent secondary luminescence the degree of polarization should fall from 60 to 28%. However, comparing the polarization spectrum with the dichroism spectrum shows that the main cause of the phenomena observed does not lie in this. In anthracene, as shown above, the dichroism falls to zero at 396 $m\mu$; if the main cause of the change in the polarization of the luminescence lay in this, then the polarization should not change at this and shorter wavelengths, and there would only be a dip in the spectral variation of the polarization of the luminescence, corresponding to the dichroism band. The results given above (obtained for excitation of the crystals from in front) clearly show that this is not so. Control experiments with excitation in transmission showed the effect of dichroism appearing in the form of a dip in the polarization spectrum at 400 $m\mu$, having the same form as for excitation from in fron (Fig. 79).

Thus dichroism cannot be the cause of the observed facts, though its possible effects must be carefully considered.

Traces of exciting light with $\lambda = 365\ m\mu$ in the case of anthracene may reliably be separated out by filters, since this wavelength lies a fair way beyond the limits of the luminescence spectrum.

An analogous sharp change of polarization at the edge of the spectrum is also found in a phenanthrene single crystal belonging to the same monoclinic system and containing four molecules in the elementary cell.

The results for phenanthrene are shown in Fig. 80. The constancy of the polarization over the whole long-wave part of the spectrum and its sharp change at the short-wave end also leads to the supposition than in this part of the spectrum we have a small trace of luminescence polarized along one of the crystallographic axes, i.e., luminescence corresponding to transitions from a free exciton state.

It is interesting to note some analogy between molecular crystals and certain Scheibe polymers. Our results show that luminescence from free excitons, characteristic of the molecular crystal as a whole and not associated with individual molecules, takes place at the short-wave edge of the spectrum, i.e., in the region of resonance luminescence. Scheibe [210] also observed luminescence characteristic of polymers (one-dimensional crystals) in the form of a resonance peak.

Thus this effect evidently appears more strongly in one-dimensional crystals.

## §4. Variation of the Polarization of Luminescence with Wavelength of the Radiation in Mixed Molecular Crystals

In view of the results obtained in studying the spectral variation of the polarization of luminescence from pure single-component molecular crystals and their interpretation, it is of great interest to conduct analogous

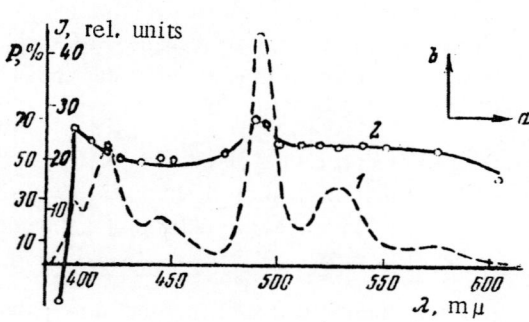

Fig. 81. 1) Luminescence spectrum and 2) spectral variation of polarization of a mixed anthracene-naphthacene crystal. Plane of the crystal ab, b-axis vertical. Naphthacene concentration $C_{naph} = 6 \cdot 10^{-5}\ g/g$.

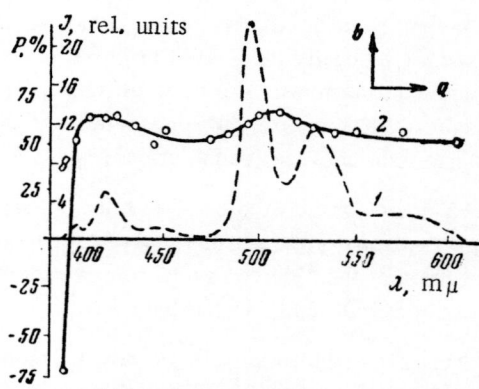

Fig. 82. Same as Fig. 81, for $C_{naph} = 2 \cdot 10^{-4}\ g/g$.

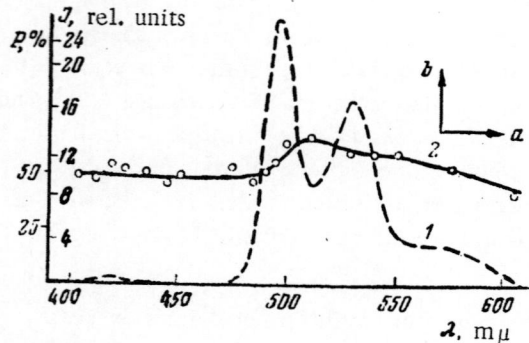

Fig. 83. Same as Fig. 81, for $C_{naph} = 10^{-3}\ g/g$.

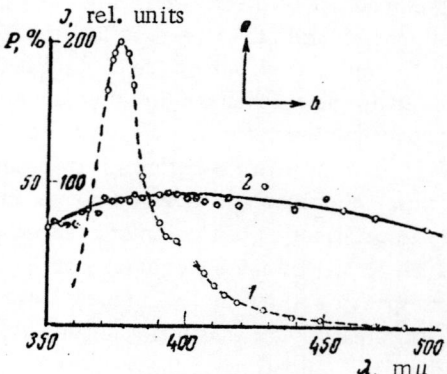

Fig. 84. 1) Luminescence spectrum and 2) spectral variation of polarization of tolan containing stilbene. Plane ab. Orientation of the axes shown by arrows.

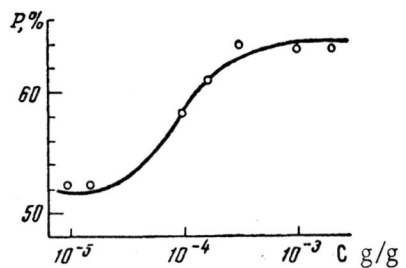

Fig. 85. Degree of polarization of the fluorescence of anthracene as a function of naphthacene concentration.

investigations of mixed crystals, i.e., crystals containing luminescent impurities. These investigations may be regarded as control experiments, in the sense that free excitons are impossible in the "interstitial" impurity lattice, especially for low concentrations.

As subjects for these experiments we selected anthracene containing naphthacene and tolan containing stilbene. The results for the anthracene-naphthacene with various naphthacene concentrations ($6 \cdot 10^{-5}$, $2 \cdot 10^{-4}$, and $10^{-3}$ g/g) are shown in Figs. 81, 82, 83.

With increasing naphthacene content, the intensity of the luminescence of the anthracene itself rapidly falls. For the greatest concentration used ($10^{-3}$ g/g), this intensity was so small that it proved impossible to measure the polarization at the extreme short-wave edge of the anthracene spectrum. For the other concentrations, a sharp change in the degree of polarization of the luminescence was observed at the edge of the anthracene spectrum, similar to that found in pure anthracene crystals.

The degree of polarization in the naphthacene impurity luminescence has approximately the same value as in the long-wave part of the anthracene spectrum. The total (not spectrally analyzed) naphthacene luminescence usually has a rather lower degree of polarization than the total anthracene luminescence, evidently because of the increase in polarization for the short-wave anthracene band. In any case, this leads to the conclusion that the orientation of the interstitial naphthacene molecules does not differ substantially from the orientation of the anthracene molecules in the main lattice. A certain slight rise of polarization is noticed in the region of the first naphthacene band. It is most interesting that no sharp changes of polarization are observed at the short-wave edge of the naphthacene spectrum for any of the concentrations studied, right up to $10^{-3}$ g/g, similar to those found in the polarization spectra of the pure crystals.

If we compare this fact with the former argument against the existence of free excitions in the naphthacene "sublattice," which is devoid of translational symmetry (especially at such low concentrations), we find support for the above suggestion that the sharp change in the polarization at the edge of the luminescence spectrum is associated with a slight amount of free exciton luminescence in this region. Analogous results were obtained for tolan containing stilbene.

As already indicated in §4 of Chapter III, it is extremely probable that the luminescence spectrum observed always belongs to the stilbene impurity, from which it is almost impossible to purify the tolan [50]. The difficulty in separating these is explained by the similarity of their molecular structure and the structure of their lattices. From the point of view of the polarization diagrams, this hypothesis cannot be refuted.

Figure 84 shows results of measuring the spectral variation of the polarization of the luminescence of a tolan single crystal. The luminescence spectrum is approximately the same as for stilbene. No sharp changes in polarization at the short-wave edge were observed, however, though the sensitivity of the apparatus enabled us to go right to the very end of the spectrum. This result is similar to that found for naphthacene in anthracene, also confirming the interpretation of the sharp changes in polarization at the end of the spectrum discussed above.

With mixed crystals of anthracene with naphthacene, yet another experimental investigation was undertaken; the results also support the interpretation in question.

If we accept, in agreement with Arganovich [28], that in activated crystals of the anthracene-naphthacene type the excitation energy transfer from the main substance to the impurity is effected by free excitons, then we may expect that the (even without this) small contribution of free exciton luminescence will be slightly reduced, whereas the relative intensity of the localized exciton luminescence will tend to rise. Hence the degree of polarization of the luminescence from the main substance should somewhat increase with increasing impurity content, approaching the value calculated from the oriented gas model.

Experiment confirms this.  Figure 85 shows the variation of the degree of polarization of anthracene luminescence with naphthacene content.  A considerable rise in polarization occurs up to a concentration of $5 \cdot 10^{-4}$ g/g, after which it slows down and tends to saturate.  This is natural, as in this region of concentrations the naphthacene fluorescent yield also attains saturation [146].

CHAPTER VI

# POLARIZATION OF RADIOLUMINESCENCE IN MOLECULAR CRYSTALS AND EFFECT OF IONIZING RADIATIONS ON THE POLARIZATION OF THEIR PHOTOLUMINESCENCE [211, 212]

Since the polarization of the luminescence of molecular crystals is independent of the anisotropy of the excitation (i.e., the polarization and direction of the exciting light), being determined by the anisotropy of the crystal itself, we should expect that the luminescence of molecular crystals would be polarized not only on photoexcitation but also for other kinds, for example, hard excitation (radioluminescence).

We conducted certain investigations into the polarization of luminescence from molecular crystals on excitation by β-rays. The β-ray source was a radioisotope of strontium $Sr^{90} \to Y^{90}$ with activity 100 mCi, the mean β-electron energy bein 0.9 MeV and the maximum 2.2 MeV. First of all, for the stilbene single crystal, we made a comparison of the polarization of luminescence with photo- and β-excitation for the integral luminescence (not spectrally analyzed). For the same crystal and the same orientation, the degree of polarization, as measured in the Cornu polarimeter, was exactly the same (within the limits of experimental error) for photoexcitation by natural light with $\lambda = 365$ mμ and for β-excitation: 32.7 and 33.1% respectively.

If the polarization of luminescence from crystals corresponded to the "oriented gas" model, i.e., were determined entirely by the structure of the lattice and the orientation of the molecules in it, the fact that the polarization is independent of the nature of the excitation would be trivial. Results given in the preceding chapters, however, show that matters are otherwise: the "oriented gas" model is not in full agreement with experimental results, polarization being associated with exciton processes in the crystal. It is therefore interesting to study in detail the polarization of luminescence from molecular crystals for various forms of excitation, in particular the spectral variation of polarization.

We studied this variation for stilbene and anthracene under β-excitation. The source was the same strontium isotope $Sr^{90}$ with activity 100 mCi. Measurement was made in the same spectropolarimetric system. These demands we were able to satisfy by going to the limit of the light power in the apparatus, by careful adjustment, placing the crystal right up to the entrance slit of the monochromator, using photoelectron multipliers of the greatest sensitivity and small dark current specially chosen from a large selection, and so forth.

It was foreseen that the apparatus could be used for successive measurements, with the same sample, of β-luminescence and photoluminescence for various exciting light wavelengths. Reliable shielding and remote control were used in the measurements. The following experimental results were obtained. In stilbene under photoexcitation, as already discussed, there was a sharp change of polarization at the short-wave end of the spectrum, associated with free exciton emission. It is important to note in this connection that for excitation by light of wavelength 365 mμ this change in polarization appeared far more significant than for shorter-wave excitation (313 and 254 mμ). We may deduce that in the latter case the part of the radiation caused by free excitons diminishes.

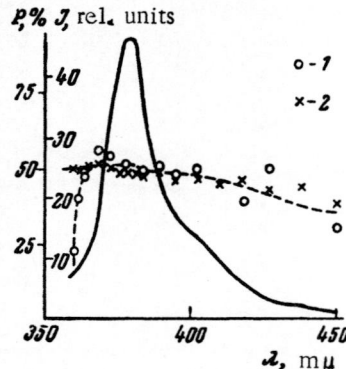

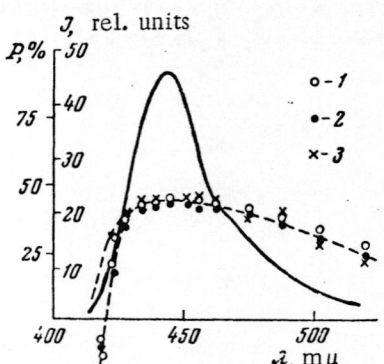

Fig. 86. Degree of polarization as a function of the wavelength of the luminescence for a stilbene crystal. 1) Excitation by light of wavelength 313 mμ; 2) β-excitation; continuous curve = luminescence spectrum. Crystal surface = ab plane (b-axis horizontal).

Fig. 87. Degree of polarization as a function of wavelength of the luminescence for an anthracene crystal. Crystal surface = ab plane (b-axis vertical). 1) Excitation by light of wavelength 365 mμ; 2) 313 mμ; 3) β-excitation; continuous curve = luminescence spectrum.

Figure 86 shows a graph of the variation of the degree of polarization with the wavelength of the luminescence from a stilbene single crystal under β-excitation (and, for comparison, under excitation by light with λ = 313 mμ). The crystal plate was cut parallel to the ab plane. It is seen that for β-excitation there is no change at all in the polarization at the end of the luminescence spectrum, i.e., the free excitons hardly radiate. In the whole long-wave part of the spectrum, however, the polarization is just the same as for photoexcitation. Several samples were measured, and the results were quite reproducible.

Figure 87 shows analogous data for anthracene. These experiments could not be performed with the usual single-crystal anthracene films obtained by sublimation; the thickness of these films was only a few microns, and the β-luminescence excited in these was of too low intensity. We used plates sawn off parallel to the ab plane from large single crystals. As with stilbene, the plates were of the order of 1 mm thick.

As already indicated, in these experiments with anthracene it was necessary to take account of the dichroism of the crystals. In order to reduce the effect of dichroism, it was distinctly better to excite the crystals from the side from which the luminescence was viewed rather than the other. However, this was experimentally impossible in the case of β-luminescence.

At the same time, the polarization of the β- and photoluminescence had to be compared in identical conditions. In the present case the difficulty was aggravated by the great thickness of the crystals. The short-wave part of the luminescence spectrum becomes considerably distorted owing to reabsorption. In order to avoid this difficulty, in Fig. 87 we show the polarization only in that region of the luminescence spectrum for which anthracene is transparent (roughly down to 415 mμ) but for which there is already a change in the polarization, since the effect under photoexcitation was studied in detail earlier, and at the moment we are simply interested in comparing it with β-excitation. It is seen from Fig. 87 that for anthracene the state of affairs appears somewhat different from that of stilbene. On passing from photo- to β-excitation, the change in polarization at the edge of the spectrum, although diminishing considerably, still remains very substantial. On changing the wavelength of the exciting light from 365 to 313 mμ there are practically no changes in polarization. In other words, in anthracene the same effect appears as in stilbene; it is much weaker, but nevertheless exists. Polarization in the long-wave part of the spectrum is the same for all forms of excitation, while the luminescence spectrum is independent of the character of the excitation.

Thus the main experimental result may be stated as follows: for β-excitation, the part played by free excitons in the luminescence of molecular crystals is less than for photoexcitation, sometimes even vanishing completely.

An analogous law, though considerably less pronounced, is found on reducing the wavelength of the exciting light. A natural hypothesis suggests linking this fact with the high probability of ionization of the molecule under β-excitation. If the subsequent recombination produces an excited molecule, we may expect with a high degree of probability that this will be a localized exciton state, since a state of ionization associated with a substantial change in the field of the molecule creates a strong local deformation of the surrounding lattice. Of course, this hypothesis needs independent confirmation.

The possibility of thus acquiring information on the role of ionization in the excitation of luminescence by hard radiation is extremely interesting in connection with the general problem of studying the mechanism of the excitation of luminescence in organic substances (including molecular crystals) by hard radiation.

The first fact which strikes us here is the low radioluminescent yield of organic substances. This is considerably less than in the case of photoluminescence. From the data of various authors, the radioluminescence energy yield of organic materials is in order of magnitude about 5%.

From this we may calculate that the expenditure of energy (for example by a fast electron) in creating one photon of luminescence is 50 to 60 eV. The processes taking place in radioluminescence may be represented in the form of the scheme proposed by Galanin [213]. This is depicted in Fig. 88.

By direct experiments in gases, the energy spent in forming one pair of ions was determined as 30 eV. This value is also supposed to be retained in condensed media. On the other hand, it is known that the ionization potential of organic molecules is approximately 10 V, and the mean excitation potential 6 V. Hence the energy of 30 eV spent in the ionization of one molecule should simultaneously lead to the excitation of three more molecules. If the processes indicated by a query (?) in the scheme have 100% probability, then in the excitation of one molecule 30/4 ~ 7 to 8 eV should be spent. In fact, the amount spent is 50 to 60 eV. This is what the low radioluminescent yield implies.

This fact may be due to one of the following two causes: either there is some extra kind of quenching peculiar to hard excitation, the mechanism being at present unknown, or else the cause lies in the properties of the hard excitation itself; the processes marked with queries may lead to radiationless energy losses, and then only the excitation of the π-electrons will give a contribution to luminescence. At the moment there are no direct experimental or theoretical proofs for one or other of these hypotheses. All the more interest attaches to all kinds of indirect data enabling some light to be thrown on the solution of this question.

Our own polarization observations which have been discussed above and the interpretation given to them agree with the first proposition and indicate that the recombination of ionized molecules leads to excited states, i.e., that the process indicated by the query in the scheme does take place.

Ionizing radiations may not only excite radioluminescence in molecular crystals; they also have a considerable effect on their photoluminescence. In a series of papers [212, 214-217] a fall in the intensity of photoluminescence from molecular crystals under the action of ionizing radiations (α- and β-particles, γ-rays, x-rays) was observed and studied. This phenomenon received the title of "degradation of luminescence." The

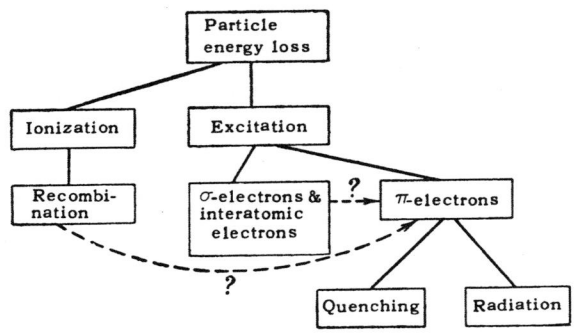

Fig. 88. Scheme of the processes taking place during radioluminescence.

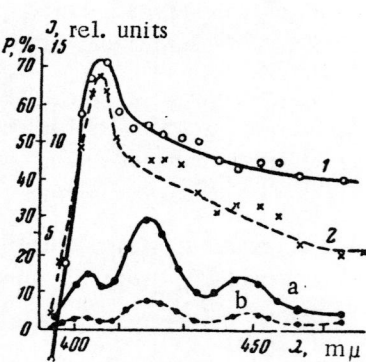

Fig. 89. a,b) Photoluminescence spectra; 1,2) spectral variation of polarization for an anthracene single crystal before (a, 1) and after (b, 2) $\beta$-radiation. Source $H^3(3.6 \cdot 10^{10}$ rad).

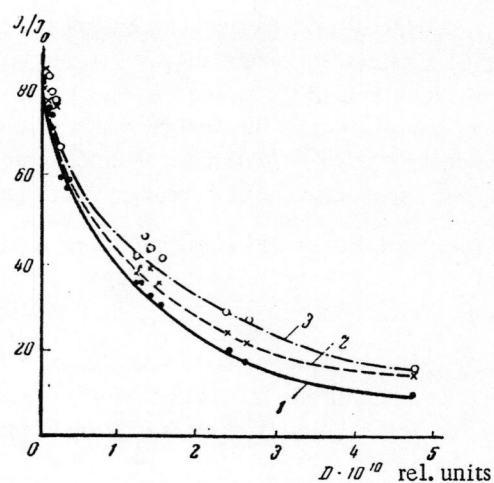

Fig. 90. Reduction in intensity (degradation) of luminescence at the maxima of the spectra for an anthracene crystal as a function of the $\beta$-radiation dose. 1) 405 m$\mu$; 2) 420 m$\mu$; 3) 445 m$\mu$.

degradation of luminescence may serve as a measure of radiation damage in molecular crystals. This degradation is irreversible or nearly so.

In the first years of the study of radioluminescence, its poor yield was associated simply with the destructive action of the radiation. However, this is evidently not the case. The irreversible degradation sets in slowly. The main processes for hard excitation, just as for other forms of the excitation of luminescence, are electronic (and this means reversible) excitation and ionization.

Let us not trouble ourselves here with questions of degradation, which will be treated specially in a paper by Khan-Magometova, but simply present certain results interesting in connection with the polarization of luminescence in molecular crystals.

We studied the effect of $\beta$-radiation on the spectral variation of the polarization of luminescence from thin anthracene single crystals obtained by sublimation. The crystals were of the order of a few microns thick. The radiation source was a tritium preparation $H^3$ of great activity, 3.4 Ci. The mean energy of the $\beta$-electrons was 5.5 keV and the maximum 18 keV. Irradiation was carried out in large doses, of the order of $10^{10}$ rad. Figure 89 gives the results of measuring the luminescence spectra and the spectral variations of polarization for an anthracene single crystal (b-axis vertical) before and after irradiation.

It must be noted that while the intensity of the anthracene luminescence was being lowered by the $\beta$-radiation (degradation) the structure of the spectrum remained unaltered; the spectrum fell as a whole, almost without distortion. The degradation of the luminescence at separate maxima of the spectrum varied with irradiation dose almost identically (Fig. 90). This indicates that the actual mechanism of luminescence in the crystal suffers no substantial changes under $\beta$-radiation.

The spectral variation of the polarization of the photoluminescence in the anthracene crystal after $\beta$-radiation also in general retained its character, including the sharp changes in polarization at the short-wave end of the luminescence spectrum, which we associated with the luminescence of free excitons. This also shows that the mechanism of luminescence in the crystal and the mechanism of exciton energy migration in the lattice are not altered as a result of $\beta$-radiation.

From these data we may conclude that the mechanism of luminescence and its excitation, on the one hand, and the degradation mechanism, on the other, are essentially different, and the small radioluminescence yield cannot be attributed to degradation. This confirms the earlier view that degradation cannot be held responsible for the low yield, on account of its slow development.

We should note a slight fall in the degree of polarization of the luminescence after $\beta$-radiation in the long-wave part of the spectrum, where it is closely bound up with the structure of the crystal lattice and the orientations of the molecules in this. These changes show that $\beta$-radiation produces some (though slight) distortion of the crystal lattice. Such changes in polarization may serve as a sensitive method of observing and studying the action of ionizing radiations on the structure of crystals.

# CHAPTER VII

# EXPERIMENTAL METHODS

In order not to disturb the continuity of the results presented, we decided to relegate the description of the apparatus and experimental methods used, to this separate chapter. In this we shall pass fleetingly over generally known methods of polarization measurements, concentrating attention on less widespread methods, and also on the original methods and apparatuses specially developed for solving the problems presented in our investigation.

## §1.  Visual Methods of Measuring the Polarization of Luminescence

Visual methods of measuring the polarization of light always finally reduce to a photometrical problem. This is evident in the classical method of Cornu, whose polarimeter is based on equating the illumination of two fields obtained by means of a Wollaston prism and corresponding to components polarized in mutually perpendicular directions.

This equalization is effected by means of a Nicol analyzer so oriented as to produce equal projections $A_1^*$ and $A_2^*$ of the electric vectors $A_1$ and $A_2$ of the two light flux components on the vibration direction N passed by the Nicol (Fig. 91). As seen from the figure, $\tan \varphi = A_2/A_1$. The ratio of intensities $I_2/I_1 = \tan^2\varphi$, and the degree of polarization

$$P = \frac{I_1 - I_2}{I_1 + I_2} = \frac{1 - \tan^2\varphi}{1 + \tan^2\varphi} = \cos 2\varphi. \qquad (VII.1)$$

It is this angle $2\varphi$ which is convenient to measure experimentally, being the angle of rotation of the Nicol analyzer between two positions giving equal illumination of the two fields.

The position stated is less obvious but no less correct in another classical method named after Savart. The main part of the Savart polariscope is the Savart plate, formed by a combination of the two crystalline quartz plates of equal thickness. Each plate is cut at an angle of 45° to the optic axis, and they are placed together so that their optic axes are in the crossed position and make an angle of 90°.

The light being studied passes through the Savart plate and an analyzer (Nicol or other polarization prism). Even if the light is partly polarized, interference fringes will appear in the field of vision. The appearance of the interference pattern indicates the presence of polarization. It is impossible, however, to measure the degree of polarization with just one Savart polariscope. For this we may use a compensation method, the compensator being, for example, a pile of glass plates. The ray refracted by the pile is partly polarized. We may place the pile at such an angle to the incident ray that the polarization created by the pile is equal in absolute magnitude and opposite in sign to the polarization of the light flux itself.

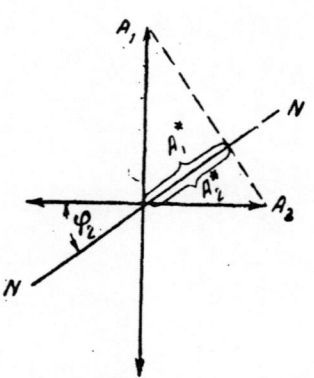

Fig. 91. Operation of the Cornu polarimeter (schematic).

In this case we may record the absence of polarization by means of the Savart polariscope, in that the interference pattern vanishes. Moreover, the maximum (light fringe) and minimum (dark fringe) brightnesses in the field of vision are proportional to $I_1$ and $I_2$ respectively, i.e., the intensities of the light flux components with polarizations along the preferred direction and perpendicular to this (see, for example, [218]).

Thus visual methods of measuring polarization reduce to balancing the illumination of two fields created by the components of the polarized flux. Hence their accuracy is restricted by the contrast sensitivity of the eye and its dependence on the brightness and spectral region of the light studied.

Below we shall estimate the limits of this accuracy and the possibility of increasing it by photoelectric methods. But first we shall present the methods of studying the polarization of luminescence from molecular crystals in which we used visual apparatus.

## §2. Systems Using the Cornu Polarimeter

The great convenience and applicability of the Cornu polarimeter are due to its portability and localzability in use. In polarimeters of the Savart type these features are lost on account of the pile of plates. These features are valuable in studying objects of small size and when selectivity of viewing direction is required.

Thus in studying the polarization of microcrystals a few microns in size we set up a system [86] in which the main components were a luminescence microscope and a Cornu polarimeter on a special stand. The polarimeter was placed above the microscope in such a way that its optic axis was vertical and coincided with the axis of the microscope. It could be moved smoothly in three directions, two horizontal and one vertical, or firmly clamped in a fixed position. Moreover, it was also possible to rotate the polarimeter in this fixed position smoothly around its vertical axis. The source of excitation was a high-pressure mercury "spot" lamp with a parabolic reflector. The 365 m$\mu$ line was mainly separated out from its spectrum by filters. The crystals under investigation were placed on the object glass. These had to be chosen as small as possible and distributed as sparsely as possible.

Measurements were carried out at magnifications from 600 to 2000 times. In the majority of cases a magnification of 600 was enough. In particular cases with the coarser crystals measurements could be made without the eyepiece. The linear dimensions of the crystals used varied between 5 and 50 m$\mu$.

Having placed the object glass on the microscope table and selected the best excitation conditions, it was necessary, by means of two mutually perpendicular micrometer screws, to choose a single crystal of adequate regularity and homogeneity in such a way that light from other crystals should not come into the field of view. One is not always successful in finding such a crystal on a given object glass. After the crystal has been selected, the polarimeter is set up over the microscope and fixed in such a position that the crystal is clearly visible through it. The image of the crystal must then be slightly defocussed so that the light flux of its fluorescence uniformly fills the entrance aperture of the polarimeter and the fields to be compared by eye are uniformly illuminated. Filters crossed with those placed in front of the excitation were set in the eyepiece.

In studying the azimuthal variation of polarization, it was experimentally inconvenient to rotate the crystal on the object stand. Instead of this, the polarimeter itself was rotated around a vertical axis, the angle of rotation being reckoned along the same graduated circle as the position of the Nicol determining the degree of polarization.

It was necessary to perform control experiments in order to check whether the optical system of the microscope itself might not introduce polarizing or depolarizing effects into the light flux being examined, as a result of multiple refractions in the lenses of the objective and eyepiece. Two experiments were made.

First, a polaroid was placed on the object glass, the filters were removed from the excitation, and visible light passed through the polaroid, acquiring linear polarization, and then through the microscope, after which its polarization was measured by the polarimeter. The azimuthal variation in this case had maximum 94%. The difference of 6% may be explained by measuring errors and possible imperfection of the polaroid. Thus we may take it that the optical system of the microscope had no significant depolarizing action.

Second, one drop of an aqueous solution of fluorescin was placed on the object glass. On excitation by natural light and observation along the direction of the exciting light, the polarization should of course be zero.

It turned out that for all orientations of the polarimeter P = 0, i.e., the microscope did not in practice polarize the light passing through it.

The polarization of luminescence from microscopic single crystals of a number of substances was measured by the method described, and measurements were also made of the variation of the degree of polarization of fluorescence from the crystals with the polarization of the exciting light. In the latter case a Nicol was placed in the way of the excitation (between the condenser mirror and the condenser of the microscope); this could be rotated around a vertical axis. The angle of rotation was measured on a graduated circle. The polarimeter and crystal were in one fixed position all the time. The degree of polarization of the fluorescence was measured as a function of the angle determining the position of the Nicol.

Another example of the merits of using the Cornu polarimeter is its application to the study of polarization diagrams. The portability of the polarimeter enabled it to be connected to a goniometer, on the graduated circle of which the stand bearing the Cornu polarimeter could slide; the polarimeter was set up so that its optic axis passed through the center of the goniometer. It was possible in this way to measure polarization in various directions in one plane. The crystal was set up so that the excited part was in the center of the goniometer. The crystals (including hemispherical ones) were fixed in a specially constructed holder which facilitated the fixing of hemispheres of different diameter and crystals of arbitrary shape and various sizes, the regulation of the diaphragm set close up against the crystal surface so as to limit the size of the excited part, and the rotation of the crystal around a horizontal axis. The latter was necessary for measuring the azimuthal variation, and also so that, by making measurements in different directions in the same horizontal plane of the goniometer (experimentally most convenient), polarization diagrams for different crystallographic planes could be obtained with various crystal orientations.

## § 3. Kavraiskii Polariscope

In a number of cases (especially in the study of the polarization of luminescence from solutions and anisotropic films) we used the polariscope proposed by Kavraiskii [219]. The Kavraiskii polariscope is essentially a development of the Savart instrument. It differs from the latter chiefly in that it uses as analyzer not a Nicol but a double image prism. For this a modified Wollaston prism is used. This differs from an ordinary Wollaston prism in that the optic axes of its parts are parallel to its surface, but make angles of ±45° with the refracting edges. The prism must satisfy the condition that its splitting angle Φ should be equal to half the angular width Λ of the parallel fringes in the interference pattern:

$$\Phi = \frac{1}{2}\Lambda.$$

The substance of the Kavraiskii method lies in this. As a result there will be superimposed on one another two interference patterns with different polarizations at corresponding points of the field. In the Savart polariscope one of these is extinguished by the Nicol. It follows that the Kavraiskii polariscope has twice the light power of the Savart model. In order to achieve this result, however, not only the analyzer but also the polariscopic section of the polariscope must be changed. In the Savart polariscope, this is the Savart plate.

The splitting of the modified Wollaston prism is constant for rays lying in a plane parallel to the edge. The fringes in the interference pattern produced by the Savart plate, however, are not quite parallel. Therefore, in order to obtain a sharp interference pattern in a wide field of vision, the Savart plate is replaced by another polariscopic piece. This consists of two quartz crystalline plates, as in the Savart plate, not crossed, but in opposition with a λ/2 plate between them. This combination gives equidistant interference fringes. Thus the Kavraiskii polariscope has two important advantages over the Savart model: double light power and considerably larger field of view.

The compensator was a pile of glass plates fixed to the base of a Cardan joint so as to be able to rotate about vertical and horizontal axes. This offered the possibility of measuring both positive and negative polarization. Calibration of the compensator was effected with respect to a light source with known polarization. For this we used the scattered light of an incandescent lamp passed through a Nicol in a definite orientation. A change in this orientation corresponded to a change in the degree of polarization, quite easy to calculate.

We checked the calibration of the pile (both in the Kavraiskii polarimeter and in all the other cases when we used a compensating pile) by the method proposed by Weber [220]. The essence of this method is as follows. Measurements are made of the polarization of luminescence from a solution (we used fluorescein in glycerine) excited by linearly polarized light in a direction perpendicular to the direction of viewing the fluorescence. For a vertical position of the electric vector in the exciting light, we call the fluorescent light components $I_{\parallel}$ and $I_{\perp}$. If the electric vector of the exciting light makes an angle of $\theta$ with the vertical, then the corresponding fluorescence components $I_{0\parallel}$ and $I_{0\perp}$ will be connected with $I_{\parallel}$ and $I_{\perp}$ in the following way:

$$I_{\theta\parallel} = I_{\parallel}\cos^2\theta + I_{\perp}\sin^2\theta; \quad I_{0\perp} = I_{\perp}.$$

Consequently

$$\frac{1}{P_\theta} = \frac{I_{\parallel}\cos^2\theta + I_{\perp}(1+\sin^2\theta)}{(I_{\parallel}-I_{\perp})\cos^2\theta} = \frac{1}{P} + \tan^2\theta\,\frac{2I_{\perp}}{I_{\parallel}-I_{\perp}}.$$

Remembering that $I_{\parallel}/I_{\perp} = (1+P)/(1-P)$, we obtain

$$\frac{1}{P_0} = \frac{1}{P} + \left(\frac{1}{P}-1\right)\tan^2\theta. \tag{VII.2}$$

Thus by varying $\theta$ we may experimentally obtain the variation of $1/P_0$ with $\tan^2\theta$ in the form of a straight line, the slope of which, together with the intercept which it makes on the axis of ordinates, should correspond to formula (VII.2) if there is no systematic error in the calibration of the compensator. In our control experiments, these condition were fulfilled within the limits of the experimental error inherent in visual measurements.

## §4. Errors in Visual Methods of Measuring the Degree of Polarization

As already noted in §1, the precision of visual methods of measuring the degree of polarization are limited by the contrast sensitivity of the eye.

The contrast sensitivity of the eye is determined by the quantity $\Delta I/I$, where I = background brightness and $\Delta I$ = brightness increment necessary to distinguish contrast. Over a wide brightness range (from 10 to 500 mL) the ratio $\Delta I/I$ changes little and remains approximately equal to 1 to 2% (mean value 1.5%). With falling brightness the contrast sensitivity worsens. For a brightness of 1 mL the photometric error reaches 2.5%, for 0.2 mL 5%, and for 0.1 mL 10%.

Differentiating the expression determining the degree of polarization

$$P = \frac{I_1-I_2}{I_1+I_2} = \frac{1-\rho}{1+\rho},$$

where

$$\rho = \frac{I_2}{I_1},$$

we obtain the following expressions for the absolute and relative errors in determining the quantity P:

$$\delta P = \pm\frac{1-P^2}{2}\frac{\delta\rho}{\rho}; \quad \frac{\delta P}{P} = \pm\frac{1-P^2}{2P}\frac{\delta\rho}{\rho}. \tag{VII.3}$$

Let the brightness of the luminescence be not less than 10 mL and the photometering error $\delta\rho/\rho = 1.5\%$. It is seen from formula (VII.3) that, even in this case, for small values of P the relative error in determining this quantity may become extremely large. Thus if P = 0.05, then $\delta P/P = 15\%$, while for P = 0.02 the relative error reaches 37%. If, however, the brightness of the luminescence under examination is less than 10 mL, the relative errors will become unacceptably large even for considerable polarizations. For example, for a brightness of the order of 0.1 mL (when the photometric error is 10%) and P = 5%, in the Savart polariscope the interference pattern will not be visible at all. These errors have a fundamental significance in visual methods and cannot be avoided.

At the same time, in studying the polarization of luminescence, just as in many other regions of science (astrophysics, meteorological optics, etc.), it is often required to measure the degree of polarization of weak light fluxes. For example, studying the polarization in individual parts of the fluorescence spectrum demanded a fundamental increase in the sensitivity and precision of experimental methods.

Apart from all that has been said, we must also bear in mind that, in studying fluorescence (especially that of molecular crystals), we often have to deal with luminescence situated in the blue and violet parts of the spectrum, where the eye has low sensitivity. Here error becomes even greater. In the ultraviolet region, of course, visual methods are out of the question. All this makes it essential to pass over to photoelectric methods of measuring the polarization of light.

## §5. Photoelectric Methods of Measuring the Polarization of Light

Whereas in visual methods the measuring accuracy is set by the limited contrast sensitivity of the eye, in photoelectric methods the accuracy depends on the receiver.

An intensity difference $I_1 - I_2$ may be measured with any previously assigned relative error if the sensitivity of the receiver is high enough.

Two possible main variations of the photoelectric polarimeter exist:

1. The intensity of the polarized light flux is modulated by some polarization method. The modulated flux is detected by a photoelectric receiver. The amplitude of the variable component of the photocurrent in this is proportional to the difference $I_1 - I_2$ ($I_1$ = intensity of component with preferred polarization, $I_2$ = that with perpendicular polarization). Measurement of the degree of polarization, i.e., the depth of modulation, may be made by various means. One of the most suitable is compensation.

2. The mutually perpendicular polarization components of the light flux are measured separately. Various modes of separation are possible; for example, successive measurements of the components through a Nicol analyzer for two mutually perpendicular orientations of the latter, or by measuring the difference in photocurrents from two receivers separately sensitive to the mutually perpendicular polarized components, which may be separated spatially, for example, by means of a doubly refracting polarization prism.

## §6. Modulation Photoelectric Polarimeter

Various means of optically modulating the polarized light flux may be used. Tumerman [221] passed the light through a rotating polaroid. This method is simple, but has substantial failings. The photocathode of every photoelectric detector is anisotropic, i.e., its sensitivity is different for different orientations of the electric vector of the incident light. This leads to additional modulation on rotating the analyzer, and hence to error. The light can be depolarized in front of the photocathode by opal glass, but this leads to a sharp fall in the measured intensity.

A number of authors have proposed methods of modulation in which the analyzer remains stationary. Wille [222], in an apparatus developed for measuring the polarization of light from the solar corona, used a modulator consisting of a rotating $\lambda/2$ plate and a stationary analyzer. Spektorov [127,223] proposed as modulator a system consisting of a rotating biquartz and a stationary analyzer. The biquartz consists of two halves: levo- and dextrorotatory quartz. The axis of rotation is displaced parallel to that of the light beam, so that the latter passes alternately through the dextrorotatory and levorotatory quartz. The direction of the vibrations passed by the stationary analyzer is set at an angle of 45° to the preferred direction of vibrations of the electric vector in the light under examination. The maximum modulation corresponds to an angle of rotation of the quartz through 45°.

Together with V. I. Gribkov, we set up an apparatus with a modulator operating on the principle just mentioned. The radiotechnical part of the system was devised by our colleague, V. I. Gribkov, and is described in detail in his paper, together with many other operating details. For measuring the degree of polarization we used a compensatory null method. The null method allowed us to avoid such complications as the nonlinearity of the photomultiplier and amplifier, and so forth.

As compensator we used a pile of glass plates. The operation of the compensator was beset by the following complication. As a result of the inhomogeneity of the photocathode, the displacement of the light beam across its surface must be avoided. But in passing through the inclined pile, the light ray is displaced parallel to itself. The displacement is different for different angles of inclination. In order to eliminate this displacement and ensure the invariability of the position of the ray in space, we used two identical piles of glass plates, rotating in opposite directions through equal angles. For measuring both positive and negative degrees of polarization, it was arranged that the axis of rotation of the two piles could be rotated by 90° around an axis coinciding with that of the light beam. The calibration of the compensator and its verification were carried out by the methods described in § 3. Details of the calibration and measurements on the apparatus are set out in V. I. Gribkov's paper.

The signal from the photomultiplier receiver passed through an amplifier to an oscillograph. The variable component was compensated by rotation of the pile. This apparatus was used to measure the polarization characteristics of the luminescence from various objects.

This system, however, has a number of failings (in particular, the need for a motor to rotate the biquartz, inevitably leading to mechanical vibration and to electromagnetic induction removable only with difficulty), and, for especially accurate and sensitive measurements, as in measuring polarization over the luminescence spectrum, we were compelled to design a system of a second kind, not using modulation.

### § 7. Spectropolarimetric Photoelectric System

Together with V. I. Gribkov we developed a static photoelectric system without modulator and compensator, based on the direct measurement of the difference between photocurrents obtained from light flux components with mutually perpendicular polarizations. This system was set up for measuring the degree of polarization of luminescence, azimuthal variations, polarization diagrams, and other characteristics as functions of the wavelength of the luminescence. Arrangements were also made for varying the wavelength of the exciting light.

Thus the apparatus was universally capable of measuring all possible polarization characteristics in the luminescence of various objects: crystals, solutions, and films. Arrangements were made to set up any required orientation of the objects and to vary this smoothly, and also to measure samples in a dewar at liquid nitrogen temperatures.

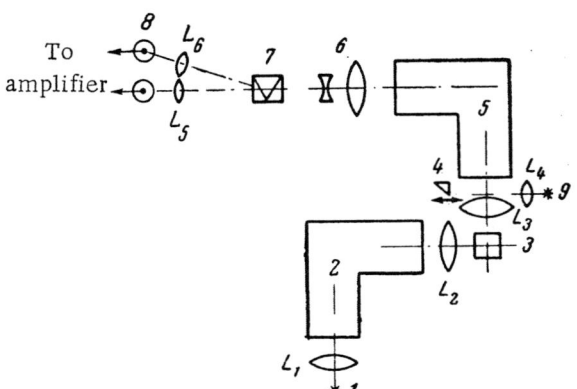

Fig. 92. Main optical scheme of the spectropolarimetric photoelectric system. 1) PRK-4; 2) quartz monochromator; 3) object being examined; 4) prism with opal glass; 5) glass monochromator; 6) objective; 7) prism; 8) photomultiplier FEU-18; 9) Incandescent lamp; $L_1$ and $L_2$ are lenses.

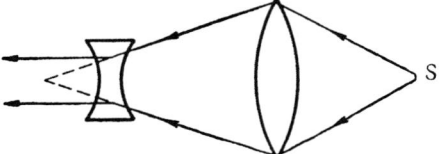

Fig. 93. Scheme of the objective in front of the polarization prism. S = glass monochromator slit.

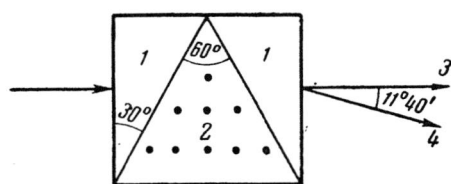

Fig. 94. Doubly refracting prism of the type proposed by Abbé. 1) Crown glass; 2) Iceland spar; 3) extraordinary ray; 4) ordinary ray.

Figure 92 shows the principal optical scheme of the system. Excitation of the luminescence was effected by the light of a mercury lamp (usually a PRK-4) passing through a quartz monochromator. It was arranged to vary the position of the monochromator so as to allow excitation of luminescence from the side, behind, or in front, depending on the object and type of problem. Light from the luminescence was collected by a condensing lens and directed into a glass monochromator. At the exit from the monochromator is placed an objective consisting of two lenses: double convex and double concave (Fig. 93). The purpose of this objective is to produce a parallel beam of small given cross section. This parallel beam then passes through a doubly refracting prism, which forms a most important element in the system.

This element is a little-used prism of the Abbé type, a description of which we found in an old German dissertation written by Grosse in 1886 [227]. On both sides of the central equilateral iceland spar prism, in which the optic axis is parallel to the refracting edge, are stuck right-angled prisms of crown glass with acute angle 30°, so that the whole forms a rectangular parallelepiped (Fig. 94). The length/breadth ratio is $2/\sqrt{3}$. The extraordinary ray passes through the prism without deviation, while the ordinary ray suffers a deviation of 11°40'. It would be possible to increase the divergence of the rays by increasing the refracting angle of the central crystalline prism. For examples for a refracting angle of 90° the divergence reaches almost 23°. However, in this case the ordinary ray may be partly cut off by the edge of the prism and the original intensity ratio of the two fluxes disrupted. A deviation angle of around 12° for the above-described construction of the prism is in fact quite adequate to separate the rays and make them fall into different detectors. This could not be done by means of a doubly refracting prism of the ordinary kind (for example, the Wollaston prism, in which the splitting angle equals only 3.4°).

Each of the light beams (with vertical and horizontal polarizations) is focused on the photocathode of an FÉU-18 photomultiplier. The photocurrents from the two photomultipliers are then amplified by a differential two-channel amplifier and recorded by a microammeter M-91 (sensitivity $1 \cdot 10^{-8}$ A/mm).

The electronic part of the apparatus was developed and constructed in our group by V. I. Gribkov and is described in detail in his paper. We shall confine ourselves to brief indications.

The main scheme of the two-channel direct current amplifier is shown in Fig. 95. The amplifier comprises two 6Zh1Zh tubes and operates in the electrometer mode. Signals from the first and second photomultipliers pass to the suppressor grids of the first and second amplifier tubes respectively. Interchangeable input resistances of 1, 10, and 100 MΩ facilitate operation in three ranges of photocurrent amplification, 10, 100, 1000 times. The photocurrents amplified in the two channels of the amplifier are set in opposition in a bridge scheme. The microammeter appears across the diagonal of the bridge. In measuring the photocurrent of just one of the two photomultipliers, the amplifier works according to the ordinary bridge scheme. On including signals from both photomultipliers simultaneously, the apparatus measures the difference between the photocurrents. In or-

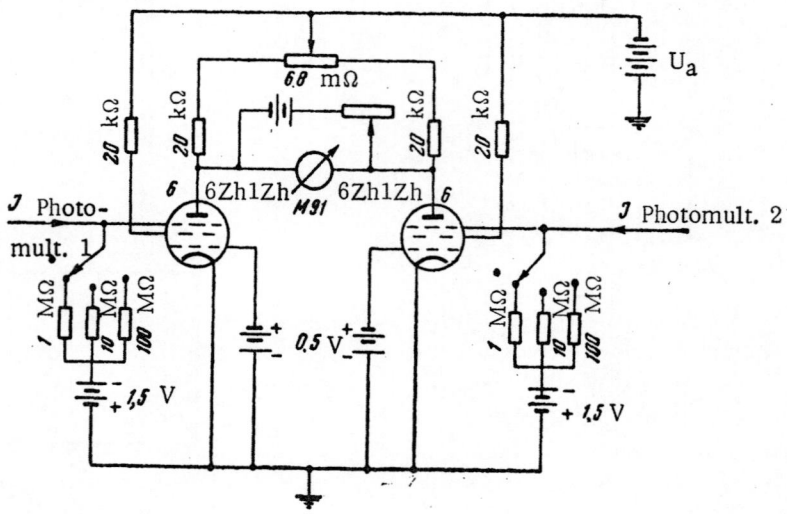

Fig. 95. Main scheme of the differential two-channel amplifier.

der to raise the sensitivity and precision of the system, it is very important to be able to measure small differences in photocurrents (i.e., small degrees of polarization) on a more sensitive range than the components themselves.

Arrangements are made in the apparatus to check the linearity of the two channels, the equality of their amplification factors, and also the possibility of regulating the two channels separately and compensating direct current from various kinds of induction and noise in the measuring system.

The maximum sensitivity of the apparatus is $1 \cdot 10^{-10}$ A. The precision of measuring the degree of polarization rises on increasing the photocurrents corresponding to the difference between the components and to the separate components, i.e., on increasing the intensity and degree of polarization of the light under examination.

For intensities corresponding to a photocurrent of $5 \cdot 10^{-8}$ A, the relative error in measuring the degree of polarization is 0.1%, for $5 \cdot 10^{-9}$ A it is 1%, and for $5 \cdot 10^{-10}$ A not more than 10%.

In working with the system, the following important correction is required. The two mutually perpendicular polarized components of the light flux under examination may fall into the apparatus under dissimilar conditions, owing to the unequal sensitivities of the photomultipliers (although photomultiplier pairs were selected from a large number so as to have the best and closest parameters, absolutely identical pairs could not be ensured), the depolarizing action of the monochromator (resulting from refraction and reflection at the boundaries of prisms and lenses), slightly differing amplification factors, and other factors leading to asymmetry of the system in respect of the vertically and horizontally polarized components.

As a result of this, clearly, there will be systematic errors in measuring the degree of polarization. In order to correct these errors, the following method was employed. Before each measurement (for each wavelength of the luminescence spectrum), in front of the entrance slit of the glass monochromator we placed a total internal reflection prism (without disturbing the setting of the whole apparatus including the sample being studied); this had opal glass on the face turned towards the slit. By means of this prism, light from an incandescent lamp set to one side (see Fig. 92) was fed into the monochromator. The lamp was so regulated that the signal arising from it should equal the measured signal from the luminescence of the sample. The system was balanced with the aid of this natural light. The opal glass completely depolarized the light, so that on switching in both photomultipliers the recording system should indicate zero. If the reading differs from zero, then it can be zeroed by slightly changing the supply voltage of the corresponding photomultiplier. After this the auxiliary prism is taken away and the required measurement made.

On the same apparatus the luminescence spectrum $I_1 + I_2 = f(\lambda)$ could also be measured. The spectral sensitivity of the system as a whole was calibrated by a band-lamp with known color temperature.

Descriptions of analogously based systems using two photomultipliers have appeared in the literature [225,226]. These, however, had markedly lower capabilities than ours. They provided for no amplification of the photocurrents, nor the spectral resolution of the radiation; the separation of the components was achieved by mirrors, leading to considerable loss in intensity, so that measurements over the spectrum could not be made owing to inadequate sensitivity, and so forth.

We were only able to study the polarization of luminescence from molecular crystals over the whole spectrum, especially at the short-wave end, where intensities are extremely small, by using our own apparatus.

In a number of cases, it was only by using our own apparatus that we could measure the variation of the polarization of luminescence from crystals with the polarization of the exciting light, especially in the case of mixed crystals, where the primary and chief difficulty was to separate out traces of exciting light when the intensity of the luminescence was very low (see §4 of chapter IV).

## §8. Preparation and Treatment of Crystals

Thin single crystal films of anthracene, anthracene-naphthacene, etc.,were obtained by sublimation in porcelain crucibles.

Large single crystals of stilbene, tolan, benzyl, etc.,were grown in a sealed test tube by the Obreimov-Shubnikov method, in a thermostatic bath by the Kiropulos method, and in other ways. Some of the crystals were grown in our group, but many were prepared for us in the Institute of Crystallography, Academy of Sciences, USSR, and in the Khar'kov Branch of the Institute of Reagents.

For the cutting of large single crystals into parts of the required size, a special saw was constructed and prepared in our group by V. I. Gribkov and A. Ya. Mitin. The crystals were cut with a silk surgical thread moistened with a weak solvent (for example, kerosene). Usually such saws were made circular, which constructionally was simplest. In this case, however, fine sawing was inevitably hindered by the knot in the thread. In order to avoid this difficulty we used a return spring in our saw. Arrangements were also made for soft but reliable fixing of the brittle crystals, smooth control of their displacement during sawing, and so forth. By means of this saw it was possible to cut crystals of various shapes and sizes in the required fashion, including extremely thin plates.

In a number of cases, spherical single crystals were required for experiments. Cut crystals of the necessary size were given spherical form by means of a rotating tube and emery. The sphere so obtained was polished by rotating in the same tube with a soft cloth moistened in kerosene or in a special chamber with chamois stuck to the inside. After this treatment we obtained completely transparent spheres with even, polished surfaces.

The orientation of the crystals was carried out using all possible direct optical methods and auxiliary phenomena and properties. Important information comes from the polarization of the luminescence itself. Thus the sublimed thin anthracene crystals with surfaces parallel to the ab-plane can easily be oriented from the azimuthal variation of polarization: the positive maximum of this corresponds to the vertical position of the b-axis.

In orienting stilbene and tolan spheres, the first stage was to determine the positions of the optic axes. The spherical form of the samples allowed Shubnikov's [150] method to be employed for observing the conoscopical picture. This method is convenient on account of its extreme simplicity. The crystal sphere is viewed between crossed polaroids, being illuminated from below by an incandescent lamp via the polarizing polaroid and opal glass (Fig. 96). The light is diffused by the opal glass. Rays in various directions pass through the lower polaroid and are converged by the sphere at various points near its surface (Fig. 97). This creates a conoscopical picture which is viewed through the upper polaroid. The conoscopical figure appears to the observer as a three-dimensional figure on the spherical surface rather than a plane. On rotating the sphere, the figure rotates with it. On rotating the upper polaroid of the apparatus, one can see changes in the conoscopical picture on the sphere, corresponding to a transition from the parallel position of the polaroids to the crossed position. If in a biaxial crystal the angle between the optic axes is small, then the points at which the two axes emerge from the sphere can be seen simultaneously. If the angle between the axes is large, then by rotating the sphere first one and then the other axis may be found and their points of emergence inked on the surface of the sphere. In

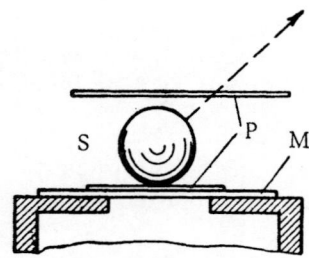

Fig. 96. Scheme of Shubnikov's apparatus for observing the conoscopical figure in a crystal sphere. S = sphere, P = polaroid, M = opal glass; dots show direction in which the conoscopical figure is viewed.

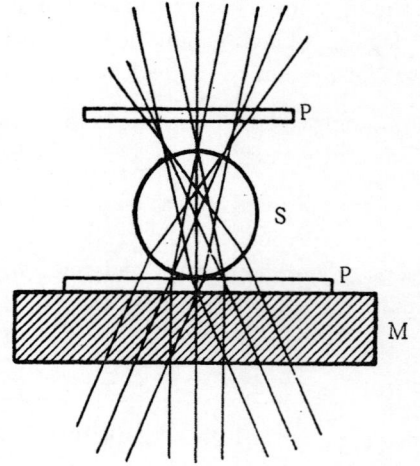

Fig. 97. Ray paths in sphere. Notation as in Fig. 96.

order to allow for the dispersion of the axes and to determine their positions for some fixed narrow spectral region, the appropriate light filters must be used. By doing this, the conoscopical picture loses brightness but gains contrast.

Apart from finding the optic axes by the method described, an important aid to the orientation of samples lies in determining the cleavage planes from the appearance of a series of cracks when the sample is cooled by flooding with a rapidly evaporating substance (for example, dichlorethane).

For orienting the crystals and finding the crystallographic axes, measurements of the polarization of luminescence are themselves extremely useful, and sometimes have a decisive influence.

Naturally one must also bear in mind all known laws relating to definite crystallographical systems and linking the crystallographical and crystal optical characteristics, such as rules determining the position of the crystallographical axes with respect to the optic axes, and so on. Specific examples of determining the positions of the crystallographic axes in the single-crystal samples used for our investigations were given in Chapter III.

For measuring the polarization characteristics of the luminescence from molecular single crystals at the temperature of liquid nitrogen, we used special dewar vessels: glass **ones** with transparent zones, and quartz ones with sealed-on windows of optical fused quartz. The crystals were placed directly in the liquid nitrogen in a plexiglass holder having poor thermal conductivity.

Great experimental difficulties in polarization measurements were created by the cracking of crystals during cooling. Although thin single crystal films obtained by sublimation could fairly easily be cooled without cracking, for large crystals this was very difficult. The appearance of cracks results in the depolarization of the luminescence. The main danger was not so much the cracks themselves as the displacement of parts of the cracked crystal relative to one another. This produced polycrystalline specimens instead of single crystals, and these emitted unpolarized or practically unpolarized luminescence.

In order to cool the crystals without cracking, cooling had to be effected very slowly. This was done by one of two means.

1. The holder bearing the crystal was let down into the liquid nitrogen by a clockwork system at approximately 3 mm/hr, a copper cold conductor welded to the part holding the crystal being first passed into the nitrogen.

2. Liquid nitrogen slowly and in small quantities was passed through a special coil with a funnel coming out directly at the bottom of the dewar. This prevented the nitrogen from falling directly on to the warm crystal during pouring, which would have led to its instant cracking.

In some cases the first method was found more convenient, and in others, the second.

Using these methods, we were able to carry out a variety of polarization measurements at low temperatures.

# CONCLUSION

Interest in molecular crystals, which has long attracted the attention of physicists, has grown greatly in recent years. This has been caused by a whole series of circumstances. Amongst these we must notice the following.

On the one hand, the retention of a certain autonomy by individual molecules allows us to separate "molecular" and "crystalline" effects and to study the latter in pure form. This approach is extremely interesting for solid state physics.

On the other hand, molecular crystals present extreme interest as simplified physical models of elementary biological objects, one of the chief features of which in contemporary thought is the ordered, oriented disposition of the molecules.

Finally, molecular crystals have important practical applications (for example as scintillators).

The basic idea of the foregoing study lies in successively examining the structure and properties of molecular crystals by a variety of quantitative methods, using the polarization of luminescence as a powerful and refined instrument for investigating the anisotropy of the medium.

Although optical methods, including luminescence, play a leading part in the study of molecular crystals, the method of polarized luminescence, especially suitable in view of the nature of the objects and the formulation of problems, has been used far too little in this realm. In particular, quantitative measurements of the various polarization characteristics have been little used. Qualitative observation of polarization has been applied rather more often.

In order to apply to the study of molecular crystals the rich arsenal of quantitative polarized luminescence methods mainly developed by the Soviet school of luminescence, though in general for isotropic media, it was necessary to take account of many specific features of molecular crystals, and accordingly to develop these methods further, as well as devising new ones.

In many cases it was required to measure the polarization of very low intensity luminescence. For this, high-sensitivity precision measuring apparatus was devised.

The main direction of development in the investigation here set out was determined by the circumstance, fundamental in molecular crystals, that the intermolecular forces in these were considerably smaller than those between the atoms in the molecules. As a result, the molecules in a lattice to a certain extent retain their individuality, and to a first approximation the crystal may be regarded as an oriented gas, neglecting the interaction of the molecules.

A first and necessary step in the investigation was the study of anisotropy in separate isolated molecules. These investigations were carried out using the polarization of luminescence from solutions of the same substances later to be examined in the crystalline state. With the help of polarization spectra and a number of supplementary data, the orientation of the radiation oscillator in the molecule was determined relative to its structural elements. Such studies were made for a considerable number of substances of various types. In a number of cases, the results obtained in liquid solutions were confirmed by additional experiments with anisotropic films colored by the same substances. A knowledge of the orientation of the radiating oscillator in the molecules is clearly needed in order to solve questions connected with the orientation of the molecule in the crystal lattice from the polarization of luminescence in the crystal.

If we start from the oriented gas model, then we can reasonably attempt to determine the orientation of the molecule in the lattice completely from the polarization of luminescence. It is clearly insufficient in this merely to determine the polarization in any one direction; we must study the space distribution of polarization (polarization diagram). By comparing the data so obtained with data on the orientation of molecules in the lattice derived in some other, independent way (for example, from x-ray analysis), we can draw conclusions regarding the validity of the oriented gas model and its limits of applicability.

These investigations were made for a number of pure and mixed molecular crystals. We had to overcome considerable difficulties associated with the effect of the double refraction of luminescence in the crystal on the polarization being studied. For this, quite heavy and complicated calculations of crystal optical corrections had to be set up in such a way that the effect of double refraction could be neglected; this called for spherical single crystals.

The main result of these investigations may be briefly formulated as follows. The experimental polarization diagrams, as far as their qualitative functional appearance was concerned, always agreed closely with diagrams calculated from the oriented gas model. Quantitatively, however, there was no agreement at all; the experimental degrees of polarization in absolute magnitude were always smaller than calculated. This discrepancy could not be explained by depolarization due to thermal vibrations or other depolarizing factors. The discrepancy in fact constituted a limit to the applicability of the oriented gas model to molecular crystals. In other words, the excited states in a crystal can to a first approximation be regarded as those of individual, isolated molecules (for the experimental and theoretical diagrams always agree qualitatively), but for quantitative evaluation the excited states of the crystal must be considered as localized excitons or deforming excitations, corresponding to certain local excited regions of the crystal. For the second case, theory immediately shows that the polarization of the luminescence must differ from that derived from the oriented gas model. But in any case the radiation does not in general proceed from excited states corresponding to free excitons.

The fact that the theoretical and experimental diagrams agree qualitatively is of great significance. It shows that the oriented gas model correctly gives many important features of the structure and properties of molecular crystals. The concept of localized excitons and deforming excitations does not annul this model, but rather makes more accurate what is already a genuine first approximation.

In a number of recent papers by various authors, the large part played by lattice defects in the luminescence of crystals has been established beyond doubt. The constancy of the polarization diagrams for different samples of the same substance and their qualitative agreement with calculation allows us to understand the role of defects more exactly. One must apparently not speak of luminescence from defects, but rather from molecules disposed around defects. The orientation of these molecules differs little from that of the molecules in the normal lattice, but the energy terms of such molecules may experience considerable displacement under the influence of the crystal lattice field distorted near a defect. The deviation from the oriented gas model and the existence of interaction between the molecules appears first of all in the transfer of excitation energy between molecules.

In molecular crystals, a fruitful way of studying this process is to examine the variation of the polarization of luminescence with that of the exciting light. Calculations not allowing for energy migration show that there should be a strong variation of the polarization of luminescence with the position of the electric vector in the exciting light. The fact that no such variation can be found in experiment may be taken as a proof of the migration of energy between molecules. In order to make sure that this proof was valid, a number of factors complicating the phenomenon were carefully taken into consideration or eliminated, in particular the effect of double refraction of the exciting light, as well as optical activity, dichroism, etc. In this way a large number of single-component and mixed crystals were examined. In all cases energy migration between the molecules of the main lattice was shown to occur.

In mixed crystals it was shown, with the help of sepcially developed and very sensitive methods, that energy migration took place between the impurity molecules. Migration occurred even for small concentrations of impurity, of the order of $5 \cdot 10^{-5}$ g/g, corresponding to a mean distance of 25 lattice constants between the impurity molecules (about 200 A). The migration in this case cannot be explained either by resonance or exciton mechanisms, and evidently has a special nature.

It must be noted that these experimental results may also be interpreted from the point of view of the hypothesis of impurity molecules. The same method was used to show the existence of migration in pure and mixed crystals at low temperatures (in liquid nitrogen).

An analogous constancy of the polarization of luminescence with respect to that of the exciting light was found in ZnS single crystals as well as molecular crystals; this may evidently be explained by the migration of holes in the crystal.

The spectral polarization of molecular crystals gave some interesting results. The extremely sensitive and accurate spectropolarimetric photoelectric systems created for this purpose enabled us to measure the degree of polarization as a function of luminescence wavelength over the whole spectrum, right up to the fall in intensity at the extreme short-wave edge, for a number of molecular crystals.

In a number of the crystals investigated (stilbene, anthracene, etc.) the following facts were noted. In the whole long-wave region of the luminescence spectrum, the polarization was approximately constant, and to the extent discussed above corresponded to the oriented gas model (in other words, to luminescence of localized excitons). At the short-wave edge of the spectrum, however, there were sharp changes in the polarization of the luminescence. These changes may be explained by the presence of weak luminescence polarized along definite crystallographic axes at the short-wave edge of the spectrum. It is natural to associate this with radiation of free excitons. From the character of the sharp changes in polarization obtained with definite orientations of the crystals, we may determine the crystallographic axes along which this luminescence is polarized. The results obtained agree with theoretical calculations based on group theory methods.

In order to make these results completely convincing, we made allowance for all possible complicating factors (traces of exciting light, dichroism of the crystals, etc.) carrying out many control experiments and correction calculations. Even with all these factors and corrections the conclusions remained valid.

The effect of lowering the temperature to that of liquid nitrogen on the polarization phenomena at the edge of the spectrum was examined.

Analogous studies were made of mixed crystals. These may be regarded as control experiments, since free excitons cannot exist in the impurity sublattice, especially at low concentrations.

The polarization of luminescence from impurities (naphthacene in anthracene, stilbene in tolan) did not show the characteristic changes of polarization at the edge of the spectrum which occurred for pure crystals.

Some interesting results were obtained also in studying the polarization of radioluminescence from molecular crystals, and the effect of ionizing radiations on the polarization of photoluminescence.

Thus it was established that the changes in polarization at the short-wave edge of the spectrum on excitation by $\beta$-rays were in some crystals (anthracene) less than for photoexcitation, while in others (stilbene) they were completely absent. This leads to the conclusion that free excitons play a smaller part under radioexcitation than under photoexcitation.

From this conclusion we may in turn make some suggestions regarding the different mechanisms of these two kinds of excitation.

In this brief conclusion we have listed only the main results of our study of the structure and properties of molecular crystals and the processes of energy migration in these by means of the polarization of luminescence. However, these quite convincingly show the fruitfulness of using quantitative polarization methods as a refined instrument of studying anisotropic media.

The work described above is the result of investigations carried out over a number of years in the S. I. Vavilov Luminescence Laboratory of the P. N. Lebedev Physical Institute, Academy of Sciences, USSR.

It is my pleasant duty to express deep gratitude to the Direct of the Laboratory, Professor V. L. Levshin for constant attention and interest in this work and for creating the conditions for its fulfilment.

I am very grateful to M. D. Galanin and V. L. Broude for useful discussions and valuable advice.

I am extremely grateful to my colleagues V. I. Gribkov, Sh. D. Khan-Magometova, and V. N. Varfolo-meeva, in joint work with whom a number of the results set out here were obtained, and also to É. Kh. Sadekov and A. Ya. Mitin for technical assistance.

I also sincerely thank L. M. Belyaev and Yu. V. Naboikin, in whose laboratories a number of crystals were grown and kindly presented to us for investigation.

# LITERATURE CITED

1. S. I. Vavilov, Microstructure of Light, Collected Works, 11:383.
2. V. L. Levshin, Photoluminescence of Solids and Liquids, Moscow, GITTL, 1951.
3. P. P. Feofilov, Polarized Luminescence of Atoms, Molecules, and Crystals, Fizmatgiz, 1959.
4. M. Born and Khuan Kun', Dynamic Theory of Crystal Lattices [Russian translation] IL, 1958.
5. E. Fermi, Molecules and Crystals [Russian translation] IL, 1947.
6. A. I. Kitaigorodskii, Organic Crystallochemistry, Izd. AN SSSR, 1955.
7. A. G. Gurvich, Theory of the Biological Field, Moscow, 1944; S. Reed, Excited Electron States in Chemistry and Biology [Russian translation IL, 1960; A. Szent-Gyorgyi, Bioenergetics [Russian translation] 1, IL, 1951; Questions of the Electron Microscopy of Tissues [Russian translation],Editor,G. M. Frank, IL, 1959; Modern Problems of Biophysics [Russian translation] 1, 2, IL, 1961.
8. Ya. I. Frenkel', Phys. Rev., 37:17 (1931); Phys. Rev., 37:1276 (1931); Zh. Eksperim.i Teor. Fiz., 6:647 (1936).
9. A. S. Davydov, Theory of the Absorption of Light in Molecular Crystals, Kiev, Izd. AN UkrSSR, 1951.
10. I. M. Robertson, Proc. Roy. Soc., A 142:674 (1933).
11. I. M. Robertson, Proc. Roy. Soc., A 140:79 (1933); A. M. Mathieson, I. M. Robertson, and V. C. Sinclair, Acta Cryst., 3:245, 251 (1950).
12. A. S. Davydov, Zh. Eksperim. i Teor. Fiz., 17:1106 (1947).
13. A. F. Lubchenko, Opt.i Spektroskopiya, 1:867 (1956).
14. I. M. Robertson and I. Woodward, Proc. Roy. Soc., A 162:568 (1937).
15. G. Ayring, J. Walter, and J. Kimball, Quantum Chemistry [Russian translation] IL, 1948.
16. D. P. Craig and P. C. Hobbins, J. Chem. Soc., 1955 (Febr.):539.
17. D. P. Craig, J. Chem. Soc., 1955 (July):2302.
18. D. P. Craig and P. C. Hobbins, J. Chem. Soc., 1955 (July):2309.
19. É. I. Rashba, Otp. i Spektroskopiya, 2:88 (1957).
20. E. I. Rashba, Opt. i Spektroskopiya, 3:568 (1957).
21. F. Seitz, The Modern Theory of Solids, New York and London, McGraw-Hill Book Co., 1940.
22. V. L. Broude, A. F. Prikhot'ko and É. I. Rashba, Usp. Fiz. Nauk, 67:99 (1959).
23. D. Fox and S. Yatsiv, Phys. Rev., 108:938 (1957).
24. S. I. Pekar, Zh. Eksperim. i Teor. Fiz., 35:522 (1958).
25. A. S. Davydov, Collection in Memory of S. I. Vavilov, Izd. AN SSSR, 1952:210.
26. V. M. Arganovich, Usp. Fiz. Nauk, 71:141 (1960).
27. U. Fano, Phys. Rev., 103:1202 (1956).
28. V. M. Arganovich, Opt. i Spektroskopiya, 4:586 (1958).
29. Yu. M. Popov and A. S. Selivanenko, Opt. i Spektroskopiya, 9:260 (1960).
30. A. F. Prikhot'ko, Izv. AN SSSR, seriya fizich., 12:499 (1948).
31. A. F. Prikhot'ko, Zh. Eksperim. i Teor. Fiz., 19:383 (1949).
32. M. T. Shpak and E. F. Sheka, Izv. AN SSSR, seriya fizich., 24:553 (1960).
33. I. V. Obreimov and A. F. Prikhot'ko, Sov. Phys., 9:48 (1936).
34. I. V. Obreimov and A. F. Prikhot'ko, Collection in Memory of S. I. Vavilova, Izd. AN SSSR, 1952:197.
35. L. E. Lyons, J. Chem. Phys., 23:1973 (1955).
36. M. S. Brodin and A. F. Prikhot'ko, Opt. i Spektroskopiya, 2:448 (1957).
37. A. F. Prikhot'ko and I. Ya. Fugol', Opt. i Spektroskopiya, 4:335 (1958).

38. I. Ya. Fugol' and S. Z. Shul'ga, Opt. i Spektroskopiya, 5:34 (1958).

39. J. W. Sidman, Phys. Rev., 102:96 (1956).

40. A. Yu. Éichis, Dissertation, Kiev, 1952.

41. A. S. Davydov, Zh. Eksperim. i Teor. Fiz., 21:673 (1951).

42. V. L. Broude, V. S. Medvedev, and A. F. Prikhot'ko, Zh. Eksperim. i Teor. Fiz., 21:665 (1951).

43. V. L. Broude and A. F. Prikhot'ko, Zh. Eksperim. i Teor. Fiz., 22:605 (1952).

44. V. L. Broude, V. S. Medvedev, and A. F. Prikhot'ko, Opt. i Spektroskopiya, 2:317 (1957).

45. V. L. Broude, Izd. AN SSSR, seriya fizich., 17:699 (1953).

46. M. S. Brodin, Opt. i Spektroskopiya, 1:876 (1956).

47. M. S. Brodin, O. S. Pakhomova, and A. F. Prikhot'ko, Opt. i Spektroskopiya, 5:122 (1958).

48. A. F. Prikhot'ko and I. Ya. Fugol', Opt. i Spektroskopiya, 7:35 (1959).

49. A. F. Prikhot'ko and M. T. Shpak, Opt. i Spektroskopiya, 4:17 (1958).

50. A. F. Prikhot'ko and M. T. Shpak, Opt. i Spektroskopiya, 4:30 (1958).

51. P. Pringsheim, Fluorescence and Phosphorescence [Russian translation] IL, 1951.

52. M. D. Galanin and Z. A. Chizhikova, Zh. Eksperim. i Teor. Fiz., 26:624 (1954).

53. G. T. Wright, Proc. Phys. Soc., B 68:701 (1955).

54. K. S. Krishnan and P. K. Seshan, Proc. Indian Acad. Sci., A 8:487 (1938).

55. K. S. Krishnan and P. K. Seshan, Current Sci., 3:26 (1934).

56. K. S. Krishnan and P. K. Seshan, Kristallogr., A 89:538 (1934).

57. K. S. Krishnan and P. K. Seshan, Acta Phys. Polon., 5:289 (1936).

58. S. C. Ganguly and N. K. Chaudhuri, J. Chem. Phys., 19:617 (1951).

59. S. C. Ganguly and N. K. Chaudhuri, J. Chem. Phys., 21:554 (1953).

60. S. C. Ganguly and N. K. Chaudhuri, Z. Phys., 135:255 (1953).

61. N. K. Chaudhuri, Z. Phys., 151:93 (1958).

62. N. K. Chaudhuri and S. C. Ganguly, Proc. Roy. Soc., 259:419 (1960).

63. S. Brodersen, Compt. Rend. Acad. Sci., 233:1094 (1951).

64. P. Pesteil, Compt. Rend. Acad. Sci., 234:2532 (1952); 235:150, 1384 (1952); 237:235 (1953); 238:75, 226, 1400, 1789 (1954).

65. P. Pesteil and M. Barbaron, J. Phys. Radium, 15:407 (1954).

66. P. Pesteil, Theses. Paris, 1954.

67. F. R. Lipsett and L. Tardif, Canad. J. Phys., 36:1438 (1958).

68. J. W. Sidman, J. Chem. Phys., 25:115 (1956).

69. J. W. Sidman, J. Chem. Phys., 25:122 (1956).

70. J. Ferguson and W. G. Schneider, J. Chem. Phys., 25:780 (1956).

71. L. E. Lyons and I. W. White, J. Chem., Soc., 1960 (December):5213.

72. P. W. Alexander, A. R. Lacey, and L. E. Lyons, J. Chem. Phys., 34:2200 (1961).

73. A. R. Lacey and L. E. Lyons, Proc. Chem. Soc., 1960 (414).

74. N. C. Wolf, Z. Naturforsch., 13a:414 (1958).

75. D. S. McClure, J. Chem. Phys., 25:481 (1956).

76. V. L. Levshin, Tr. Fiz. Inst. Akad. Nauk, 1:19 (1938).

77. F. Perrin, J. Phys. Radium, 7:390 (1926).

78. A. Yablonskii, Z. Phys., 96:236 (1935).

79. H. Sponer and E. Teller, Rev. Mod. Phys., 13:75 (1941).

80. A. N. Terenin, Photochemistry of Dyes, Izd. AN SSSR, 1947.

81. H. Kautsky and A. Hirsch, Chem. Ber., 65:401 (1932).

82. A. Yablonskii, Nature, 133:140 (1934); Acta Phys. Polon., 3:421 (1934); 4:371, 389 (1935).

83. P. Pringsheim, Acta Phys. Polon., 4:331 (1935).

84. P. P. Feofilov, Zh. Eksperim. i Teor. Fiz., 12:328 (1942).

85. P. P. Feofilov, Izv. AN SSSR, seriya fizich., 9:317 (1945).

86. N. D. Zhevandrov, Tr. Fiz. Inst. Akad. Nauk, 6:121 (1955).

87. J. M. Preston and P. C. Tsien, J. Soc. Dyers and Colorists, 66:261, 357 (1950).

88. J. H. Scharf, Z. Naturforsch., 10b: 355 (1955).

89. W. Kuhn, H. Dührkopf, and H. Martin, Z. Phys. Chem. (B), 45:121 (1939).

90.   S. Nikitin, Compt. Rend. Acad. Sci., 205:124, 1058 (1937).

91.   J. Grailich, Krystallographisch-Optische Untersuchungen. Wien, 1858.

92.   F. Weigert, Verh. d. Deut. Phys. Ges. (3), 1:100 (1920).

93.   S. I. Vavilov and V. L. Levshin, Z. Phys., 16:135 (1923).

94.   V. L. Levshin, Zh. Rus. Fiz. Khim. Obshch., ch. fiz., 57:283 (1925). Z. Phys. 32: 307 (1925).

95.   F. Perrin, Ann. de Phys. 12:169 (1929); Acta Phys. Pol., 5:335 (1936).

96.   P. P. Feofilov, Dokl. Akad. Nauk, SSSR, 57:343 (1947).

97.   A. N. Sevchenko, G. P. Gurinovich, and K. N. Solov'ev, Dokl. Akad. Nauk SSSR, 128:510 (1959).

98.   V. L. Levshin, Z. Phys., 26:274 (1924).

99.   E. Gaviola and P. Pringsheim, Z. Phys., 24:24 (1924).

100.  F. Weigert and G. Käppler, Z. Phys., 25:99 (1924).

101.  M. D. Galanin, Tr. Fiz. Inst. Akad. Nauk, 12:3 (1960).

102.  Th. Förster, Fluoreszenz Organischer Verbindungen. Gottingen, 1951.

103.  M. D. Galanin, Tr. Fiz. Inst. Akad. Nauk, 5:341 (1950).

104.  V. L. Levshin, Zh. Rus. Fiz. Khim. Obshch., 56:235 (1924); Z. Phys., 26:274 (1924).

105.  G. P. Gurinovich and A. N. Sevchenko, Izv. AN SSSR, seriya fizich., 22:1407 (1958).

106.  A. N. Sevchenko, G. P. Gurinovich, and A. M. Sarzhevskii, Dokl. Akad. Nauk SSSR, 126:979 (1959).

107.  A. N. Sevchenko, G. P. Gurinovich, and A. M. Sarzhevskii, Dokl. Akad. Nauk SSSR, 127:1191 (1959).

108.  S. I. Vavilov, Izv. AN OMEN:1451 (1932). S. I. Vavilov and E. M. Brumberg, Phys. Z. Sowjetunion, 3:103 (1933). S. I. Vavilov, Dokl. Akad. Nauk SSSR, 17:459 (1937).

109.  S. I. Vavilov, Zh. Eksperim. i Teor. Fiz., 10:1363 (1940).

110.  P. P. Feofilov, Dokl. Akad. Nauk SSSR, 55:407 (1947).

111.  J. Ketskemety and L. Szalay, Acta Phys. Acad. Sci. Hung., 5:305 (1955).

112.  A. N. Sevchenko and G. P. Gurinovich, Dokl. Akad. Nauk SSSR, 117:798 (1957).

113.  P. P. Feofilov, Dokl. Akad. Nauk SSSR, 92:545, 743 (1953).

114.  P. P. Feofilov, Opt. i Spektroskopiya, 1:131 (1956).

115.  P. P. Feofilov, Dokl. Akad. Nauk SSSR, 99:975 (1954).

116.  P. Frohlich, Z. Phys., 35:193 (1926).

117.  S. I. Vavilov, Z. Phys., 55:690 (1929).

118.  A. N. Sevchenko, G. P. Gurinovich, and K. N. Solov'ev, Dokl. Akad. Nauk SSSR, 133:564 (1960).

119.  L. A. Kravtsov and V. V. Gruzinskii, Dokl. Akad. Nauk Belorus. SSR, 1:103 (1957).

120.  P. P. Feofilov, Zh. Eksperim. i Teor. Fiz., 12:328 (1942).

121.  E. Laffitte, Compt. Rend. Acad. Sci., 232:812 (1951); 234:424 (1952); 235:36 (1952); 236:680 (1953); J. Phys. Radium, 15:375 (1954); Ann. de Phys., 10:71 (1955).

122.  R. Stupp and H. Kuhn, Helv. Chim. Acta, 35:2469 (1952).

123.  I. C. Goedheer, Nature, 176:928 (1955).

124.  G. P. Gurinovich, I. N. Ermolenko, A. N. Sevchenko, and K. N. Solov'ev, Material of the Tenth Conference on Spectroscopy, I, L'vov, 1957:375.

125.  N. D. Zhevandrov, Dokl. Akad. Nauk SSSR, 74:25 (1950).

126.  V. I. Gribkov and N. D. Zhevandrov, Dokl. Akad. Nauk SSSR, 98:565 (1954).

127.  L. A. Spektorov, Dokl. Akad. Nauk SSSR, 65:485 (1949).

128.  A. A. Shishlovskii, Zh. Eksperim. i Teor. Fiz., 7:1252 (1937).

129.  C. A. Coulson, Proc. Phys. Soc., 60:257 (1948).

130.  M. V. Vol'kenshtein and L. A. Vorovinskii, Dokl. Akad. Nauk SSSR, 85:737 (1952).

131.  K. N. Jones, Chem. Rev., 41:353 (1947).

132.  N. D. Zhevandrov, V. L. Levshin, and K. K. Mozgova, Izv. AN SSSR, seriya fizich., 13:49 (1949).

133.  I. V. Obreimov and A. F. Prikhot'ko, Izv. AN SSSR, seriya fizich., 14:550 (1950).

134.  R. Williams, J. Chem. Phys., 26:1186 (1957).

135.  N. D. Zhevandrov, Izv. AN SSSR, seriya fizich., 20:570 (1956).

136.  W. Kuhn, Helv. Chim. Acta, 31:1780 (1948).

137.  N. D. Zhevandrov, Izv. AN SSSR, seriya fizich., 20:553 (1956).

138.  I. V. Obreimov, A. F. Prikhot'ko, and I. V. Rodinkova, Zh. Eksperim. i Teor. Fiz., 18:409 (1948).

139.  M. Born, Optics [Russian translation], Moscow, ONTI, 1937:306.

140. R. Pockels, Lehrbuch. der Kristalloptik. Lepzig-Berlin, 1906.

141. M. D. Galanin and Z. A. Chizhikova, Zh. Eksperim. i Teor. Fiz., 26:624 (1954).

142. G. Lorentz, Lectures on Theoretical Physics [Russian translation] ONTI, 1935.

143. G. S. Landsberg, Optika, Moscow-Leningrad, GTTI, 1954:711.

144. N. E. Nechaev, A. N. Faidysh, and O. P. Kharitonova, Tr. Inst. Fiz. AN UkrSSR, (2):81 (1952).

145. E. I. Bowen, J. Chem. Phys., 13:306 (1945).

146. E. L Bowen, E. Mikiewicz, and F. W. Smith, Proc. Phys., Soc., A 62:26 (1949).

147. M. D. Galanin and Z. A. Chizhikova, Opt. i Spektroskopiya, I:175 (1956).

148. V. N. Varfolomeeva and N. D. Zhevandrov, Dokl. Akad. Nauk SSSR, 115:1115 (1957).

149. V. N. Varfolomeeva and N. D. Zhevandrov, Opt. i Spektroskopiya, 5:571 (1958).

150. N. M. Melankholin and S. V. Grum-Grzhimailo, Methods of Studying the Optical Properties of Crystals, Izd. AN SSSR, 1954:170.

151. J. M. Robertson and I. Woodward, Proc. Roy. Soc., A 164:436 (1938).

152. I. V. Obreimov, A. F. Prikhot'ko, and K. G. Shabaldas, Zh. Eksperim. i Teor. Fiz., 6:1062 (1936).

153. P. Pesteil and M. Barbaron, J. Phys. Radium, 15:92 (1954).

154. E. G. McRae, J. Chem. Phys., 33:932 (1960).

155. P. Pesteil and M. Barbaron, Compt. Rend. Acad. Sci., 236:1763 (1953); 237:884 (1953).

156. F. Perrin, Theses. Paris, 1929.

157. Grisebach, Z. Phys., 101:20 (1936).

158. W. Kuhn and Freudenberg, Handb. u. Jahrbuch d. Chem. Phys., 8, Part 3:47 (1932).

159. A. Cotton, Ann. Chim. Phys., 8:360 (1896).

160. S. Chandrasekhar, Proc. Indian Acad. Sci., A 39:243 (1954).

161. V. M. Arganovich, Dokl. Akad. Nauk SSSR, 47:797 (1954); Opt. i Spektroskopiya, I:338 (1956); 2:738 (1957).

162. B. N. Samoilov, Zh. Eksperim. i Teor. Fiz., 18:1030 (1948).

163. M. S. Brodin and Ya. O. Dovgii, Opt. i Spektroskopiya, 12:285 (1962).

164. O. Neunhoeffer and H. Ulrich, Z. Elektrochem., 59:122 (1955).

165. F. Veigert, Optical Methods of Chemistry, Goskhimtekhizdat, 307 (1933).

166. F. Perrin, Ann. Phys., 21:169 (1929); 17:73 (1937).

167. D. L. Dexter, J. Chem. Phys., 21:836 (1953).

168. S. I. Vavilov, M. D. Galanin, and F. M. Pekerman, Izv. AN SSSR, seriya fizich., 13:18 (1949).

169. B. Ya. Sveshnikov and P. P. Feofilov, Zh. Eksperim i Teor. Fiz., 10:1372 (1940).

170. A. N. Sevchenko, Dokl. Akad. Nauk SSSR, 42:349 (1944).

171. S. I. Vavilov, Dokl. Akad. Nauk SSSR, 16:263 (1937).

172. A. N. Faidysh, Dokl. Akad. Nauk UkrSSR, (3):215 (1953).

173. V. M. Arganovich and A. N. Faidysh, Opt. i Spektroskopiya, 1:983 (1956).

174. V. M. Arganovich and A. N. Faidysh, Opt. i Spektroskopiya, 1:885 (1956).

175. A. N. Faidysh, I. Ya. Kucherov, and A. A. Terskii, Opt. i Spektroskopiya, 1:403 (1956).

176. I. Ya. Kucherov, A. N. Faidysh, and Z. N. Fesenko, Opt. i Spektroskopiya, 2:462 (1957).

177. A. N. Faidysh and I. Ya. Kucherov, Ukr. Fiz. Zh. 2:68 (1957).

178. I. Ya. Kucherov and A. N. Faidysh, Izv. AN SSSR, seriya fizich., 22:29 (1958).

179. L. S. Cohen and A. Weinreb, Phys. Rev., 93:1117 (1954); Bull. Res. Concl. Israel, 3:442 (1954); Proc. Phys. Soc., B 69:593 (1956).

180. T. P. Belikova and M. D. Galanin, Opt. i Spektroskopiya, 1:168 (1956).

181. T. P. Belikova, M. D. Galanin, and Z. A. Chizhikova, Izv. AN SSSR, seriya fizich., 20:384 (1956).

182. H. Kallmann and M. Furst, Phys. Rev., 79:857 (1950); 81:853 (1951); 85:816 (1952).

183. N. D. Zhevandrov, Dokl. Akad. Nauk SSSR, 83:677 (1952).

184. N. D. Zhevandrov, Dokl. Akad. Nauk SSSR, 100:455 (1955).

185. N. D. Zhevandrov, Izv. SSSR, seriya fizich., 26:67 (1962).

186. N. D. Zhevandrov, V. I. Gribkov, and Sh. D. Khan-Magometova, Opt. i Spektroskopiya, 11:629 (1961).

187. N. D. Zhevandrov, V. I. Gribkov, and Sh. D. Khan-Magometova, Opt. i Spektroskopiya, 13:96 (1962).

188. N. C. Allen, Philos. Mag., 3:1037 (1927).

189. K. Banerjee and K. L. Sinha, Indian J. Phys., 11:409 (1937).

190. M. S. Brodin and A. F. Prikhot'ko, Material of the Tenth Conference on Spectroscopy, I, L'vov, 16 (1957).

191. V. L Gribkov, N. D. Zhevandrov, and Sh. D. Khan-Magometova, Izv. Akad. Nauk SSSR, seriya fizich., 24:740 (1960).

192. O. Simpson, Proc. Roy. Soc., 238:402 (1957).

193. D. Birks, Scintillation Counters [Russian translation] IL, 74, 1955.

194. V. L. Broude and M. L Onoprienko, Contribution to the 13th Conference on Spectroscopy, in Leningrad, July, 1960. Summaries, p. 22, Izd. AN SSSR, 1960.

195. V. L. Broude and É. I. Rashba, Contribution to the Ukrainian Conference on Physical Optics, Kiev, February, 1961.

196. T. Förster, Ann. d. Phys., 2:55 (1948); Z. Naturforsch., 4a:321 (1949).

197. M. D. Galanin, Zh. Eksperim. i Teor. Fiz., 28:485 (1955).

198. V. M. Arganovich, Opt. i Spektroskopiya, 9:113 (1960).

199. L. E. Lyons and J. W. White, J. Chem. Phys., 29:447 (1958).

200. V. I. Gribkov and N. D. Zhevandrov, Opt. i Spektroskopiya, 8:275 (1960).

201. M. N. Alentsev and E. I. Panasyuk, Opt. i Spektroskopiya, 5:207 (1958).

202. P. P. Feofilov, Contribution to the Second Conference on the Physics of Alkali Halide Crystals, 1961.

203. G. B. Bokii, Introduction to Crystal Chemistry, Izd. MGU, 1954:105, 357.

204. D. S. Belyankin and V. P. Petrov, Crystal Optics, Gosgeolizdat, 1951.

205. A. Lempicki, Phys. Rev. Letters, 2:155 (1959).

206. V. V. Antonov-Romanovskii, J. Phys. Radium, 17:661 (1956).

207. N. D. Zhevandrov, Izv. AN SSSR, seriya fizich., 22:1332 (1958).

208. N. D. Zhevandrov, V. L Gribkov, and V. N. Varfolomeeva, Izv. AN SSSR, seriya fizich., 23:57 (1959).

209. G. N. Lewis, T. T. Magel, and D. Lipkin, J. Amer. Chem. Soc., 62:2973 (1940).

210. G. Scheibe, Angew. Chemie, 50:51 (1937).

211. V. L Gribkov, N. D. Zhevandrov, and Sh. D. Khan-Magometova, Opt. i Spektroskopiya, 10:549 (1961).

212. Sh. D. Khan-Magometova, N. D. Zhevandrov, and V. L Gribkov, Izv. AN SSSR, seriya fizich., 24:561 (1960).

213. M. D. Galanin, Contribution to the Tenth Conference on Luminescence, Moscow, 1961.

214. J. B. Birks and F. A. Black, Proc. Phys. Soc., A 64:511 (1951).

215. F. A. Black, Philos. Mag., 44:263 (1953).

216. J. H. Schulman, H. W. Etzel, and J. G. Allard, J. Appl. Phys., 28:792 (1957).

217. H. B. Rosenstock and J. H. Schulman, J. Chem. Phys., 30:116 (1959).

218. S. L Vavilov, Technical Encyclopedia, Article "Polariscope," 17:325.

219. V. V. Kavraiskii, Zh. Eksperim. i Teor. Fiz., 20:619 (1950).

220. G. Weber, Biochem. J., 51:145 (1952).

221. L. A. Tumerman, Dokl. Akad. Nauk SSSR, 58:1945 (1947).

222. H. Wille, Optik, 9 (2):84 (1952).

223. L. A. Spektorov, Dissertation, Lenin MGPI, 1949.

224. W. Grosse, Über Polarisationsprismen. Dissertation. Hannover, 10:1886.

225. G. Weber, J. Opt. Soc. America, 46:962 (1956).

226. J. Ketskemety, L. Gargya, and E. Salkonts, Acta Phys. Chem. Szeged, 3:1 (1957).

# VIBRATIONAL SPECTRA AND STRUCTURE
# OF CERTAIN OXIDES IN THE CRYSTALLINE
# AND GLASSY STATES

## V. P. Cheremisinov

# INTRODUCTION*

Among the physical methods for studying the structure of matter in various stages of aggregation, an important position is held by investigations with the aid of Raman and infrared absorption spectra, which together provide an indication of the vibrational spectra of the substance studied. Important conclusions concerning the structure of a substance can be drawn from such characteristics of the vibrational spectrum as the number of lines, their position in the spectrum, and line intensity and polarization. However, study of the vibrational spectrum does not always yield a unique solution to the problem, especially in the case of multiatomic molecules or coordination compounds with multiatomic unit cells. Therefore other methods (chemical, etc.) and especially x-ray structural analysis have to be applied to the problem of structure determination. Of course, other methods are also useful, but it must be pointed out that the decisive part in solving the problem of structure is played by the x-ray structural and spectroscopic methods which supplement each other and can in many cases offer an entirely unambiguous answer.

From the structural standpoint, solid inorganic substances can be divided into three classes (determined by the specific nature of spatial bond distribution): substances with molecular lattice structure, substances with coordination lattice structure, and substances differing in structure from the substances of the first two classes. They include substances with layer and chain lattice structures. This classification of substances in accordance with lattice structures induces differences in the approach to studies of their vibrational spectra. Studies of the vibrational spectra of substances with a molecular structure are confined to vibrations of the atoms within a single molecule, as in such substances the intermolecular forces are very much weaker than the forces within the molecule, and the vibrational spectrum is therefore determined by the vibrations of the atoms in the molecule.

The molecule as a unit cannot be identified in substances with a nonmolecular structure. Therefore studies of the vibrational spectra of such substances are usually based on an examination of the vibrations of the atoms in the unit cell, i.e., of the least volume from which the whole lattice is built up by translation. It should

---

*Dissertation for the scientific degree of Candidate of Physicomathematical Sciences. Presented at the Physics Faculty of the Moscow State University, November 14, 1962.

Vladimir Petrovich Cheremisinov, a scientist of the P. N. Lebedev Institute of Physics, Academy of Sciences of the USSR, died on May 3,1964, in Canada, in his thirty-fourth year. A tragic accident cut short his brief, bright, and promising life. He had worked for ten years in the Institute of Physics. During this time his colleagues grew to know and love him. His scientific interests lay in investigations of the vibrational spectra and structure of glassy oxides. He carried out a series of researches in this field with the aid of his own original technique. His Candidate's Thesis, successfully presented in November 1962, was concerned with the same subjects. In the last year of his short life he studied absorption spectra of diatomic molecules. His interests and activities were wide and diverse. He combined his work in the Institute of Physics with studies in the conservatory and sporting activities, and did extensive social work. Vladimir Petrovich **Cheremisinov was a man of high** culture, perfect honesty, and a kind heart. The workers of the P. N. Lebedev Institute of Physics, Academy of Sciences of the USSR, have lost in him a splendid comrade, a worthy citizen, and a talented young investigator.

Staff of the Laboratory
of Low-Temperature Plasma Optics

be pointed out that some authors [1] base the interpretation of the vibrational spectra of such substances not on the unit cell but on a definite atomic group which is bonded in accordance with the valences of the atoms. However, this approach fails to explain a number of experimental facts [2].

A theoretical examination of a series of oxides led Zachariasen [3] to the conclusion that oxides such as $SiO_2$, $GeO_2$, $P_2O_5$, $As_2O_5$, $P_2O_3$, $As_2O_3$, $B_2O_3$, $Sb_2O_3$, $Sb_2O_5$, $V_2O_5$, $Nb_2O_5$, and $Ta_2O_5$ should form glasses.* According to this hypothesis, all glass-forming oxides in the crystalline and glassy states should have a coordination lattice structure but, in contrast to a crystal lattice, the glass lattice is a random, nonperiodic, and nonsymmetrical network. On the other hand, Lebedev's crystallite hypothesis [5] assumes the existence of ordered regions in glass.

In recent years the views of the supporters of the two hypotheses have begun to converge [6]. The result of this convergence may be summarized as follows: glass is a system which has both crystalline and random features.

It must be pointed out that the above list of glass-forming oxides is incomplete. It should probably also include $TeO_2$; although it has not been obtained as a pure one-component glass, tellurium glasses are obtained in presence of even very small amounts of other substances [7].

Together with the coordination theory of glass-forming oxides and their glasses, the molecular structure theory has developed increasingly in recent years.

For example, Sidorov [4] carried out a detailed study of the infrared absorption spectra of glass-forming oxides in the 2-24 $\mu$ region. However, the structure could be established only for the oxides with known Raman spectra ($P_2O_3$, $P_2O_5$, $B_2O_3$). Detailed studies of the vibrational spectra of these oxides [4, 8-12] showed that they have a molecular lattice structure. No definite conclusions could be drawn concerning the structure of the other glass-forming oxides, the Raman spectra of which had not been studied.

The main object of the present investigation was to study the Raman spectra, and also the infrared absorption spectra as far as 36 $\mu$, of a number of oxides, and to use the data obtained on the vibrational spectra for a further examination of the structure of glass-forming oxides not only in the crystalline state but also in the form of glasses whenever possible.

The glass-forming oxides chosen for the investigation included oxides which, according to various indications and inferences, could be regarded as having a molecular structure, and also oxides to which a coordination structure could be assigned. It would have been interesting to investigate substances with structures differing from both these. Unfortunately, none could be found among the glass-forming oxides. Therefore the substance taken to typify this class was a glassy polyphosphate with a chain structure, formed by dehydration of orthophosphoric acid.

In the choice of substances it was also taken into account whether a given oxide had been obtained in the glassy state.

The following abbreviations are used in the tables:

| 1) vw | very weak | 7) RS | Raman spectrum |
|-------|-----------|-------|----------------|
| 2) w | weak | 8) IR | infrared spectrum |
| 3) m | medium | 9) pol | polarized |
| 4) s | strong | 10) sh | shoulder [band on the |
| 5) vs | very strong | | wing (shoulder) of a |
| 6) max | maximum | | neighboring band] |

---

* Sidorov [4], who investigated a number of these oxides, called them glass-forming. We shall use this term here.

CHAPTER I

# EXPERIMENTAL

## §1. Apparatus

The infrared absorption spectra were investigated with the double-beam instrument designed in the Institute of Physics of the Academy of Sciences by Malyshev, Markov, and Shubin [13-15]. The design and characteristics of the instrument and the special techniques in its use are not discussed here, as good and accurate accounts are given in the papers cited above and in [4]. It may be added that, in addition to NaCl and KBr prisms which were used for studying infrared absorption spectra in the range from 2 to 24 $\mu$, we used the KRS-5 prism (a mixture of thallium bromide and iodide crystals) for obtaining absorption spectra in the far infrared region, from 24 to 36 $\mu$. It should be pointed out that, apart from its mechanical weakness, this prism has another very significant fault: the reflection coefficient is very large. The cesium iodide prism has better optical characteristics than KRS-5, but as it was not available all our investigations were conducted with the KRS-5 prism.

As the atomic weight of tellurium is high, it was to be expected that some of the fundamental vibration frequencies in the tellurium dioxide spectrum may be in the long-wave region. Therefore the infrared absorption spectrum of $TeO_2$ was investigated in the 53-100 $\mu$ region. The absorption spectrum was obtained with the aid of the single-beam spectrophotometer for investigations in the far infrared region.* The design of this instrument is described in [16].

The Raman spectra were recorded by means of the ISP-51 instrument with a camera having a focal length of 270 mm. The exciting line for the Raman spectra was the mercury line of $\lambda = 4358$ A in most cases. The line of $\lambda = 4047$ A was used in a few experiments. The light source was a low-pressure mercury vapor lamp [17], almost free of continuous background.

When it was necessary to find lines in the Raman spectrum close to the exciting line, the prisms of the instrument were taken out of the position of minimum deviation in order to increase dispersion. This reduced considerably the background in the spectrum due to light scattering by the optical components of the instrument. The main disadvantage of the powder method, used by us, for studying Raman spectra is the long exposure time. To reduce this undesirable effect, a low-pressure mercury vapor lamp with an end optical window was constructed. A diagram of this type of lamp is shown in Fig. 1, where a is mercury, b are molybdenum electrodes, c are side tubes for cooling the mercury by means of water, and d is the optical window. The end position of the window not only had the advantage of shortening the exposure time (as the light from the entire arc is used), but also greatly increased the life of the lamp. Experiments showed [18] that low-pressure mercury vapor lamps of the usual type lose 20% light intensity within 100 hr of operation owing

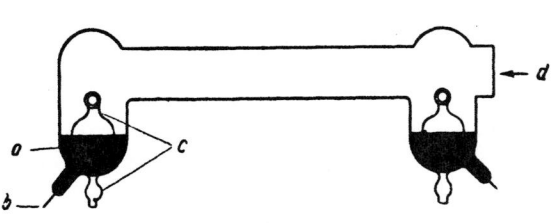

Fig. 1. Low-pressure mercury vapor lamp.

*The author thanks A. I. Demeshina and V. K. Murzin, members of the Semiconductor Laboratory, for their assistance for the use of the instrument.

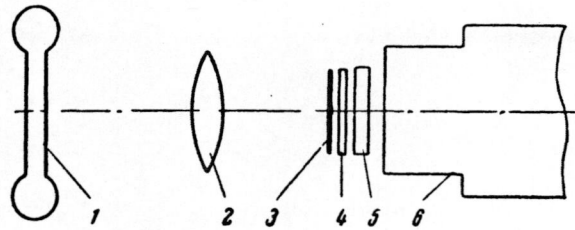

Fig. 2. Diagram of apparatus for investigating the Raman spectra of powders.

to blackening of the glass. With the window in the end position, near to a region of effective cooling, there is no blackening even after 1000 hr.

The apparatus used for qualitative investigation of the degree of depolarization of the Raman spectrum lines of certain oxides is described in [19].*

## §2. Investigation Technique

The infrared absorption spectra were investigated by the powder method, the powders being finely ground in paraffin oil. A layer of the mixture was applied onto a window made from KBr and KRS-5. The KBr windows were used for spectral investigations in the 2-24 $\mu$ region; and KRS-5 in the 24-36 $\mu$ region. The layer thickness was chosen experimentally. The same method was used for investigating the absorption spectrum of $TeO_2$ in the 53-100 $\mu$ region. A previous investigation of the transmission of paraffin oil showed good transmission (about 95%) without absorption bands in this region. In this case the mixture of powder and oil was applied onto a teflon support about 1 mm thick. Unfortunately, because of the low light flux, wide spectrometer slits had to be used, and this ultimately influenced the quality of the spectra. We used a slit width of 6 mm.

Raman spectra were obtained by two methods. If the specimens were sufficiently large, the scattered light was observed at an angle of 90° to the exciting light. For substances available as powders the exciting light was passed through a layer of the substances for investigation of the Raman spectra. Specimens which were too small for investigation by the first method were powdered and investigated by the transmission technique.

Attempts to investigate Raman spectra of powders by the reflection method were unsuccessful, mainly because (despite filtration) a strongly overexposed exciting line was obtained, giving strong scattering on the optical components of the ISP-51 instrument and masking by its background the Raman lines, especially in the low-frequency region. The high intensity of the exciting line is probably attributable to superposition of Fresnel scattering from the surface of the substance of the Rayleigh scattering. We therefore chose the transmission method for our investigation.

The choice of the light source played an important part in simplifying the equipment generally used for powder investigations. The usual procedure for investigation of powders by the transmission method is to isolate the exciting light by means of a high-transmission monochromator, to pass it through a layer of powder and filters, and to direct the scattered light onto the slight of the spectrograph used for recording the Raman spectrum. Of course, the transmission of such equipment is very low. Therefore the investigations were conducted without a monochromator, in view of the absence of background in our low-pressure mercury vapor lamp, in order to simplify the equipment and increase its transmission. A diagram of the apparatus is shown in Fig. 2, where 1 is a low-pressure mercury vapor lamp, 2 is a condenser, 3 is a filter for separating out the exciting line, 4 is a layer of the substance to be investigated, 5 is a filter for weakening the exciting line, and 6 is the spectrograph. The earlier investigations of Raman spectra were carried out with a low-pressure mercury vapor lamp of the usual type, the central optical region of the lamp being projected (as shown in Fig. 2) by the condensing lens onto the layer of material. Subsequently, in order to decrease the exposure time further, we used an end-window lamp placed directly in front of the filter. This arrangement is shown in Fig. 3.

It must be pointed out that, apart from the light source, the following factors play an important part in influencing quality of the spectra when this technique is used: selection of light-sensitive photographic plates, optimum slit width, choice of "crossed" filters and of the exciting line and, most important, the optimum thickness of the powder layer.

---

*The author is very grateful to M. M. Sushchinskii for providing the apparatus and for assistance in determination of the polarization spectra of glassy $TeO_2$ and polyphosphate.

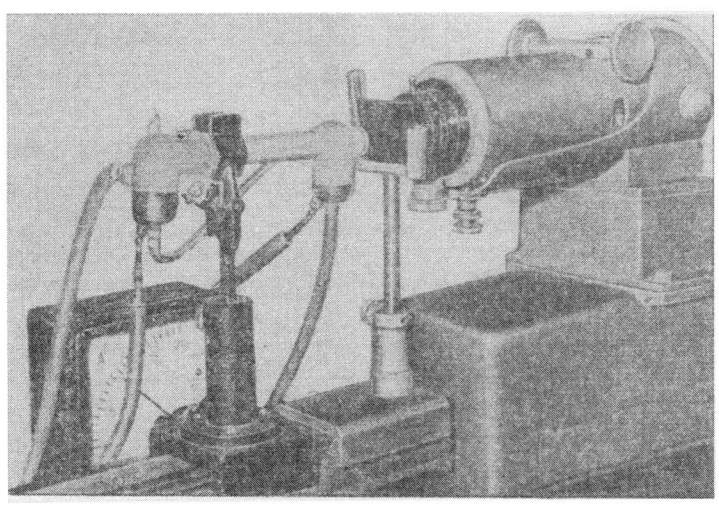

Fig. 3. Apparatus for investigating the Raman spectra of powders.

The thickness of the powder layer is very important in determination of Raman spectra by the powder method. The optimum thickness, which depends on the size and transparency of the grains, was found experimentally. Various thicknesses, from 0.25 to 2 mm, were used in our investigations.

If the thickness is unsuitable, the intensity ratio of the exciting line and the Raman spectrum lines becomes unfavorable. At thicknesses less than the optimum the intensity of the exciting line is so high that its background overlaps nearly all the Raman lines. Thicker layers result in considerable absorption not only of the exciting line but also of the Raman lines. Admittedly, it is possible to obtain a Raman spectrum in the latter case, but excessively long exposures are needed. Consequently, selection of the optimum layer thickness is very important in investigations of the Raman spectra of powders.

It should be noted that although this method has a number of advantages (it is suitable for substances without single large crystals, and small amounts of material can be used), it also has a serious defect—it is impossible to obtain polarization spectra, so that interpretation of the vibrational spectra becomes considerably more difficult.

Plates with maximum sensitivity in the 4000-5000 A region (Agfa Orthochrome Raman Plates) were used for recording the Raman spectra. The plates were developed in metol—hydroquinone developer.

Two filters were used for filtering the light from the mercury vapor lamp: the Hg 436 filter transmitted the exciting line of $\lambda = 4358$ A (the Hg 405 filter was used for the line of $\lambda = 4047$ A), and a filter consisting of a 0.003 M solution of $K_3Fe(CN)_6$ weakened the exciting line and transmitted the Raman lines. The latter filter was used only in conjunction with the Hg 436 filter, and only for substances for which a good intensity ratio between the exciting line and the Raman lines could not otherwise be obtained.

§3. Preparation and Purity of the Specimens

The crystalline oxides $As_2O_3$, $As_2O_5$, $Sb_2O_3$, and $TeO_2$ were obtained as the analytical grades, $GeO_2$ as the chemically pure grade, and $Sb_2O_5$ as the pure grade. $As_2O_3$ glass was prepared at our request in the Institute of Glass* by fusion of the cubic form of $As_2O_3$. Glassy $GeO_2$ and the crystalline insoluble form of $GeO_2$ were prepared in our laboratory [20].

A special high-temperature furnace with six Silit rods arranged vertically around its periphery was constructed for preparation of glassy germanium dioxide. The furnace temperature was measured by means of an

---

*I take this opportunity to thank N. I. Chutkina of the Institute of Glass for providing the crystalline $As_2O_3$ and preparing the glass $As_2O_3$.

optical pyrometer. The starting material was a white crystalline powder of the soluble form of germanium dioxide with a very low impurity content ($SiO_2$ < 0.0004%). Because of the considerable shrinkage and intense evaporation of germanium dioxide, several trial meltings had to be carried out in order to obtain a specimen of the required size. A platinum test tube was used as the crucible. Although at about 1400°C germanium dioxide is a fairly liquid mass, prolonged exposure at this temperature was not possible because of intense evaporation of the dioxide. Therefore the specimens contained a few gas bubbles. The specimens prepared in this manner were annealed at a cooling rate of 0.2 degree/min. Three specimens were obtained in all: two prisms 10 × 10 × 20 mm, and one cylinder 16 mm in diameter and 60 mm ling. The specimens were quite suitable for determination of Raman spectra.

The white, fine-grained, powdered germanium dioxide supplied by our industry consists mainly of the soluble form with a small admixture of the insoluble modification. The pure soluble form was prepared by crystallization of an aqueous solution of the commercial $GeO_2$ powder. The insoluble crystalline form was obtained by the hydrothermal method [21] from the soluble form. The catalytic action of steam was used for converting the one form into the other. Ten grams of the soluble form and 10 g of distilled water were put in a platinum crucible with a tightly fitting lid. The crucible was put in a stainless steel bomb sealed by means of a screw stopper. The bomb was heated in a furnace, kept for 120 hr at 480°, and cooled rapidly to room temperature. The product was freed from iron oxides in a bath with steam from 6 N aqueous hydrochloric acid. The residue was washed with warm distilled water to remove traces of acid and heated at 400°C for about 12 hr. There was no appreciable loss of weight when the powder was washed with HCl; this indicated complete conversion of the soluble into the insoluble form. In addition to the solubility test, the different forms of $GeO_2$ were identified by their refractive indices. The refractive indices found were in good agreement with the data reported in [22].

Glassy tellurium dioxide [7] was prepared from white crystalline $TeO_2$ powder, prepared by chemical synthesis, which was shown by x-ray diffraction to have a tetragonal lattice. The tellurium dioxide content of the powder was 99.0-99.5%; i.e., the starting material contained 0.5-1% impurities. This crystalline (tetragonal) form was the starting material for preparation of glass tellurium dioxide. The powdered tellurium dioxide was placed in an alundum crucible into a TG-2 furnace, heated to 800-850°, and kept at that temperature for 30-40 min until fusion was complete. The melt was then cooled to 400° at the rate of 100 degree/hour and kept for about an hour at that temperature. The glass was then cooled in the closed furnace with the current cut off. A transparent specimen was obtained in the form of a cylinder 10 mm in diameter and 10 mm long. The glass was greenish in color and was found by analysis to contain up to 6% $Al_2O_3$, which entered the specimen from the crucible walls during the melting.

The glassy polyphosphate was prepared by thermal dehydration. The starting material was orthophosphoric acid, $H_3PO_4$, in one case, and ammonium dihydrogen phosphate, $(NH_4)H_2PO_4$, in the other. The required amount of substance was put in a crucible and held at a definite temperature until the melt was quite free from gas bubbles. $H_3PO_4$ was dehydrated at about 650°, while $(NH_4)H_2PO_4$ required a temperature of about 800°C. The melt was poured out into a previously heated mold and cooled rapidly. As a protection against moisture, the glassy specimens were handled in an air-tight chamber in an argon atmosphere. Absolute alcohol was used in the grinding and polishing operations. The glasses prepared by this method were in the from of transparent prisms 100 × 15 × 15 mm. As the specimens were highly hygroscopic, they were placed in a special cell with a drying agent ($P_2O_5$) and were used in this form for studies of the Raman spectra.

CHAPTER II

# VIBRATIONAL SPECTRA OF OXIDES WITH MOLECULAR LATTICE STRUCTURE

Studies of the vibrational spectra of the oxides $P_2O_3$ and $P_2O_5$ in the crystalline state, and of $B_2O_3$ in the glassy state, and comparison of the results with the results of other physicochemical investigations showed [4, 8-12] that these oxides have a molecular structure. It follows that among the glass-forming oxides, all of which should have a coordination structure according to the Zachariasen theory, several have a molecular structure. However, it was found that several more oxides, according to conjectures or results of various investigations, may be regarded as having a molecular structure. These are the oxides $As_2O_3$, $Sb_2O_3$, $As_2O_5$. It was shown by x-ray analysis [23] that $As_2O_3$ and $Sb_2O_3$ have a molecular structure. With regard to $As_2O_5$ there is merely a statement that, by analogy with $P_2O_5$, it should also have a molecular structure. According to the Zachariasen theory, all these oxides should form glasses. Unfortunately, we obtained only one ($As_2O_3$) in the glassy state. No data could be found in the literature concerning its structure in that state, apart from [4] where it is suggested from the analogy between the infrared absorption spectra of the crystalline and glassy forms that the structure is molecular.

Thus, studies of the vibrational spectra of these oxides should show whether the results of our investigations agree with the views cited above concerning their structure.

## §1. Arsenious Anhydride [24]

Solid arsenious anhydride exists in two states, crystalline and glassy [25]. There are two forms of crystalline arsenious anhydride, cubic and monoclinic. The low-temperature (cubic) crystalline form is completely converted into the high-temperature (monoclinic) form at 275°.

Electron diffraction studies of arsenious anhydride in the gas phase [26] showed that the vapor consists of $As_4O_6$ molecules having symmetry of the point group $T_d$. It was shown by x-ray diffraction [23] that the cubic form has a molecular lattice. The structure of the monoclinic form has not been definitely established, and there are only some conjectures on this subject [27].

It is suggested in [27] that the glassy form has "a disordered progressive atomic lattice," which appears to confirm the Zachariasen hypothesis. However, the fact that when the glassy form crystallizes it passes into the cubic modification may itself indicate similarity of the two structures. In fact, a study of the infrared absorption spectrum of glassy arsenious anhydride [4] in the region from 2 to 24 μ showed a resemblance to the infrared absorption spectrum of the cubic form. This similarity of the infrared spectra is a further justification for the assertion that the glassy and low-temperature forms of arsenious anhydride have a similar structure. Therefore we may expect that a study of the structure of the different forms of arsenious anhydride based on data on its vibrational spectrum would not only confirm x-ray diffraction and chemical data, but would help in the study of the structure of glassy arsenious anhydride, as the x-ray diffraction method does not give reliable results in this case.

Discussion of Results. Before our results, only the infrared spectrum [28] of arsenious anhydride had been obtained. We therefore investigated not only the infrared absorption spectrum, with the range extended to 36 μ,

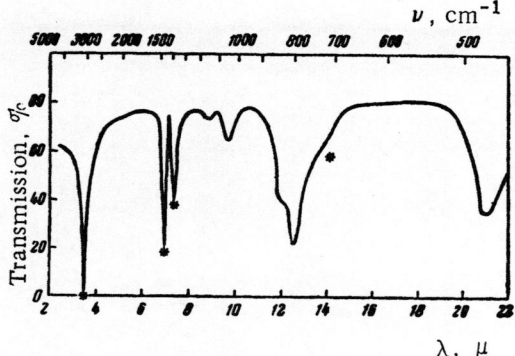

Fig. 4. Infrared absorption spectrum of crystalline arsenious anhydride in the 2-22 μ region. Bands due to the oil are indicated by asterisks in all the figures.

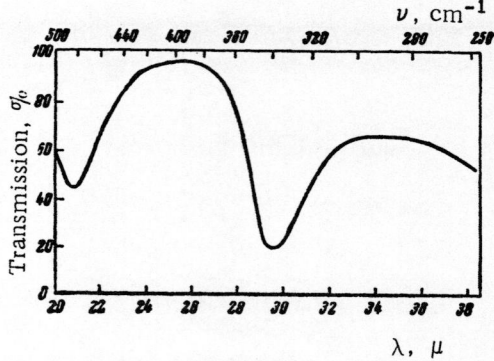

Fig. 5. Infrared absorption spectrum of crystalline arsenious anhydride in the 20-30 μ region.

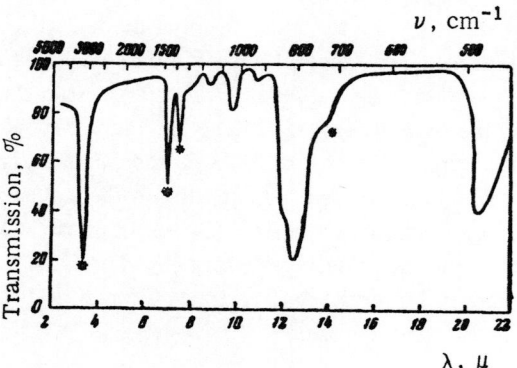

Fig. 6. Infrared absorption spectrum of glassy arsenious anhydride in the 2-22 μ region.

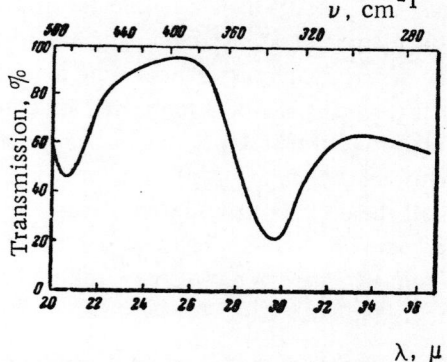

Fig. 7. Infrared absorption spectrum of glassy arsenious anhydride in the 20-36 μ region.

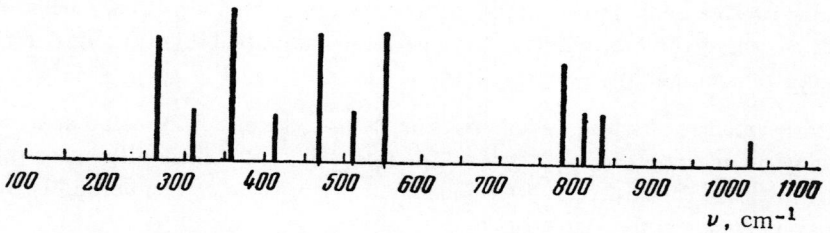

Fig. 8. Raman spectrum of crystalline arsenious anhydride.

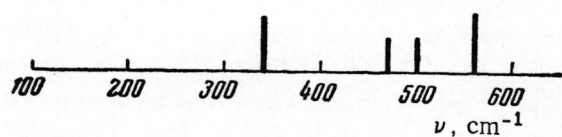

Fig. 9. Raman spectrum of glassy arsenious anhydride.

36 $\mu$, but also the Raman spectrum. The infrared absorption and Raman spectra of the crystalline cubic and glassy forms of arsenious anhydride are given in Figs. 4-9. In later work, the infrared absorption spectrum [29] and both the infrared and Raman spectra [30] were obtained. All the data on the vibrational spectrum of arsenious anhydride are collected in Table 1. The penultimate column of the table gives average values based on all the literature data and our results. In general, the agreement between all the results is good. However, the estimate of the 175 cm$^{-1}$ line intensity in [30] appears to be too high, as such an intense line should have been recorded in our measurements.

It should also be noted that our assessment of the intensity of the 842 cm$^{-1}$ line in [24] is in all probability incorrect. As this line is next to the very strong 801 cm$^{-1}$ line, its intensity may be too high. We therefore used the estimate given in [30]. The last column of the table gives results for the vibrational spectrum of glassy arsenious anhydride. The data on the infrared absorption spectrum in the 2-24 $\mu$ region for this form of arsenious anhydride are in good agreement with the results of [4], where the spectrum was investigated down to 24 $\mu$. It should be noted that in this case the measurements of the intensities are very weak. There is good agreement between the lines of the Raman spectrum of the glassy form and the strongest lines of the Raman spectrum of the cubic crystalline form of arsenious anhydride, confirming the conclusion [4] that their structures are similar. It is interesting to note that the lines of the Raman spectrum of the glassy form are fairly narrow, and there is no significant change of line breadth in the transition from the crystalline to the glassy form.

Table 1 shows that eight absorption bands were found in the infrared absorption spectrum of arsenious anhydride; four of these ($\nu = 252, 345, 482, 801$ cm$^{-1}$) are estimated as strong and the rest as weak. It is difficult to assign the band at 840 cm$^{-1}$ to this group, as its intensity is masked by the strong band at 801 cm$^{-1}$. Evidently the frequencies of the most intense bands should be assigned to fundamental vibration frequencies, and those of the weak bands to component frequencies or overtones. Thus, we have four or five strong bands in the infrared absorption spectrum.

It should be pointed out that three strong bands in the infrared spectrum coincide within the limits of experimental error with lines in the Raman spectrum ($\nu = 262, 476$ and 815 or 785 cm$^{-1}$). The fourth strong line of the infrared absorption spectrum at $\nu = 345$ cm$^{-1}$ is not found in the Raman spectrum. Later we shall consider the suggestion that the line at $\nu = 245$ cm$^{-1}$ is either lacking in the Raman spectrum or is very weak.

In addition to the three above-named frequencies, all the Raman frequencies assessed as strong and medium should be taken as fundamental. The possibility is not excluded that certain weak frequencies may also be fundamental. Thus, in the Raman spectrum we have seven to nine fundamental vibration frequencies. From a comparison with the infrared absorption spectrum a possible model of crystalline arsenious anhydride may be suggested.

The peculiarities which follow from a comparison of the two spectra should first be noted.

1. The number of frequencies in the Raman spectrum (7-9) is considerably greater than the number of fundamental frequencies in the infrared absorption spectrum (4). This indicates that the molecule cannot have symmetry of the point groups $C_2, C_3, C_8, C_{3v}$ and $C_{\infty v}$, as these have active frequencies in both spectra.

2. Some frequencies are found in both spectra. This indicates that the molecule cannot belong to point groups with a center of symmetry or to point groups for which the Raman frequencies are forbidden in the infrared spectrum or vice versa.

The following symmetry point groups are therefore excluded: $C_i, C_{2h}, D_{2h} \equiv V_h, S_6, C_{4h}, C_{6h}, D_{3d} \equiv S_{6v}, D_{4d} \equiv S_{8v}, D_{4h}, D_{5h}, D_{\infty h}, O_h$.

3. We assume that the line at 345 cm$^{-1}$ is present only in the infrared absorption spectrum. The molecule must then belong to one of the following symmetry groups: $D_3, D_4, D_5, D_6, C_{3h}, C_{5h}, D_{3h}$.

Table 2 gives calculations, analogous to those for $B_2O_3$ [10], of the numbers of frequencies active in the infrared absorption and Raman spectra for each of these point symmetry groups. For each symmetry group the molecular models selected are only those giving the number of frequencies closest to the experimental data.*

---

*As Herzberg's book [31] does not give the derivation of the number of vibrations for point group symmetry types $D_5$ and $C_{5h}$, we derived the appropriate formulas for computing the number of vibrations.

TABLE 1. Vibration Frequencies of $As_2O_3$, $cm^{-1}$

| Our work | | [28] | [4] | [29] | [30] | | Vibrational spectrum of crystal (average) | | Vibrational spectrum of glass | |
|---|---|---|---|---|---|---|---|---|---|---|
| RS | IR | IR | IR | IR | RS* | IR | RS | IR | RS | IR |
| | | | | | 83 (4) | | 83 (vs) | | | |
| | | | | | 110 (1/2) | | 110 (w) | | | |
| | | | | | 131 (1/2) | | 131 (w) | | | |
| | | | | | 175 (2) | | 175 (m) | | | |
| 270 (s) | | | | | 255 (1) | 252 (m) | 262 (m) | 262 (m) | | |
| | | | | | | 278 (w) | | 278 (w) | | |
| 306 (w) | | | | | 297 (1/2) | | 302 (w) | | | |
| | | | | | 332 (1/2) | | 332 (w) | | | |
| | 340 (s) | | | 349 (vs) | | 347 (m) | | 345 (s) | 340 (w) | 333 (m) |
| 365 (vs) | | | | | 368 (3) | | 366 (vs) | | | |
| 415 (w) | | | | | 430 (1) | | 422 (w) | | | |
| 473 (s) | 479 (s) | | 488 (s) | 480 (s) | 480 (1/2) | 488 (s) | 476 (m) | 482 (s) | 470 (vw) | 488 (m) |
| 515 (w) | | | | | 532 (1) | | 532 (w) | | 500 (vw) | |
| 550 (s) | | | | | 561 (2.5) | | 555 (vs) | | 560 (w) | |
| | | | | | 629 (0) | | 629 (vw) | | | |
| | | | | | 676 (0) | | 676 (vw) | | | |
| | | | | | 710 (0) | | 710 (vw) | | | |
| 780 (m) | 802 (vs) | 800 | 802 (vs) | 805 (vs) | 790 (1) | 800 (s) | 785 (m) | 801 (vs) | | 802 (vs) |
| 815 (m) | | | | | | | 815 (w) | | | |
| 840 (m) | 842 (s) | 840 | 842 (s) | 840 (sh) | | 838 (sh) | 840 (m) | 840 (sh) | | 842 (sh) |
| 1037 (vw) | 1041 (w) | 1040 | 1041 (w) | | 1038 (0) | 1038 (w) | 1037 (vw) | 1040 (vw) | | |
| | 1152 (vw) | | 1152 (vw) | | | | | 1152 (vw) | | |

*Intensities given in parentheses are on the arbitrary scale of [30].

138

TABLE 2. Selection Rules for Various Models of the Arsenious Anhydride Molecule with Symmetry Groups $D_3$, $D_4$, $D_5$, $D_6$, $C_{3h}$, $C_{5h}$, $D_{3h}$

| Group | Formula of molecule | No. of frequencies active in IR only | Total frequencies active in IR | No. of frequencies active in RS |
|---|---|---|---|---|
| $D_3$ | $As_4O_6$ | 3 | 11 | 13 |
| $D_4$ | $As_6O_9$ | 5 | 15 | 24 |
| $D_5$ | $As_8O_{12}$ | 6 | 22 | 30 |
| $D_6$ | $As_8O_{12}$ | 4 | 13 | 23 |
| $C_{3h}$ | $As_4O_6$ | 3 | 8 | 11 |
| $C_{5h}$ | $As_8O_{12}$ | 7 | 14 | 27 |
| $D_{3h}$ | $As_2O_3$ | 1 | 3 | 5 |
| $D_{3h}$ | $As_4O_6$ | 2 | 7 | 12 |
| Experiment | | 1 | 4—5 | 7—9 |

Table 2 shows that models of the symmetry groups $D_{3h}$ give the best agreement with experiment, but even these differ considerably from the experimental data in the number of frequencies. Therefore these models are hardly acceptable. The only remaining assumption is that the frequency 345 cm$^{-1}$ is also present in the Raman spectrum but either it is weak or the 332 cm$^{-1}$ frequency in the Raman spectrum should also be assigned to the same vibration.

Therefore all four fundamental frequencies of the infrared spectrum must be assumed present in the Raman spectrum. Therefore we have 8-10 fundamental vibrations in the Raman spectrum of arsenious anhydride. Accordingly, the arsenious anhydride molecule must belong to one of the following point groups: $C_{2v}$, $D_2 \equiv V$, $C_4$, $C_5$, $C_6$, $S_4$, $C_{4v}$, $C_{5v}$, $C_{6v}$, $D_{2d} \equiv V_d \equiv S_{4d}$, T and $T_d$. In the case of boric anhydride, Sidorov and Sobolev [10] calculated the most probable molecular model for these point groups. Table 3, which gives the selection rules for various structural models of the arsenious anhydride molecule, shows that the closest agreement with experiment is given by the model of the $As_4O_6$ molecule having symmetry of the point group $T_d$. The model of the $As_4O_6$ molecule is shown in Fig. 10.

Interpretation of the Vibrational Spectrum Frequencies of the Molecule. It has been shown that the model of the $As_4O_6$ molecule belonging to the $T_d$ point group is in good agreement with experiment. The selection rules for the vibrations of a molecule with this symmetry point group are given in Table 4 [31].

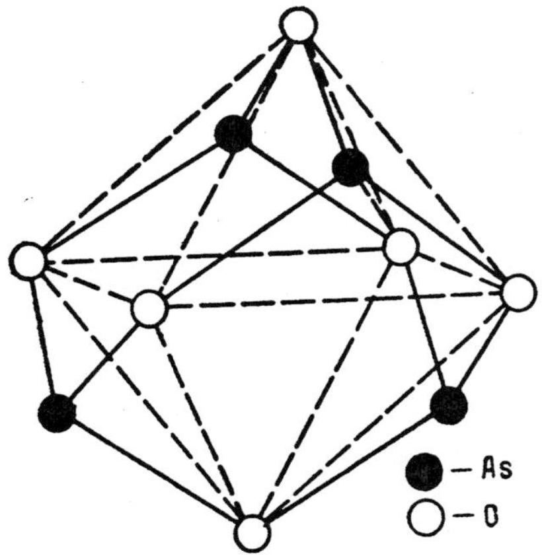

It follows from Table 4 that the $As_4O_6$ molecule with point group symmetry $T_d$ must have two totally symmetric vibrations of type $A_1$, active only in the Raman spectrum, two doubly degenerate vibrations of type E, active only in the Raman spectrum, two triply degenerate vibrations of type $F_1$, forbidden in both spectra, and four triply degenerate vibrations of type $F_2$, active in both the Raman and the infrared spectrum.

Lack of line-polarization data precludes interpretation of the $As_4O_6$ vibrational spectrum. However, assignment of the fundamental vibration frequencies is possible. As the totally symmetric vibrations have high intensity in the Raman spectrum, evidently the 366 and 555 cm$^{-1}$ frequencies 252, 482, and 801 cm$^{-1}$ should be assigned to triply degenerate vibrations of type $F_2$, as they are also active in the

Fig. 10. Structure of the $As_4O_6$ molecule.

● — As
○ — O

139

TABLE 3. Selection Rules for Various Structural
Models of the Arsenious Anhydride Molecule

| Symmetry group | Formula of molecule | No. of frequencies active (IR) | No. of frequencies active (RS) |
|---|---|---|---|
| $C_{2v}$ | $As_2O_3$ | 8 | 9 |
| $D_2 = V$ | $As_2O_3$ | 7 | 9 |
| $C_4$ | $As_4O_6$ | 12 | 18 |
| $C_5$ | $As_4O_6$ | 12 | 15 |
| $C_6$ | $As_4O_6$ | 10 | 13 |
| $S_4$ | $As_4O_6$ | 12 | 17 |
| $C_{4v}$ | $As_4O_6$ | 11 | 17 |
| $C_{5v}$ | $As_4O_6$ | 12 | 15 |
| $C_{6v}$ | $As_4O_6$ | 10 | 13 |
| $D_{2d}$ | $As_4O_6$ | 10 | 17 |
| $T$ | $As_8O_{12}$ | 13 | 23 |
| $T_d$ | $As_4O_6$ | 4 | 8 |
| $T_d$ | $As_8O_{12}$ | 9 | 17 |
| Experiment | | 4—5 | 8—10 |

TABLE 4. Selection Rules for the Point Group $T_d$

| Vibration type | Selection | | Number of vibrations |
|---|---|---|---|
| | RS | IR | |
| $A_1$ | Allowed | Forbidden | 2 |
| $A_2$ | » | » | 0 |
| $E$ | » | » | 2 |
| $F_1$ | » | » | 2 |
| $F_2$ | » | Allowed | 4 |

infrared spectrum. The frequency 345 cm$^{-1}$ should also be assigned to $F_2$, because for the $T_d$ point group only $F_2$ vibration frequencies should be present in the infrared spectrum. However, this method of comparing the activities and intensities of frequencies can hardly be conclusive for finding vibrations of type E; we therefore calculated the vibration frequencies of the $As_4O_6$ molecule having symmetry of the point group $T_d$. For this purpose we first used secular equations derived with the assumption of a valence force model for $X_4O_6$ molecules having $T_d$ symmetry [30].

The potential energy expression for this model contains three force constants: the valence constant $f$ for the As—O bond and the deformation constants d and g for the OAsO and AsOAs angles; it may be assumed that $d \approx g$, as both the angles in $As_4O_6$ are 109.5°. As deformation constants are usually smaller than valence constants by an order of magnitude, we put $d = g = \frac{1}{10} f$. However, it was found that this model is not suitable, because substitution of class $A_1$ frequencies into the equation for totally symmetric vibrations gives complex values for the force constants.

In view of the fact that the distance between the nonbonded As—As atoms (3.28 A) is considerably less than the sum of the Van der Waals radii (4.00 A), we introduced an additional force constant $f_s$ between these atoms, i.e., we used a mixed valence-force and central-force model and, as above, applied the secular equations of [30]. In calculation of the frequencies of the molecular skeleton of arsenious anhydride two lines with frequencies of 366 and 555 cm$^{-1}$ were taken for determination of the force constants.

These frequencies gave the following values for the force constants of the $As_4O_6$ molecule:

$$f = 2.72 \cdot 10^5 \text{ dynes/cm}; \qquad d = 0.272 \cdot 10^5 \text{ dynes/cm};$$
$$g = 0.272 \cdot 10^5 \text{ dynes/cm}; \qquad f_s = 0.70 \text{ dynes/cm}.$$

TABLE 5. Experimental and Calculated Vibration
Frequencies of the $As_4O_6$ Molecule, $cm^{-1}$

| Vibration type | Interpretation | Experiment | Calculation |
|---|---|---|---|
| $A_1$ | $\nu_1$ | 555 | 555 |
| | $\nu_2$ | 366 | 366 |
| $E$ | $\nu_3$ | 532 | 570 |
| | $\nu_4$ | 175 | 210 |
| $F_2$ | $\nu_5$ | 785 | 700 |
| | $\nu_6$ | 476 | 530 |
| | $\nu_7$ | 345 | 370 |
| | $\nu_8$ | 252 | 260 |

The other fundamental vibration frequencies of the $As_4O_6$ molecule, given in Table 5, were found from these force constants. Frequencies of type $F_1$ were not calculated, as by the selection rules they should not be active either in the Raman or in the infrared absorption spectrum.

Table 5 shows that calculations not only confirm the above assignment of the frequencies for type $A_1$ and $F_2$ vibrations, but make it possible to find frequencies for type E vibrations. The remaining frequencies can be interpreted as combination frequencies or overtones, or as vibration frequencies of type $F_1$ which appear owing to disregard of the selection rules.

The above discussion shows that the vibrational spectrum of arsenious anhydride can be satisfactorily interpreted on the assumption that the compound consists of $As_4O_6$ molecules having $T_d$ symmetry. This is in good agreement with the x-ray structural investigations reported in [23].

The great similarity between the vibrational spectra of crystalline and glassy arsenious anhydride, indicated by the data in Table 1, shows that glassy arsenious anhydride also has a molecular structure and consists of $As_4O_6$ molecules having symmetry of the point group $T_d$.

## §2. Arsenic Pentoxide [33]

Arsenic pentoxide exists in two forms, crystalline and glassy [34]. In contrast to arsenious anhydride, arsenic pentoxide has only one crystalline modification, the physical and chemical properties of which have been studied very little. With regard to the structure of the crystalline modification it is stated in [23] that, by analogy with phosphoric anhydride, its lattice should have a molecular structure, consisting of $As_4O_{10}$ molecules. Our discussion of the experimental results obtained in investigations of its vibrational spectrum is based on this statement.

Discussion of Results. The infrared absorption spectrum of arsenic pentoxide in the range from 2 to 36 $\mu$ (Figs. 11 and 12) has seven absorption bands, of which only one is weak while the others are of approximately equal intensity. The band in the region of about 3 $\mu$ corresponds to vibration of the OH group, indicating the presence of a small amount of water in the specimen.*

The Raman Spectrum of arsenic pentoxide (Fig. 13) contains 13 lines. A larger number of lines is possible, because in the region up to 360-400 $cm^{-1}$ determination of weaker lines is difficult owing to the strong influence of the background due to the exciting mercury line of $\lambda = 4358$ A.

*The suggestion that we obtained the spectrum of arsenic acid rather than that of its anhydride can be rejected, as our spectrum differs greatly from the spectrum of the acid. Moreover, when the powder specimen was prepared in the oil the anhydride was in contact with the air for only a short time (less than 30 sec), insufficient for the acid to form.

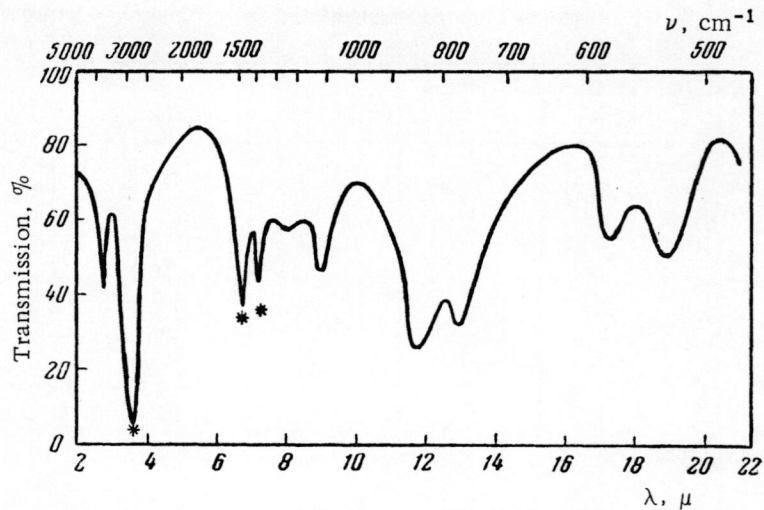

Fig. 11. Infrared absorption spectrum of crystalline arsenic pentoxide in the 2-22 μ region.

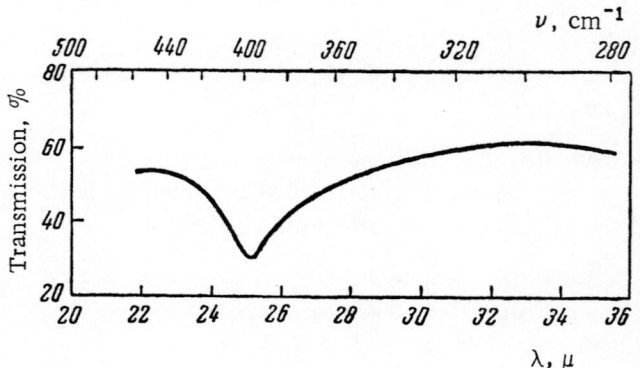

Fig. 12. Infrared absorption spectrum of crystalline arsenic pentoxide in the 20-36 μ region.

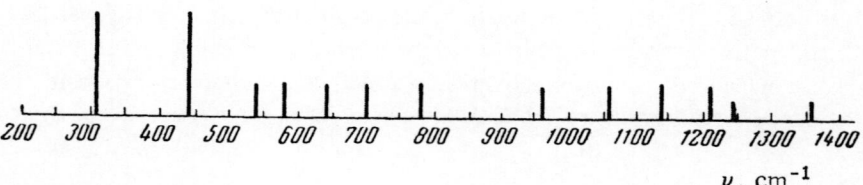

Fig. 13. Raman spectrum of crystalline arsenic pentoxide.

The line frequencies of the Raman and infrared absorption spectra and approximate assessments of their intensities are given in Table 6.

It follows from Table 6 and Figs. 11-13 that the infrared absorption spectrum contains six bands with intensities assessed higher than weak. Evidently, the frequencies of these bands must be taken to be fundamental vibration frequencies.

The Raman spectrum contains 11 lines assessed as "weak" and "very weak," and two lines of higher intensity. As the molecular weight of arsenic pentoxide is considerable, lines with frequencies above 1200 cm$^{-1}$ need hardly be taken into consideration in determining the fundamental vibration frequencies, as they are too high. Therefore all the frequencies in the Raman spectrum below this limit may be taken as fundamentals, be-

TABLE 6. Vibrational Spectrum of Arsenic Pentoxide, cm$^{-1}$

| RS | IR | RS | IR |
|----|----|----|----|
| 305 (m) |          | 954 (w)   |          |
|          | 395 (m)  | 1060 (w)  |          |
| 440 (m)  |          | 1140 (w)  | 1131 (m) |
| 535 (w)  | 527 (m)  | 1205 (s)  |          |
| 580 (w)  | 574 (m)  | 1245 (vw) | 1243 (w) |
| 640 (w)  |          | 1358 (vw) |          |
| 700 (w)  |          |           |          |
| 777 (w)  | 775 (m)  |           |          |
|          | 858 (s)  |           |          |

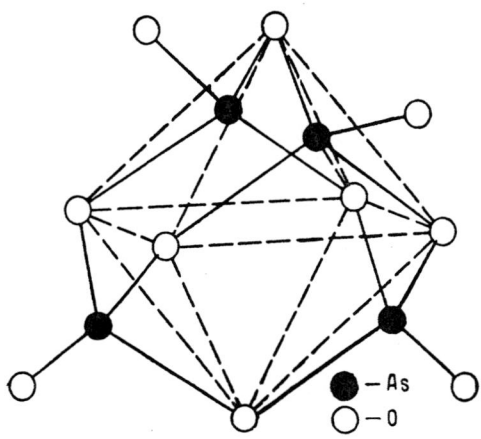

Fig. 14. Structure of the $As_4O_{10}$ molecule.

cause they are all of roughly the same intensity. We thus have 10-11 fundamental vibration frequencies appearing in the Raman spectrum. The 1205 cm$^{-1}$ frequency, at the boundary of the fundamental frequency region, is also assigned to the fundamental vibrations.

The following conclusions may be drawn from the Raman and infrared absorption spectra and from the activities of the frequencies in these spectra.

1. The arsenic pentoxide molecule cannot belong to the point groups $C_2$, $C_3$, $C_5$, $C_{3v}$, $C_{5v}$, as it is found experimentally that the number of Raman frequencies (10-11) is considerably greater than the number of infrared absorption frequencies (6).

2. The molecule cannot belong to the point groups $D_{\infty h}$, $C_i$, $C_{2h}$, $S_6$, $C_{4h}$, $C_{6h}$, $D_{3d}$, $D_{4d}$, $D_{4h}$, $D_{5h}$, $D_{6h}$, $O_h$, as some frequencies (535, 580, 777, and 1140 cm$^{-1}$) are found in both spectra.

3. Some frequencies in the infrared absorption spectrum are not found in the Raman spectrum (395 and 858 cm$^{-1}$). We calculated the numbers of frequencies for point groups having infrared frequencies forbidden in the Raman spectrum. These are the following point groups: $D_3$, $D_4$, $D_5$, $D_6$, $C_{3h}$, $C_{5h}$, $D_{3h}$. The results are given in Table 7, with experimental data shown in the bottom line. The possible models were chosen so as to give numbers of frequencies close to the experimental.

Numbers of frequencies were estimated for different positions of the atoms in a molecule belonging to a given point group but, as Table 7 shows, this has very little influence on the number of frequencies. The model with $D_{3h}$ point group symmetry gives the closest agreement, but even this deviates from the experimental data, while the other groups show even greater deviations. Therefore these models are hardly acceptable and the remaining assumption is that the frequencies of 395 and 858 cm$^{-1}$ are allowed by the selection rule in the Raman spectrum although they were not observed experimentally. Apparently these lines are difficult to detect by the powder method because of their low intensity, and the 395 cm$^{-1}$ frequency is also influenced by the background due to the exciting line.

Thus, if we assume that all the infrared frequencies are also present in the Raman spectrum, the latter should contain 12-13 frequencies rather than the 10-11 fundamental frequencies observed experimentally.

A molecule with all its infrared frequencies also present in the Raman spectrum should belong to one of the following point groups: $C_{2v}$, $D_2 \equiv V$, $C_4$, $C_5$, $C_6$, $S_4$, $C_{4v}$, $C_{5v}$, $C_{6v}$, $D_{2d}$, $T$ and $T_d$. The numbers of fre-

TABLE 7. Selection Rules for Models of Arsenic Pentoxide Molecules with Symmetry $D_3$, $D_4$, $D_5$, $D_6$, $C_{3h}$, $C_{5h}$, $D_{3h}$

| Symmetry group | No. of frequencies in RS | No. of frequencies in IR | No. of frequencies active in both spectra | Molecular model |
|---|---|---|---|---|
| $D_3$ | 18 | 18 | 12 | $As_4O_{10}$ |
| | 19 | 17 | 12 | $As_4O_{10}$ |
| $D_4$ | 23 | 13 | 9 | $As_4O_{10}$ |
| | 22 | 14 | 10 | $As_4O_{10}$ |
| $D_5$ | 19 | 11 | 8 | $As_4O_{10}$ |
| | 18 | 12 | 8 | $As_4O_{10}$ |
| $D_6$ | 25 | 20 | 13 | $As_6O_{15}$ |
| | 26 | 19 | 13 | |
| $C_{3h}$ | 18 | 13 | 7 | $As_4O_{10}$ |
| | 18 | 12 | 6 | $As_4O_{10}$ |
| | 19 | 17 | 7 | $As_2O_5$ |
| $C_{5h}$ | 18 | 8 | 4 | |
| $D_{3h}$ | 8 | 5 | 3 | |
| Experiment | 10—11 | 6 | 4 | |

TABLE 8. Selection Rules for Models of Arsenic Pentoxide Molecules with Symmetry $C_{2v}$, $D_2 \equiv V$, $C_4$, $C_5$, $C_6$, $S_4$, $C_{4v}$, $C_{5v}$, $C_{6v}$, $T$, $T_d$, $C_{2d}$

| Symmetry group | No. of frequencies in RS | No. of frequencies in IR | Frequencies active in both spectra | Frequencies active only in RS | Molecular model |
|---|---|---|---|---|---|
| $C_{2v}$ | 15 | 14 | 14 | 1 | $As_2O_5$ |
| $D_2 \equiv V$ | 15 | 11 | 11 | 4 | $As_2O_5$ |
| $C_4$ | 27 | 18 | 18 | 9 | $As_4O_{10}$ |
| $C_5$ | 22 | 16 | 16 | 6 | $As_4O_{10}$ |
| $C_6$ | 18 | 12 | 12 | 6 | $As_4O_{10}$ |
| $S_4$ | 11 | 7 | 7 | 4 | $As_2O_5$ |
| $C_{4v}$ | 11 | 8 | 8 | 3 | $As_2O_5$ |
| $C_{5v}$ | 9 | 6 | 6 | 3 | $As_2O_5$ |
| $C_{6v}$ | 21 | 18 | 18 | 3 | $As_4O_{10}$ |
| $D_{2d}$ | 11 | 7 | 7 | 4 | $As_2O_5$ |
| $T$ | 23 | 15 | 15 | 8 | $As_6O_{15}$ |
| $T_d$ | 12 | 6 | 6 | 6 | $As_4O_{10}$ |
| Experiment | 12—13 | 6 | 6 | 6—7 | |

quencies for molecular models of these groups are calculated in Table 8. The bottom line gives experimental data, which are in best agreement with the $As_4O_{10}$ molecule having symmetry of the point group $T_d$. A model of the $As_4O_{10}$ molecule is shown in Fig. 14.

<u>Interpretation of the Vibrational Spectral Frequencies of the $As_4O_{10}$ Molecule.</u> The selection rules for the $As_4O_{10}$ molecule belonging to the $T_d$ point group require the presence of three totally symmetric vibrations of

TABLE 9. Interpretation of the Vibrational Spectrum of the
Arsenic Pentoxide Molecule

| RS, cm$^{-1}$ | IR, cm$^{-1}$ | Interpre-tation | Calcula-ted fre-quencies, cm$^{-1}$ | Symmetry type of highest vi-bration state |
|---|---|---|---|---|
| 305 (m) | | $\nu_3$ | | $A_1$ |
| | 395 (m) | $\nu_{15}$ | | $F_2$ |
| 440 (m) | | $\nu_2$ | | $A_1$ |
| 535 (w) | 525 (m) | $\nu_{14}$ | | $F_2$ |
| 580 (w) | 574 (m) | $\nu_{13}$ | | $F_2$ |
| 640 (w) | | $\nu_5 - \nu_3$ | 649 | $E$ |
| 695 (w) | | $\nu_6$ | | $E$ |
| 777 (w) | 775 (m) | $\nu_{12}$ | | $F_2$ |
| | 858 (s) | $\nu_{11}$ | | $F_2$ |
| 954 (w) | | $\nu_5$ | | $E$ |
| 1060 (w) | | $\nu_4$ | | $E$ |
| 1140 (w) | 1131 (m) | $\nu_{10}$ | | $F_2$ |
| 1205 (w) | | $\nu_1$ | | $A_1$ |
| 1245 (vw) | 1243 (w) | $\nu_{11} + \nu_{15}$ | 1253 | $A_1 + E + F_1 + F_2$ |
| 1358 (vw) | | $\nu_3 + \nu_4$ | 1365 | $E$ |

type $A_1$ active in the Raman spectrum, three doubly degenerate type E vibrations active in the Raman spectrum, three triply degenerate type $F_1$ vibrations forbidden in both spectra, and six triply degenerate type $F_2$ vibrations permitted in both spectra.

The vibrational spectral frequencies of arsenic pentoxide can be interpreted from the activities and intensities of the frequencies.

Since totally symmetric vibrations are active only in the Raman spectrum and are of high intensity, the most intense lines 305 and 440 cm$^{-1}$ must be assigned to type $A_1$ vibrations. The frequencies 395, 535, 580, 777, 858, and 1131 cm$^{-1}$ should be assigned to type $F_2$ vibrations, as they are active in both spectra.

By analogy with phosphoric anhydride, which has a similar structure [4], the highest frequency of 1205 cm$^{-1}$, lying at the boundary of the fundamental frequencies, should be assigned to type $A_1$ totally symmetric vibrations. The frequencies 954 and 1060 cm$^{-1}$ are assigned to type E vibrations, as they give a satisfactory interpretation of the 640 and 1358 cm$^{-1}$ frequencies and account for the absence of these frequencies in the infrared absorption spectrum. The 700 cm$^{-1}$ frequency should probably be assigned to a third fundamental vibration of type E. The weak band at 1245 cm$^{-1}$, active in both spectra, is satisfactorily interpreted with the aid of the frequencies 395 and 858 cm$^{-1}$.

The interpretation of all the frequencies in the vibrational spectrum of arsenic pentoxide is given in Table 9. It must be pointed out that owing to the lack of polarization data the interpretation for some frequencies cannot be regarded as unique (type E vibrations).

The data on the vibrational spectrum of arsenic pentoxide and their interpretation show that they agree best with a structure of the $As_4O_{10}$ molecule belonging to the $T_d$ point group.

We did not investigate glassy arsenic pentoxide, but as crystalline arsenic pentoxide exists in only one form it may be assumed that the glassy form, like the crystalline, has a molecular structure with $As_4O_{10}$ as the structural unit. This assumption, which of course should be verified, is based on the analogy between the vibrational spectra of crystalline and glassy forms of glass-forming oxides, such as arsenious anhydride.

In conclusion it must be pointed out that although spectroscopic data confirm Ormont's view that crystalline arsenic pentoxide has a molecular structure, a thorough x-ray structural investigation is needed for a final answer to the problem.

## §3. Antimonous Anhydride

Antimonous anhydride exists in three forms: two crystalline (rhombic and cubic), and glassy. According to x-ray data [23], the structure of the cubic form is molecular and that of the rhombic, nonmolecular. Vapor density determinations showed that antimonous anhydride vapor consists of $Sb_4O_6$ molecules having $T_d$ symmetry [36].

Antimonous and arsenious anhydrides have much in common, both in their position in Mendeleev's periodic table and in their physicochemical properties. However, cubic antimonous anhydride has a higher melting point than the cubic form of arsenious anhydride. The higher melting point is apparently associated with a relative decrease of the intermolecular distance in comparison with the intramolecular. In the cubic form of arsenious anhydride the distance between arsenic and oxygen atoms of neighboring molecules is 2.78 A, and within the molecule it is 2.01 A; in the cubic form of antimonous anhydride the distance between Sb and O atoms is 2.61 A in neighboring molecules, while within the molecule it is 2.22 A according to [37], and 2.00 A according to other data [38,39]. This fairly substantial relative decrease of the intermolecular distance makes the melting point of antimonous anhydride (652°C) higher than that of arsenious anhydride (272°C). The question of which of the x-ray values for the bond length is to be preferred is discussed below, in subsection b, where calculation of the fundamental vibration frequencies of the antimonous anhydride molecule will require a knowledge of the kinematic coefficients, while for calculation of the latter it is necessary to know the numerical values of the molecular parameters.

The structure of the antimonous anhydride molecule can be determined from data on the Raman and infrared absorption spectra.

Discussion of Results. The infrared absorption spectrum and the Raman spectrum of the cubic form of antimonous anhydride are given in Figs. 15 and 16 respectively. The results are summarized in Table 10, which also contains infrared spectrum data from [4,40]. All the data on the infrared spectrum are in agreement, apart

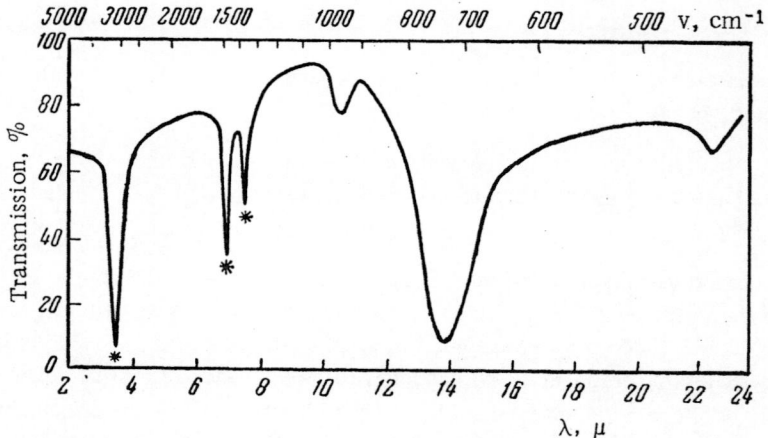

Fig. 15. Infrared absorption spectrum of crystalline antimonous anhydride in the 2-24 μ region.

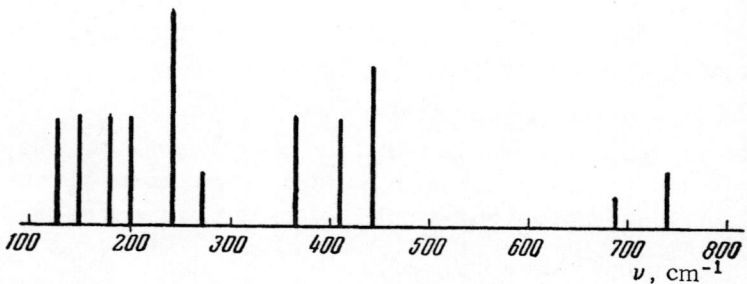

Fig. 16. Raman spectrum of crystalline antimonous anhydride.

TABLE 10. Vibration Frequencies of $Sb_2O_3$, $cm^{-1}$

| Our work | | [40] IR | [4] IR | [41] IR | Average data | | $As_2O_3$ | |
|---|---|---|---|---|---|---|---|---|
| RS | IR | | | | RS | IR | RS | IR |
| | | | | | | | 84 (vs) | |
| | | | | | | | 110 (w) | |
| 127 (m) | | | | | 127 (m) | | 131 (w) | |
| 145 (m) | | | | | 145 (m) | | 175 (m) | |
| 180 (m) | | | | 179 | 180 (m) | 179 | 262 (m) | 252 (m) |
| 197 (m) | | | | | 197 (m) | | 302 (w) | |
| | | | | | | | 332 (w) | |
| | | | | | | | | 345 (s) |
| 242 (vs) | | | | | 242 (vs) | | 366 (vs) | |
| 271 (w) | | | | | 271 (w) | | 422 (w) | |
| 365 (m) | | | | 357 | 365 (m) | 357 | 476 (m) | 482 (s) |
| 419 (m) | | | | | 419 (m) | | 532 (w) | |
| 444 (s) | 450 (w) | | 467 (s) | 435 | 444 (s) | 450 (w) | 555 (vs) | |
| | | | 547 (s) | | | | 629 (vw) | |
| | | | 588 (s) | | | | 676 (vw) | |
| 688 (w) | | 690 (w) | 695 (s) | | 688 (w) | 690 (w) | 710 (vw) | |
| 741 (w) | 732 (vs) | 740 (vs) | | | 741 (w) | 736 (vs) | 785 (m) | 801 (vs) |
| | | | | | | | 815 (w) | |
| | | 950 (vw) | | | | 950 (vw) | 840 (w) | 840 (sh) |
| | | | | | | | 1037 (vw) | 1040 (vw) |
| | | | | | | | | 1152 (vw) |

from the data in [4], which apparently refer to another modification of $Sb_2O_3$. In [41] the infrared spectrum was studied by the residual ray method; therefore some discrepancies in the line frequencies may be expected between the results of this work and absorption data. This is found to be the case for the band at 435 $cm^{-1}$, which has the absorption value of 450 $cm^{-1}$. In Table 10, which gives average values for the infrared spectrum of antimonous anhydride, we used data from [41] for the low-frequency region. For comparison the table also gives the vibrational spectrum of the cubic form of arsenious anhydride, the line frequencies being taken from Table 1. It is seen that there is considerable similarity between the spectra of the two oxides both in the number of the most intense lines and in the intensity distribution of these lines over the spectrum.

We will discuss experimental data on the vibrational spectrum of antimonous anhydride on the basis of the molecular structure hypothesis. The infrared absorption spectrum contains six bands, three of which (450, 690, and 950 $cm^{-1}$) are assessed as weak. Evidently the three remaining bands [41] must be assigned to fundamental vibration frequencies. They are also present in the Raman spectrum. In addition to these three frequencies, the Raman spectrum contains six frequencies assessed as not below "medium." Thus, we have nine fundamental vibration frequencies in the Raman spectrum and three in the infrared absroption spectrum.

The most probable model for the cubic form of crystalline antimonous anhydride can be chosen from a comparison and examination of all the fundamental vibration frequencies in the infrared absorption and Raman spectra.

Taking the numbers and activities of the frequencies into account, we can exclude a number of point groups.

1. The molecule cannot belong to the point groups $C_2$, $C_3$, $C_s$, $C_{3\nu}$, $C_{\infty\nu}$, as for these groups all the frequencies are active in both spectra.

2. The point groups $C_i$, $C_{2h}$, $D_{2h} \equiv V_h$, $S_6$, $C_{4h}$, $C_{6h}$, $D_{3d} \equiv S_{6\nu}$, $D_{4d} \equiv S_{8\nu}$, $D_{4h}$, $D_{6h}$, $D_{\infty h}$, $O_h$ are also excluded, as they should not have frequencies active in both spectra.

TABLE 11. Selection Rules for Various Structural
Models of the Antimonous Anhydride Molecule

| Symmetry group | Formula of molecule | Frequencies active in IR | Frequencies active in RS |
|---|---|---|---|
| $C_{2v}$ | $Sb_2O_3$ | 8 | 9 |
| $D_2 \equiv V$ | $Sb_2O_3$ | 7 | 9 |
| $C_4$ | $Sb_4O_6$ | 12 | 18 |
| $C_5$ | $Sb_4O_6$ | 12 | 15 |
| $C_6$ | $Sb_4O_6$ | 10 | 13 |
| $S_4$ | $Sb_4O_6$ | 12 | 17 |
| $C_{4v}$ | $Sb_4O_6$ | 11 | 17 |
| $C_{5v}$ | $Sb_4O_6$ | 12 | 15 |
| $C_{6v}$ | $Sb_4O_6$ | 10 | 13 |
| $D_{2d}$ | $Sb_4O_6$ | 10 | 17 |
| $T$ | $Sb_8O_{12}$ | 13 | 23 |
| $T_d$ | $Sb_4O_6$ | 4 | 8 |
| $T_d$ | $Sb_8O_{12}$ | 9 | 17 |
| Experiment | | 3 | 9 |

3. No frequency in the infrared absorption spectrum is inactive in the Raman spectrum. Therefore the following point groups should be excluded: $D_3$, $D_4$, $D_5$, $D_6$, $C_{3h}$, $C_{5h}$, $D_{3h}$.

Thus, the following point groups must be considered: $C_{2v}$, $D_{2d} \equiv V$, $C_4$, $C_5$, $C_6$, $S_4$, $C_{4v}$, $C_{5v}$, $C_{6v}$, $D_{2d} \equiv V_d \equiv S_{4v}$, $T$ and $T_d$.

Table 11 gives the calculated numbers of frequencies active in both the infrared and the Raman spectrum for these point groups. The end of the table gives experimental results, which agree best with the model for $Sb_4O_6$ having symmetry of the point group $T_d$. It is thus seen that the results given by an investigation of the vibrational spectrum of antimonous anhydride are in good agreement with x-ray structural data.

Interpretation of the Frequencies of the Vibrational Spectrum of the $Sb_4O_6$ Molecule. According to the selection rules, for the $Sb_4O_6$ molecule having symmetry of the point group $T_d$ we should have two totally symmetric vibrations of type $A_1$ active only in the Raman spectrum, two doubly degenerate vibrations of type $E$ active only in the Raman spectrum, two triply degenerate vibrations $F_1$ forbidden in both spectra, and four triply degenerate $F_2$ vibrations active both in the Raman and in the infrared spectrum.

The results show that totally symmetric vibrations of type $A_1$ are very easy to find, because they are stronger than the other lines. These are the lines at 242 and 444 cm$^{-1}$. It is also very easy to identify three out of the four $F_2$ vibrations by their activity in both spectra. These are the lines with frequencies of 197, 365, and 741 cm$^{-1}$. As it is very difficult to assign the remaining lines to definite vibration types, we were forced to calculate the vibration frequencies of the $Sb_4O_6$ molecule. Secular equations derived with the assumption of a valence-force model for a molecule of the type $X_4O_6$ with $T_d$ symmetry [30] were first used. However, the calculations require a knowledge of the numerical values of the kinematic coefficients, and to calculate the latter it is necessary to know the masses of the atoms in the molecules and the bond angles. To find the angles, we must know the bond lengths, Sb—O in this case; several numerical values are given in the literature, 2.00 A according to some sources [38,39] and 2.22 A according to others [37].

To find the length of the Sb—O bond we use the similarity between the physicochemical properties and lattice structures of the crystalline cubic forms of antimonous and arsenious anhydrides. In arsenious anhydride vapor, consisting of $As_4O_6$ molecules, the length of the As—O bond is 1.80 A, which is less than the sum of the

TABLE 12. Physical Characteristics of the $Sb_4O_6$ Molecule

| Parameter | Value |
|---|---|
| $r$ (Sb — O) | 2.22 A |
| $r$ (Sb — Sb) | 3.62 A |
| $r$ (O — O) | 3.62 A |
| $\delta$ (O — Sb — O) | 109°30′ |
| $\beta$ (Sb — O — Sb) | 109°30′ |
| Space symmetry group | $O_h$ |
| Number of molecules in unit cell | 2 |
| Melting point | 652° C |
| Distance between Sb and O of neighboring molecules | 2.61 A |
| Symmetry point group | $T_d$ |

covalent radii of the atoms (1.87 A), indicating a certain degree of double bonding. At the same time, conden-sation of $As_4O_6$ molecules into the cubic crystalline form results in interesting changes in the molecular parame-ters. The As—O distance greatly increases in the crystalline state, reaching 2.00 A, which indicates not only that this bond is of the single type in the crystal molecules but also that the distance is greater than the sum of the two covalent radii. This experimental fact is due to considerable intermolecular interaction between $As_4O_6$ molecules (as indicated by the fairly high melting point, 272°C).

The Sb—O distance in the antimonous anhydride molecule, calculated from the covalent radii of the atoms, is 2.07 A. This value differs considerably from the experimental values found by x-ray diffraction (2.00 and 2.22 A), and it is therefore difficult to decide which is to be preferred. However, the higher melting point of antimonous anhydride (652°C) indicates even greater intermolecular interaction in this compound. It is there-fore to be expected that the Sb—O distance in the $Sb_4O_6$ molecule with a crystalline lattice structure should be greater than the distance found from the covalent radii. Accordingly, the value of 2.22 A was taken for the length of the Sb—O bond in the antimonous anhydride molecule. Using the Sb—O and O—O bond lengths we can find the angle $\delta$ = O—Sb—O, and with the additional length Sb—Sb we can determine the angle $\beta$ = Sb—O—Sb. All the parameters found for the $Sb_4O_6$ molecule are given in Table 12.

The angles thus found, and also the atomic masses, were used for calculating the kinematic coefficients with the aid of formulas from [42]. Substituting, as in the case of the $As_4O_6$ molecule (see §1) all the parame-ters of the $Sb_4O_6$ molecules and the totally symmetric vibration frequencies ($\nu_1 = 242$ cm$^{-1}$ and $\nu_2 = 444$ cm$^{-1}$), we find the following values for the force constants:

$$f = 1.73 \cdot 10^5 \text{ dynes/cm;} \qquad d = 0.173 \cdot 10^5 \text{ dynes/cm;}$$
$$g = 0.173 \cdot 10^5 \text{ dynes/cm;} \qquad f_s = 0.52 \cdot 10^5 \text{ dynes/cm.}$$

The remaining vibration frequencies of the other types were calculated with the aid of these force con-stants. The calculated and experimental frequencies are given in Table 13. Among the type $F_2$ frequencies, 319 cm$^{-1}$ occurs twice; i.e., two fundamental vibrations of this type have the same frequency according to cal-culations. If the experimental frequency of 357 cm$^{-1}$ is assigned to one vibration of this type, the other fre-quency should, by analogy with the $As_4O_6$ spectrum, be in the region of 250-300 cm$^{-1}$ (see Table 10). However, it is not found in the spectrum. Absence of this $F_2$ frequency is apparently due to strong intermolecular interac-tion which infringes the selection rules.

An analogous effect, but in a weaker form because the intermolecular interaction is weaker, is found in the spectrum of the $As_4O_6$ molecule. There the 345 cm$^{-1}$ of the same vibration as in $Sb_4O_6$ appears only in the infrared spectrum and is absent from the Raman spectrum, whereas in theory the frequencies of this type should be active in both spectra.

| Vibration type | Interpretation | Experiment | Calculated |
|---|---|---|---|
| $A_1$ | $\nu_1$ | 444 | 444 |
|      | $\nu_2$ | 242 | 242 |
| $E$  | $\nu_3$ | 419 | 440 |
|      | $\nu_4$ | 145 | 150 |
|      | $\nu_5$ | 736 | 600 |
| $F_2$ | $\nu_6$ | 357 | 319 |
|      | $\nu_7$ | — | 319 |
|      | $\nu_8$ | 199 | 184 |

Table 13 shows that, despite the fairly crude calculation model, there is otherwise fairly satisfactory agreement between the calculated and experimental fundamental vibration frequencies of the $Sb_4O_6$ molecule. The remaining frequencies of the vibrational spectrum of the $Sb_4O_6$ molecules are either component frequencies or overtones.

The following experimental fact is interesting. In the Raman spectrum the low-frequency lines in the 100-200 $cm^{-1}$ region are doublets; the frequencies of the components of one doublet are 180 and 197 $cm^{-1}$, and of the other 127 and 145 $cm^{-1}$, and the intensities of the lines are equal. Calculations show (see Table 13) that one of the components of each doublet is assigned to the fundamental vibration frequencies. The second components of the doublets may be regarded as combination frequencies the intensity of which increases owing to Fermi resonance. The average values of each doublet are $\nu_4 = 136$ and $\nu_8 = 188$ $cm^{-1}$; the first frequency is of type E and the second of type $F_2$.

The combination frequencies $\nu_2 - \nu_4$ and $\nu_6 - \nu_8$ not only have numerical values close to the fundamental frequencies $\nu_4$ and $\nu_8$, but also vibrational levels of the same type. Therefore Fermi resonance between these fundamental and component frequencies is quite possible, leading to the appearance of doublets consisting of bands of equal intensity.

CHAPTER III

# VIBRATIONAL SPECTRA OF GLASS-FORMING
# OXIDES WITH COORDINATION LATTICE STRUCTURE

## §1. Antimonic Anhydride

The higher oxide of antimony has as yet been found only in the crystalline state. Studies of its structure showed that this crystalline form has cubic structure with $O_h$ symmetry [43]. Antimonic anhydride has not yet been obtained in the glassy state, although by the Zachariasen theory this compound should form a glass.

The positions of the elements in the periodic table suggest considerable similarity between the physico-chemical properties of arsenic pentoxide and antimonic anhydride.

In fact, there is some similarity, but there is also a difference which relates to their structure. Whereas in the case of arsenic pentoxide it was reported (and subsequently confirmed by our investigations of its Raman spectrum) that it has a molecular lattice structure and consists of $As_4O_{10}$ molecules belonging to the point group $T_d$, we did not find any such indications in the literature of a molecular structure in the case of antimonic anhydride. However, because of the similarity in properties to $As_2O_5$, it is possible to postulate that is has a molecular structure and consists of $Sb_4O_{10}$ molecules having symmetry of the point group $T_d$ and then to investigate whether this model gives agreement with the results of Raman spectroscopy.

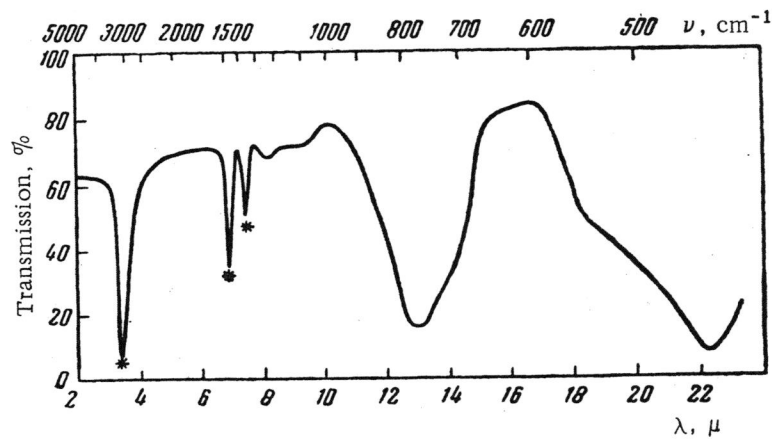

Fig. 17. Infrared absorption spectrum of crystalline antimonic anhydride.

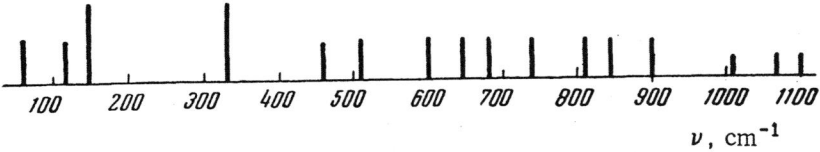

Fig. 18. Raman spectrum of crystalline antimonic anhydride.

TABLE 14. Vibration Frequencies of $Sb_2O_5$, $cm^{-1}$

| Our work | | [40] IR | [4] IR |
| IR | RS | | |
|---|---|---|---|
| | 58 (w) | | |
| | 117 (w) | | |
| | 147 (m) | | |
| | 332 (m) | | |
| 450 (vs) | | | |
| | 460 (w) | | Band |
| | 510 (w) | | 417—827 |
| 550 (m) | | | |
| | 600 (w) | | |
| | 649 (w) | | 659 (vw) |
| | 612 (w) | 685 (vw) | |
| 700 (sh) | | | 704 (s) |
| | 745 (w) | 740 (s) | |
| 770 (vs) | | | |
| | 814 (w) | | |
| | 847 (w) | | |
| | 902 (w) | | |
| | 1019 (vw) | | |
| | 1085 (vw) | | |
| | 1102 (vw) | | |
| | 1250 (vw) | 3225 (w) | |

The infrared absorption and Raman spectra are given in Figs. 17 and 18, and the results are summarized in Table 14. Data from [4,40] on the infrared absorption spectrum are also included. The table shows that our results on the infrared spectrum differ from the literature data. As in the case of $As_2O_5$, the Raman lines are somewhat weak. The Raman spectrum was found to contain 16 lines of which only two (as in the case of $As_2O_5$) have higher intensity than the others. Frequencies above 1000 $cm^{-1}$ need hardly be taken into consideration, because they are too high for fundamental vibration frequencies.

Thus, in view of the equal intensity of all the lines, all the frequencies below this limit in the Raman spectrum may be taken as fundamental vibration frequencies. In the infrared absorption spectrum there are six frequencies; of these, four should be taken as fundamentals as they are the most intense. There are no corresponding frequencies in the Raman spectrum for any of the frequencies in the infrared absorption spectrum. The consequence of this inactivity of the infrared frequencies in the Raman spectrum is that the vibrational spectrum must be determined by the symmetry group which satisfies the condition of inactivity of Raman frequencies in the infrared absorption spectrum and vice versa.

This contradicts the $Sb_4O_{10}$ model having symmetry of the point group $T_d$, as here all the frequencies of the infrared absorption spectrum should be active in the Raman spectrum. However, the experimental data on the vibrational spectrum do not contradict a unit cell with space symmetry group $O_h$, found by x-ray diffraction. The above condition of inactivity of the infrared frequencies in the Raman spectrum is satisfied for this group, which has a center of symmetry. Therefore it may be assumed that the vibrational spectrum of crystalline $Sb_2O_5$ is determined by a unit cell belonging to the $O_h$ space group in a coordination lattice. Thus, antimonic anhydride has the structure of a coordination compound and differs sharply from $As_2O_5$, which has a molecular structure, in this respect. As crystalline $Sb_2O_5$ exists only as the cubic form, it is to be expected that the structure of glassy $Sb_2O_5$ is analogous to that of the crystalline form. Further and more detailed studies of the structure of antimonic anhydride with the aid of data on its vibrational spectrum are possible only with single crystals, which are difficult to prepare at present.

TABLE 15. Physicochemical Properties of Different Forms of Germanium Dioxide

| Parameters | Insoluble form | Soluble form | Glass |
|---|---|---|---|
| Crystal form | Tetragonal | Hexagonal | Amorphous |
| Crystal structure type | Rutile | $\alpha$-quartz | – |
| Density, g/ml at 25°C | 6.239 | 4.228 | 3.637 |
| Refractive index $\{$ $n_0$ | 1.99 | 1.695 | – |
| $\quad$ $n_e$ | 2.05-2.10 | 1.735 | 1.607 |
| Inversion point, °C | 1033 ± 10 | 1033 ± 10 | – |
| Melting point, °C | 1086 ± 5 | 1116 ± 4 | – |
| Solubility, in 100 g water | Insoluble | 0.453 (25°C) | 0.5184 (30°C) |
| Action of hydrofluoric acid | No action | Reacts to form $H_2GeF_6$ | |
| Action of hydrochloric acid | " " | Reacts to form $GeCl_4$ | |
| Action of caustic soda at 100°C | Dissolves very slowly | Dissolves very rapidly | |

## §2. Germanium Dioxide [20,42]

By the Zachariasen theory, germanium dioxide (like silicon dioxide) is an outstanding example of a glass-forming oxide with a coordination structure. Both substances pass readily into the glassy state, both have a coordination structure, and satisfy all the main postulates of the Zachariasen theory. The physicochemical properties of these oxides are also very similar [22,44]. Therefore a similarity in their vibrational spectra is also to be expected. From this standpoint, study of the vibrational spectra of the different forms of germanium oxide is not only of independent interest but makes an additional contribution to studies of silicon dioxide.

Germanium dioxide exists in several forms: two crystalline (soluble and insoluble), and glassy. According to Goldschmidt's physicochemical data on $GeO_2$ [45] and to Zachariasen's x-ray data [46], germanium dioxide obtained by crystallization from aqueous solution has a crystal lattice analogous to the lattice of low-temperature $\alpha$-quartz. It was shown by Müller's x-ray investigations [47] that crystalline $GeO_2$ obtained by another method, devitrification of fused $GeO_2$, has the same crystalline structure. This crystalline form is partially soluble in water and is known as the soluble form of $GeO_2$.

Müller and Bland [48] discovered a second crystalline form of $GeO_2$, which they called insoluble, because it is almost insoluble in water. This form of $GeO_2$ is obtained by a hydrothermal method from the soluble form; it was shown by Morton [22], who used the x-ray equipment described in [49], to have a lattice of the rutile type.

Apart from the structural difference, these two forms have entirely different physicochemical properties. This is illustrated by Table 15 [22]. The table also includes data on the glassy form of $GeO_2$, formed by cooling of a melt of the soluble crystalline form.

The table shows considerable similarities between the properties of the soluble crystalline form and the glassy form of $GeO_2$. The properties of the insoluble form are entirely different.

Discussion of Results and Interpretation of the Vibrational Spectrum Frequencies. The infrared absorption and Raman spectra of the different forms of germanium dioxide are given in Figs. 19-26. The results are summarized in Table 16. This table also includes infrared spectral data from [50].

Table 16 also shows very good agreement between our data for the soluble form and the data in [50]. For the glassy form the author of [50] found only the strongest band at 895 cm$^{-1}$. There is no agreement for the insoluble form.

The vibrational spectra of $GeO_2$ have the following characteristics.

1. It follows from an examination of the spectra of the soluble crystalline and glassy forms of germanium dioxide that the lines of the vibrational spectrum of the glassy form of $GeO_2$ are considerably broader than the corresponding lines of the soluble form, and neighboring lines even merge to give one broad band. For example, it is clear from the spectra shown in Figs. 19 and 23 that the 515, 551, and 587 cm$^{-1}$ frequencies of the spectrum

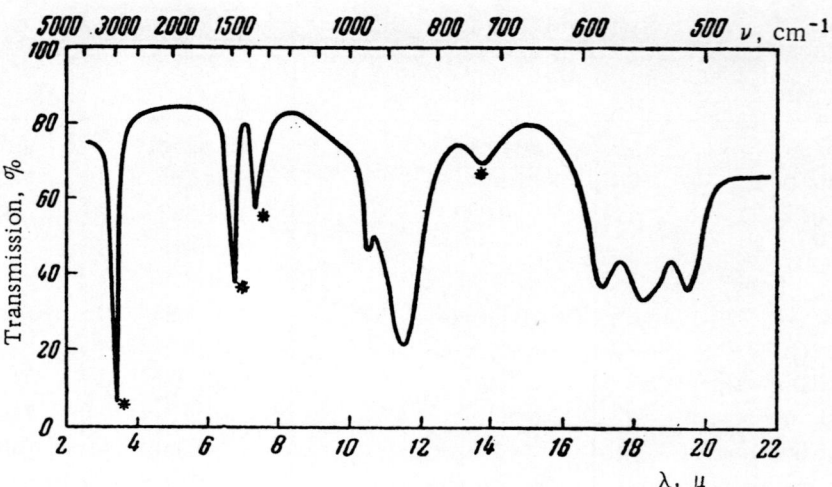

Fig. 19. Infrared absorption spectrum of the crystalline soluble form of germanium dioxide in the 2-22 μ region.

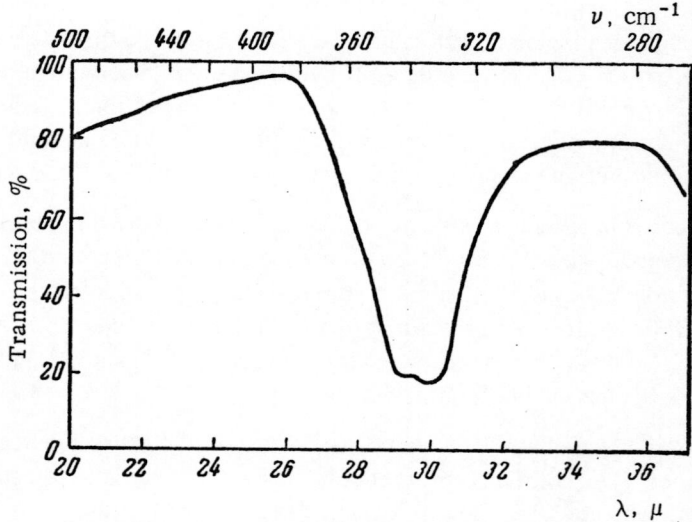

Fig. 20. Infrared absorption spectrum of the crystalline soluble form of germanium dioxide in the 20-30 μ region.

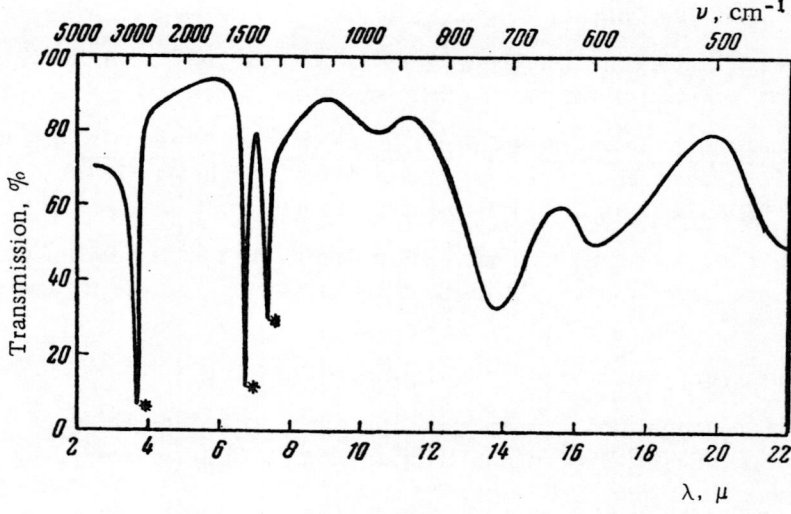

Fig. 21. Infrared absorption spectrum of the crystalline insoluble form of germanium dioxide in the 2-22 μ region.

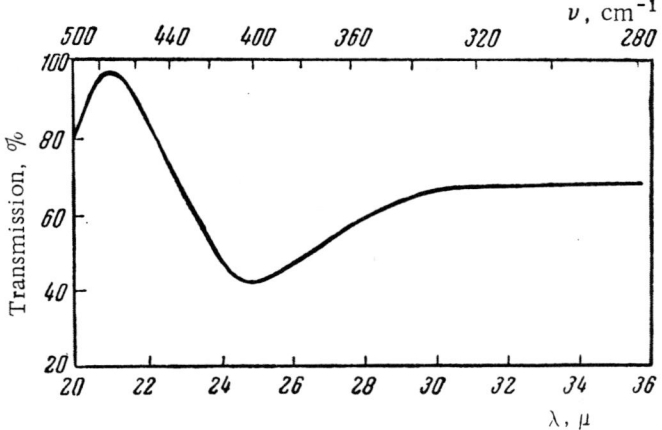

Fig. 22. Infrared absorption spectrum of the crystalline insoluble form of germanium dioxide in the 20-36 μ region.

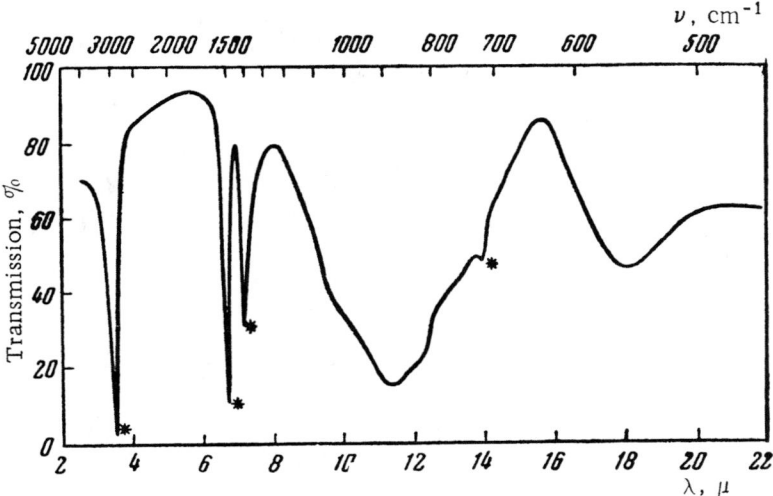

Fig. 23. Infrared absorption spectrum of glassy germanium dioxide in the 2-22 μ region.

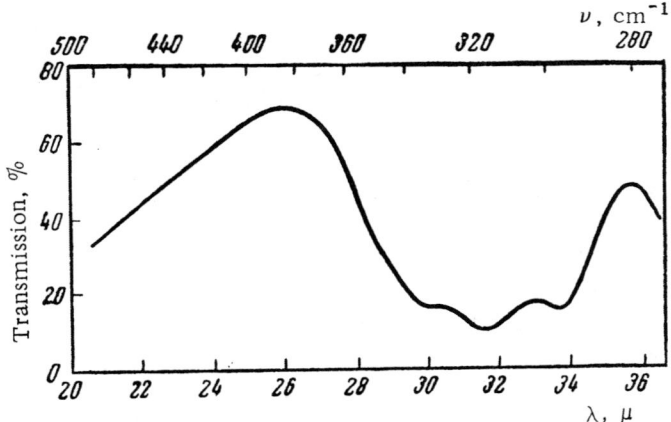

Fig. 24. Infrared absorption spectrum of glassy germanium dioxide in the 20-36 μ region.

TABLE 16. Vibration Frequencies of $GeO_2$, cm$^{-1}$

| Crystalline soluble form | | | Crystalline insoluble form | | | Glassy form | | |
|---|---|---|---|---|---|---|---|---|
| RS | IR | IR [50] | RS | IR | IR [50] | RS | IR | IR [50] |
| 340 (m) | 333 (s) | 332 | | | | 193 (vw) | | |
| | 343 (s) | 345 | | | | 250 (vw) | 295 (sh) | |
| 408 (m) | | | | 407 (vs) | 435 | 315 (m) | 315 (s) | |
| 440 (s) | | | | | | 342 (m) | 334 (sh) | |
| 508 (m) | 515 (m) | 515 | Spectrum not found | | 513 | 373—461 (vs) | | |
| 560 (w) | 551 (m) | 540 | | | 515 | 520—615 (m) | 564 (s) | |
| | 587 (m) | 585 | | | | | | |
| 640 (w) | | | | 606 (m) | | | | |
| | | | | 709 (vs) | | | | |
| 888 (vw) | 872 (vs) | | | | | 825—894 (w) | 888 (vs) | 895 |
| 959 (vw) | 955 (sh) | 963 | | 950 (vw) | | 935—973 (w) | 966 (sh) | |

of the crystalline form broaden and merge into a single band with a maximum at 564 cm$^{-1}$ in the infrared absorption spectrum of the glassy form; frequencies in the Raman spectrum of the soluble form (408 and 440 cm$^{-1}$) merge to give one broad band at 373-461 cm$^{-1}$ in the spectrum of the glassy form. Further, a characteristic feature of the Raman spectrum of glassy $GeO_2$ is the existence of a continuous spectrum adjacent to the exciting line of $\lambda = 4358$ A and extending to approximately 600 cm$^{-1}$.

2. With line broadening taken into account, there is reasonable agreement between the Raman spectra of the soluble and glassy forms of $GeO_2$, both in the positions and in the activities of the frequencies. This agreement between the Raman frequencies was to be expected from the great similarity of the physicochemical properties of the two forms, discussed above. At the same time, the insoluble form differs greatly in physicochemical properties from the other two forms, and therefore the vibrational spectrum was found to be entirely different. The distinction is especially prominent in the Raman spectrum. The insoluble form did not give a Raman spectrum at all; this seems to indicate considerable bond ionicity in this compound.

Thus, the spectral lines of the glassy form are approximately in the positions of the sharp lines of the soluble crystalline form. This experimental fact suggests that the structure of the glassy form of $GeO_2$ is similar to that of the soluble crystalline form. Therefore study of the structure of the soluble form sheds light on the structure of glassy $GeO_2$.

As already noted, there are considerable similarities between germanium and silicon compounds. Investigations of the soluble crystalline form of germanium dioxide [22] showed that it is similar both in lattice structure and in physicochemical properties to one of the crystalline forms of silicon dioxide—low-temperature $\alpha$-quartz. On the vibrational spectrum of $\alpha$-quartz a considerable amount of data has been obtained, which were

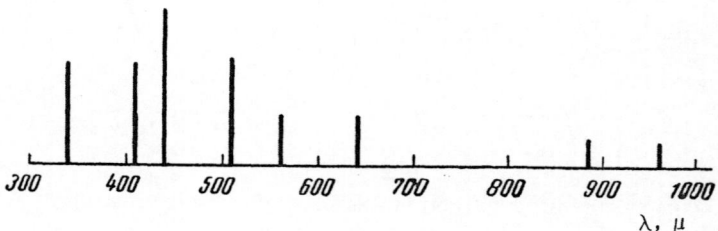

Fig. 25. Raman spectrum of the crystalline soluble form of germanium dioxide.

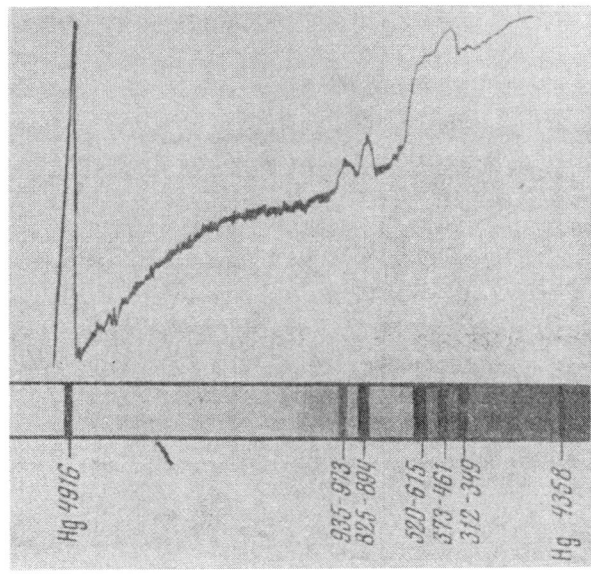

Fig. 26. Raman spectrum of glassy germanium dioxide.

used for detailed analysis of the spectral frequencies [51,52]. A similar analysis of the vibrational spectrum frequencies can be performed for the soluble form of germanium dioxide, which has a lattice similar to the $\alpha$-quartz lattice. For a unit cell consisting of three $GeO_2$ groups and belonging to space symmetry group $D_3$ the selection rules require 24 fundamental vibration frequencies, including four type A vibrations active in the Raman spectrum only, four type B vibrations active in the infrared absorption spectrum only, and eight doubly degenerate type E vibrations active in both spectra.

On the basis of a study of the activities of frequencies in the Raman and infrared absorption spectra of germanium and silicon dioxides, the frequencies were assigned to vibration types and compared with the corresponding frequencies in the $SiO_2$ spectrum of $\alpha$-quartz (Table 17). It was taken into account that the frequencies in the $GeO_2$ spectrum should be lower than the corresponding frequencies in the $SiO_2$ spectrum because of the higher atomic weight of germanium. In the comparison of the frequencies of the spectra of the two oxides it was also taken into account that the data on the Raman spectrum of $\alpha$-quartz [52] are more complete because the spectrum was excited by the mercury resonance line of $\lambda = 2537$ A and was obtained from a single crystal, whereas the spectrum of germanium dioxide was investigated by the powder method with the mercury lines of $\lambda = 4358$ and 4047 A so that, of course, not all the frequencies of the Raman spectrum were revealed. Further, it was taken into account that the infrared absorption spectrum was investigated only up to 36 $\mu$. For

TABLE 17. Comparison of Vibrational Spectra of $GeO_2$ and $SiO_2$

| Vibration type | Vibration spectrum frequencies of silicon dioxide, cm$^{-1}$ | Vibration spectrum frequencies of germanium dioxide, cm$^{-1}$ | Vibration type | Vibration spectrum frequencies of silicon dioxide, cm$^{-1}$ | Vibration spectrum frequencies of germanium dioxide, cm$^{-1}$ |
|---|---|---|---|---|---|
| A | 206 | — | | 128 | — |
| | 356 | — | | 265 | — |
| | 466 | 440 | | 394–404 | 343 |
| | 1082 | 640 | E | 452 | 408 |
| | 145 | — | | 696 | 515 |
| B | 509 | 333 | | 795–806 | 551 |
| | 780 | 587 | | 1063 | 872 |
| | 1046 | 872 | | 1160 | 955 |

157

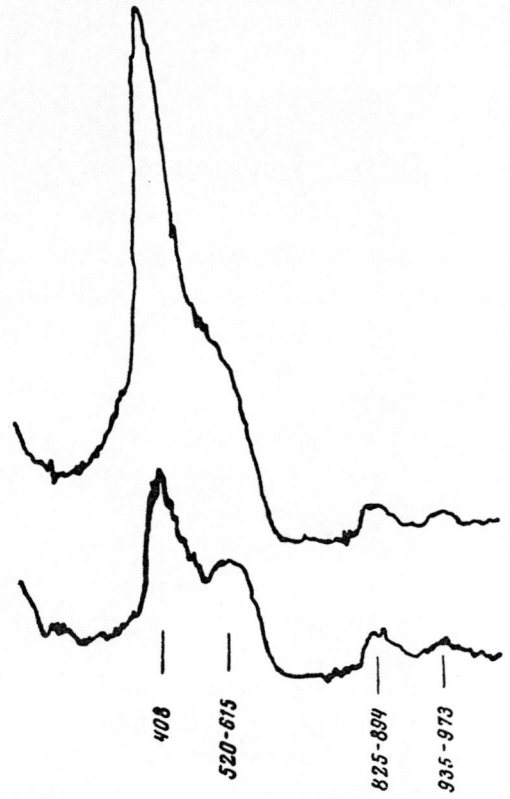

these reasons analogs of the low-frequency lines of the silicon dioxide spectrum could not be found in the vibrational spectrum of germanium dioxide. The other lines of the vibrational spectra show fairly good analogy apart from the frequency 408 cm$^{-1}$. The frequencies 440 and 640 cm$^{-1}$ of the germanium dioxide spectrum were assigned to type A, as they are active only in the Raman spectrum. According to the selection rules, the frequency 408 cm$^{-1}$ should also be assigned to this vibration type. However, in view of the fact that two totally symmetric vibrations cannot usually have such close frequencies, the frequency 408 cm$^{-1}$ was assigned to type E, where it corresponds to the frequency 452 cm$^{-1}$ of the SiO$_2$ spectrum. The frequencies 343, 515, 551, 872, and 955 cm$^{-1}$ were also assigned to type E, as they are active in both spectra. The frequencies 333 and 587 cm$^{-1}$ were assigned to type B, as they appear only in the infrared absorption spectrum. The frequency 872 cm$^{-1}$ was also assigned to this type despite the fact that it is active in both spectra. It corresponds to the frequency 1046 cm$^{-1}$ of the same type in the spectrum of SiO$_2$, which is very close to the frequency 1063 cm$^{-1}$ of type E. Apparently the two vibrations of type E and B in the spectrum of germanium dioxide are so close that we must assign the same frequency of 872 cm$^{-1}$ in both.

Fig. 27. Polarization Raman spectrum of glassy germanium dioxide. Intensities of scattered light polarized in two mutually perpendicular directions are given.

With these comments taken into consideration, the agreement between the results for the vibrational spectrum of the soluble form of germanium dioxide and the results of x-ray diffraction studies may be regarded as fairly good. Further, more detailed study of the vibrational spectrum of GeO$_2$ would be possible only if a single crystal of germanium dioxide can be obtained.

In view of the fact that the glassy forms of GeO$_2$ and SiO$_2$ correspond to the crystalline forms discussed above, similarity between the spectra of the glassy forms is also to be expected. As the characteristics of the vibrational spectrum of glassy SiO$_2$ have been thoroughly studied, to make the comparison complete we investigated the degree of depolarization of the lines in the Raman spectrum of GeO$_2$. The polarization spectrum is shown in Fig. 27.

This figure shows that the bands of the Raman spectrum at 825-894, 935-973, and 520-615 cm$^{-1}$ are depolarized. The strongest band, 373-461 cm$^{-1}$ with a maximum at 408 cm$^{-1}$, is polarized.

With the depolarization data and the band positions and intensities in the spectra taken into account, Fig. 18 gives a comparison of the spectra of the glassy forms of SiO$_2$ and GeO$_2$, although the comparison is somewhat hindered by the fact that the lines in the spectra of glasses are greatly broadened. Nevertheless, a considerable similarity between the spectra of the two oxides in the glassy state can be seen. Of course, it was taken into account that the bands in the vibrational spectrum of GeO$_2$ should have lower frequencies than the corresponding bands for SiO$_2$ because of the greater atomic weight of Ge. It is interesting to note that both substances give a continuous background in the Raman spectrum, adjacent to the exciting line and extending to about 600 cm$^{-1}$.

Thus, the results show fairly close resemblance between the soluble form of GeO$_2$ and low-temperature α-quartz. This provides additional confirmation that this form of GeO$_2$ should contain three GeO$_2$ groups in the unit cell which belongs to the space group D$_3$. Interpretation of the spectral frequencies gives satisfactory agree-

TABLE 18. Vibrational Spectra of Glassy Silicon and Germanium
Dioxides, cm$^{-1}$

| Glassy SiO$_2$ | | Glassy GeO$_2$ | |
|---|---|---|---|
| [53] RS | [4] IR | RS | IR |
| 30—120 (vs) | | | |
| | | 193 (vw) | |
| 285 (s) | | 250 (vw) | 295 (sh) |
| 370 (vs) | | 342 (m) | 334 (sh) |
| 430 (vs) | | 373—461 (vs) | |
| 495 (vs) | 476 (vs) | 315 (m) | 315 (s) |
| 635 (vw) | | 520—615 (m) | |
| 660 (vw) | | | 564 (s) |
| 775—845 (s) | 785 (vs) | | |
| 1022—1098 (s) | 1066 (vs) | 825—894 (w) | 888 (vs) |
| 1140—1246 (w) | 1147 (sh) | 935—973 (w) | 966 (sh) |

ment with this conclusion. The vibrational spectrum of glassy GeO$_2$ is apparently determined by the same structural lattice unit as that of the corresponding crystalline form.

## §3. Tellurium Dioxide [54]

In Zachariasen's theory, tellurium dioxide was not classified as a glass-forming oxide. At that time tellurium glasses were unknown and had not been obtained ouwing to difficulties in preparation. Recently the production of tellurium glasses, based on tellurium dioxide, has started. Pure one-component TeO$_2$ glass has not yet been obtained, although in presence of very small amounts of other oxides (not glass formers) a glass is relatively easy to obtain. Since tellurium glasses are based mainly on tellurium oxide, this oxide should probably also be regarded as a glass-forming oxide.

Two forms of crystalline tellurium dioxide are known, tetragonal and rhombic [55,56]. The rhombic form of tellurium dioxide occurs very rarely as the native mineral tellurite. X-ray diffraction studies showed that the unit cell of the rhombic form contains eight TeO$_2$ groups and belongs to the D$_{2h}$ space group [31]. The tetragonal form of tellurium dioxide can only by obtained by chemical methods [57,58]. There is no agreed view among x-ray crystallographers with regard to its structure. Goldschmidt [56] and later Ormont [23] concluded that the unit cell of the tetragonal form of tellurium dioxide contains two TeO$_2$ groups and belongs to the D$_{4h}$ space group. Czech scientists [57] concluded from their investigations that the unit cell of this form belongs to the D$_4$ symmetry group and contains four TeO$_2$ groups.

The glassy form of tellurium dioxide is obtained by fusion of the tetragonal crystalline form. Brady [59] concluded from the results of x-ray diffraction studies that the unit cell of glassy tellurium dioxide contains eight TeO$_2$ groups and belongs to the D$_{2h}$ symmetry group. Thus, according to Brady, the lattice structure of the glassy form is similar to the structure of the rhombic crystalline form of tellurium dioxide.

Discussion of Results. The infrared absorption and Raman spectra of the crystalline (tetragonal) and glassy forms of tellurium dioxide are given in Figs. 28-34. The frequencies and intensity assessments are given in Table 19.

The table and figures show that there is, in general, good agreement between the vibrational spectra of crystalline and glassy tellurium dioxide. It should be noted, however that in the spectrum of the glassy form all the lines are weaker (therefore the weak lines of the vibrational spectrum of the crystalline form are not found in the spectrum of the glassy form) and also broader, so that close lines merge into a borad band. Thus, the Raman spectrum of the glassy form has a band at 380-500 cm$^{-1}$ with individual lines at 399, 456, and 491 cm$^{-1}$

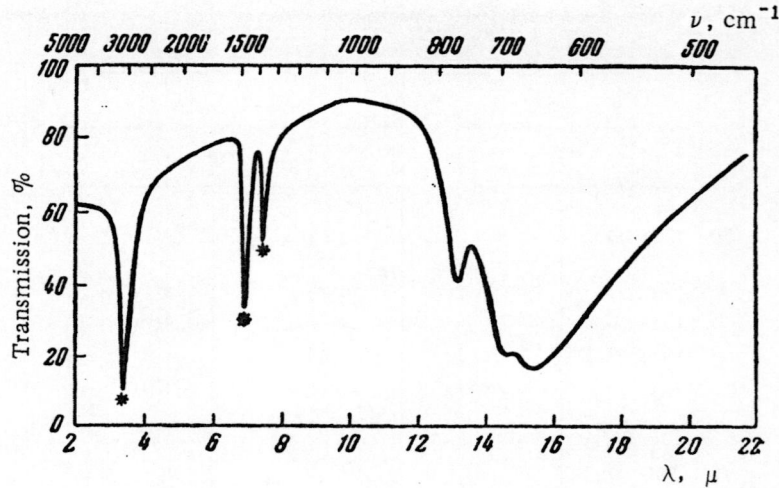

Fig. 28. Infrared absorption spectrum of crystalline tellurium dioxide in the 2-22 μ region.

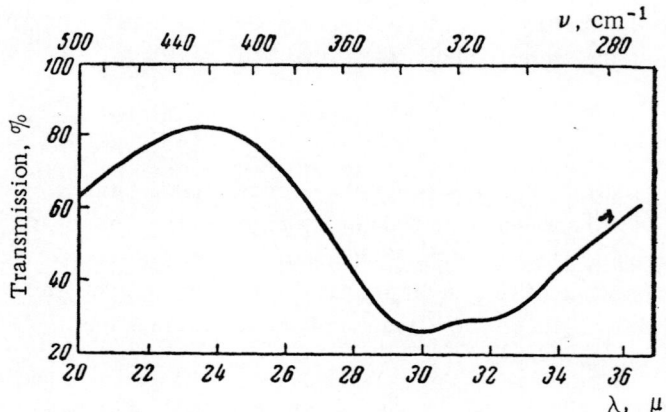

Fig. 29. Infrared absorption spectrum of crystalline tellurium dioxide in the 20-36 μ region.

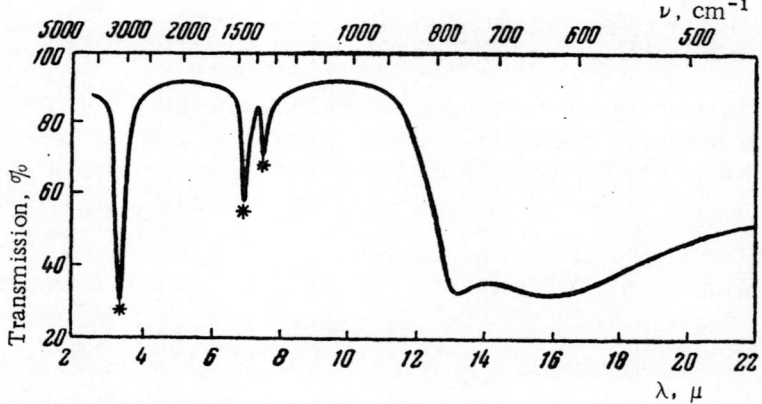

Fig. 30. Infrared absorption spectrum of glassy tellurium dioxide in the 2-22 μ region.

TABLE 19. Vibration Frequencies of $TeO_2$, $cm^{-1}$

| Crystalline form | | Glassy form | |
| --- | --- | --- | --- |
| RS | IR | RS | IR |
| 58 (vs) | | | |
| 89 (s) | | | |
| | 113 (m) | | |
| 121 (vs) | 119 (m) | 120 (s) | |
| | 137 (m) | | |
| 146 (vs) | | 157 (s) | |
| | 174 (vs) | | |
| 204 (m) | | | |
| 266 (w) | | | |
| | 314 (vs) | | max 330 (m) |
| | 334 (vs) | | |
| 343 (w) | | | |
| 365 (vw) | | | |
| 392 (s) | | 399 (m) | |
| 409 (m) | | 456 (w) | |
| 495 (w) | | 491 (w) | |
| 590 (m) | | | |
| 648 (vs) | 648 (vs) | max 643 (s) | max 646 (m) |
| 667 (m) | | | |
| 717 (m) | 714 (vs) | | |
| 766 (m) | 760 (m) | max 775 (m) | max 768 (m) |

emerging through the background, and a band at 590-830 $cm^{-1}$ with two maxima, at 643 and 775 $cm^{-1}$. Similar broadening is observed in the infrared absorption spectrum. It is interesting to note that the broadening does not affect the low-frequency lines in the Raman spectrum of the glassy form (120 and 157 $cm^{-1}$). As was pointed out earlier, a similar boradening effect is observed in the transition from the crystalline to the glassy form in the case of $GeO_2$.

As already noted, the spectra of the glassy and crystalline (tetragonal) forms of tellurium dioxide are very similar in the positions and activities of the lines; hence it may be concluded that the unit cells determining their vibrational spectra are also similar. It is therefore incorrect to assert [59] that the structure of glassy tellurium dioxide is similar to that of the rhombic form. This conclusion is also confirmed by the fact that by the selection rules the rhombic form, which has the $D_{2h}$ symmetry group for the unit cell, should not have frequencies active in both the Raman and the infrared absorption spectrum [31]. Yet the experimental results show that certain frequencies are active in both spectra. Therefore the conclusions of [59] are contradicted by experimental facts. Accordingly, when we determine the unit cell and its space group for the crystalline (tetragonal) form of tellurium dioxide, we also determine the structure of glassy tellurium dioxide.

We now pass to a discussion of the results obtained for tetragonal tellurium dioxide. As already noted, different conclusions have been drawn from x-ray diffraction data concerning the lattice structure of this form. We calculated and compared the numbers of fundamental vibration frequencies for unit cells determined by x-ray diffraction with our results. This comparison made it possible to select the unit cell model closest to our results. The calculated numbers of fundamental vibration frequencies in relation to their activities are given in Table 20. The bottom line gives the numbers of frequencies found experimentally. In view of the fact that the vibrational spectrum of crystalline tellurium dioxide was investigated by the powder method, and also that strong absorption occurs over the entire region investigated, all the frequencies found were taken to be fundamental vibration frequencies in the Raman spectrum. In the infrared spectrum all the frequencies were assessed

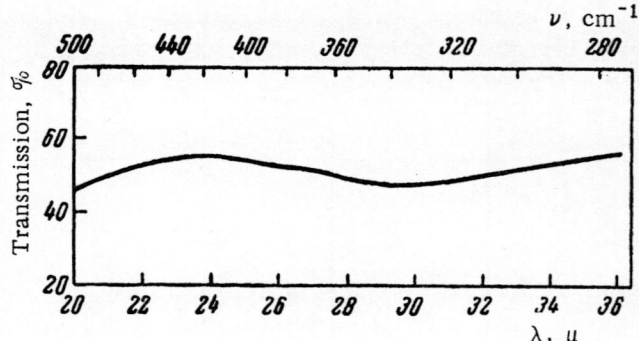

Fig. 31. Infrared absorption spectrum of glassy tellurium dioxide in the 20-36 μ region.

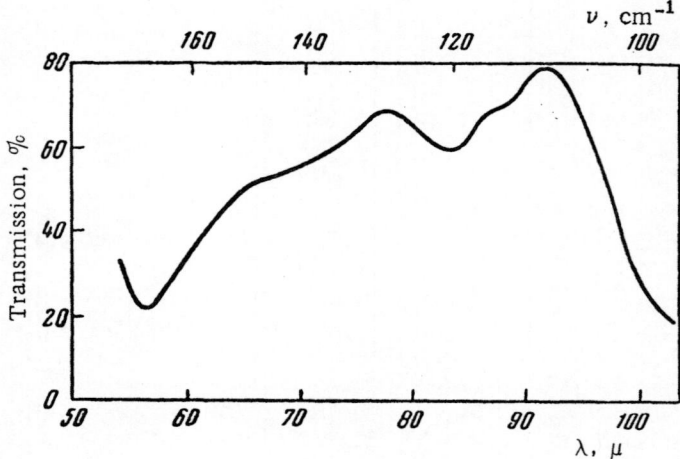

Fig. 32. Infrared absorption spectrum of crystalline tellurium dioxide in the 53-100 μ region.

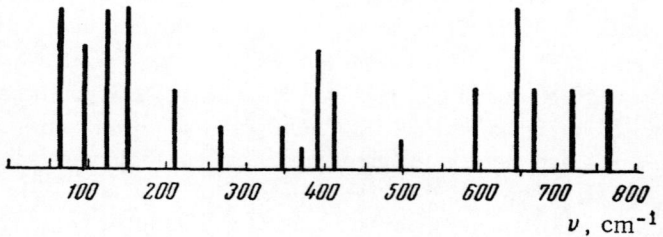

Fig- 33. Raman spectrum of crystalline tellurium dioxide.

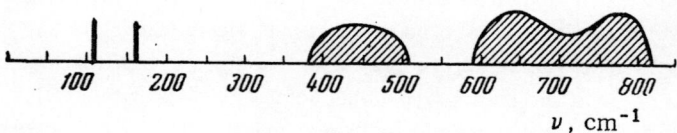

Fig. 34. Raman spectrum of glassy tellurium dioxide.

TABLE 20. Selection Rules for Symmetry Groups $D_{2h}$, $D_4$, and Comparison with Experimental Data

| Symmetry group | No. of groups in unit cell, Z | No. of RS frequencies | No. of IR frequencies | Frequencies active in IR only | Frequencies active in IR and RS |
|---|---|---|---|---|---|
| $D_{2h}$ [23,56] | 2 | 6 | 3 | 3 | None |
| $D_4$ [57] | 4 | 21 | 12 | 4 | 8 |
| Experiment | | 16 | 9 | 4—5 | 4—5 |

as not below "medium" in intensity, and since fundamental frequencies have the highest, all the infrared lines found were taken to be fundamental vibration frequencies.

Comparison of the calculated and experimental results shows that, by the number and activity of the frequencies, the experimental data are in best agreement with calculations for a unit cell with $Z = 4$, belonging to space group $D_4$. Therefore this result will be used in the subsequent interpretation of the vibrational spectrum of the tetragonal form of tellurium dioxide.

<u>Structure of Tetragonal Tellurium Dioxide.</u> It has been shown that the lattice of the tetragonal crystalline form of tellurium dioxide contains a unit cell belonging to space group $D_4$. The unit cell of space group $D_4$ has one fourfold screw axis and four twofold screw axes perpendicular to it. The cell contains four tellurium atoms and eight oxygen atoms. The positions of the atoms in the unit cell of tetragonal tellurium dioxide were determined by x-ray structural analysis [57]: Te–(1,2,3,4)–[x,x,0],[$\bar{x}$,$\bar{x}$,$\frac{1}{2}$],[$\frac{1}{2}$–x,$\frac{1}{2}$ + x,$\frac{1}{4}$],[$\frac{1}{2}$ + x,$\frac{1}{2}$–x,$\frac{3}{4}$] with parameter x = 0.030;

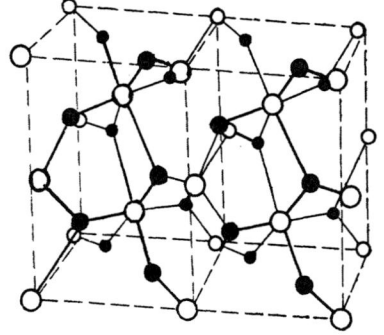

Fig. 35. Model of the structure of crystalline tellurium dioxide.

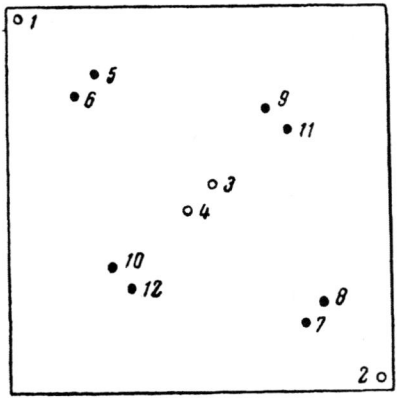

Fig. 36. Projection of the atoms in the unit cell of the crystalline tellurium dioxide lattice onto the basal plane.

$$O - (5, 6, 7, 8, 9, 10, 11, 12) - [x, y, z],$$

$$[\bar{x}, \bar{x}, \tfrac{1}{2}], \ [x, \bar{y}, \tfrac{1}{2} + z], \ [\bar{y}, \bar{x}, \tfrac{1}{2} - z],$$

$$[\tfrac{1}{2} - y, \tfrac{1}{2} + x, \tfrac{1}{4} + z], \ [\tfrac{1}{2} + x, \tfrac{1}{2} - y, \tfrac{3}{4} - z],$$

$$[\tfrac{1}{2} - x, \tfrac{1}{2} + y, \tfrac{1}{4} - z], \ [\tfrac{1}{2} + y, \tfrac{1}{2} - x, \tfrac{3}{4} + z]$$

with parameters x = 0.177; y = 0.227; z = 0.217. The digits in parentheses represent the numbering of the atoms in the unit cell. A model of the lattice structure of the tetragonal form of tellurium dioxide, shown in Fig. 35, was constructed from the calculated coordinates.

Figure 35 shows only two unit cells with the nearest atoms of neighboring cells; white circles represent tellurium atoms, and black, oxygen atoms. Each tellurium atom is surrounded by six oxygen atoms, and each oxygen atom by three tellurium atoms; this agrees with the coordination numbers of these components. It should be noted that the six oxygen atoms surrounding a tellurium atom form an irregular, distorted octahedron.

Figure 36 shows a projection of the $Te_4O_8$ unit cell onto a plane perpendicular to the fourfold axis. The numbering of the atoms on the projection corresponds to the numbering given for the coordinates.

<u>Interpretation of the Vibrational Spectrum Frequencies of Tellurium Dioxide.</u> By considering the unit cell consisting of 12 atoms, we can determine the elements of symmetry characteristic of symmetry group $D_4$. The symmetry elements of this group are the following.

1. Identity.
2. A double fourfold screw axis $2C_4$.
3. A twofold screw axis coinciding with the fourfold axis $C_2'' \equiv C_4^2$.
4. Four twofold axes perpendicular to the fourfold axis $2C_2'$ and $2C_2$, these being screw axes.

Applying the operations of symmetry to the atoms of the tellurium dioxide cell, we can write the transpositions of the atoms which take place. These transpositions are given below; the numbers of the atoms undergoing transposition in the operation of symmetry are in parentheses.

1. I—each atom coincides with itself

2. $2C_4$ — $\begin{cases} (1, 3, 2, 4)\ (11, 8, 10, 6)\ (5, 9, 7, 12) \\ (4, 2, 3, 1)\ (6, 10, 8, 11)\ (12, 7, 9, 5) \end{cases}$

3. $C_2''$ — $(1, 2)\ (3, 4)\ (11, 10)\ (8, 6)\ (5, 7)\ (9, 12)$

4. $2C_2$ — $\begin{cases} (1, 3)\ (2, 4)\ (5, 10)\ (6, 12)\ (11, 7)\ (8, 9) \\ (1, 4)\ (3, 2)\ (11, 5)\ (9, 6)\ (10, 7)\ (8, 12) \end{cases}$

5. $2C_2'$ — $\begin{cases} (3)\ (4)\ (1, 2)\ (5, 8)\ (7, 6)\ (11, 9)\ (10, 12) \\ (1)\ (2)\ (3, 4)\ (5,6)\ (8, 7)\ (9, 10)\ (11, 12) \end{cases}$

Taking into account the positions of the atoms and the transposition resulting from symmetry operations, we calculated the number of fundamental vibration frequencies for each class of this symmetry group. The results are given in Table 21. The selection rules for the vibration frequencies were taken from Herzberg's book [31]. They are also given in Table 21.

Thus, for the unit cell of 12 atoms we have four totally symmetric vibrations of class $A_1$ active only in the Raman spectrum, four class $A_2$ vibrations, antisymmetric with respect to the twofold axis, active only in the infrared absorption spectrum, nine vibrations of class $B_1$ and $B_2$, antisymmetric with respect to the fourfold axis, active in the Raman spectrum, and eight doubly degenerate class E vibrations active in both spectra.

Knowing the number of fundamental vibrations and the selection rules for the symmetry group $D_4$, we can now use the results for assignment of the frequencies in the vibrational spectrum of tetragonal tellurium dioxide to vibration classes. As data on the degree of depolarization cannot be obtained by the powder method, the degree of depolarization of certain lines in the Raman spectrum of glassy $TeO_2$ was estimated qualitatively. Unfortunately, owing to the influence of the background this could be done only for the band at 590-830 cm$^{-1}$.

Because of the analogy between the unit cell structures of the glassy and tetragonal forms, it is possible to apply the data on the degree of depolarization of the band in the spectrum of the glassy form to the corresponding lines in the Raman spectrum of tetragonal tellurium dioxide. Therefore it may be stated qualitatively that the lines from 590 cm$^{-1}$ upward are depolarized.

In addition to the degree of depolarization, we will use such important characteristics as intensities and activities. It is well known that totally symmetric vibrations are not only Raman-active but usually have the highest intensities. Therefore frequencies which are active in the Raman spectrum only and are the most intense and polarized should be assigned to totally symmetric vibrations of class $A_1$. These frequencies in the vibrational spectrum are 89, 146, and 329 cm$^{-1}$. The intense frequencies 121 and 648 cm$^{-1}$ cannot be assigned to this symmetry type, because they are also present in the infrared absorption spectrum and, moreover, the 648 cm$^{-1}$ frequency is depolarized.

Table 21. Selection Rules for Symmetry Group $D_4$

| Vibration type | Selection rules | No. of fundamental vibration frequencies |
|---|---|---|
| $A_1$ | RS | 4 |
| $A_2$ | IR | 4 |
| $B_1$ | RS | 5 |
| $B_2$ | RS | 4 |
| E | $\begin{cases} \text{RS} \\ \text{IR} \end{cases}$ | 8 |

TABLE 22. Assignment of Frequencies to Vibration Classes

| RS, cm$^{-1}$ | IR, cm$^{-1}$ | Symmetry class of the highest vibration state | RS, cm$^{-1}$ | IR, cm$^{-1}$ | Symmetry class of the highest vibration state |
|---|---|---|---|---|---|
| 58 (vs) | | ? | 343 (w) | 334 (vs) | $E$ |
| 89 (s) | | $A_1$ | 365 (vw) | | $B_1$ or $B_2$ |
| | 113 (m) | $A_2$ | 392 (s) | | $A_1$ |
| 121 (vs) | 119 (m) | $E$ | 409 (m) | | $B_1$ or $B_2$ |
| | 137 (m) | $A_2$ | 495 (w) | | $B_1$ or $B_2$ |
| 146 (vs) | | $A_1$ | 590 (m) | | $B_1$ or $B_2$ |
| | 174 (vs) | $A_2$ | 648 (vs) | 648 (vs) | $E$ |
| 204 (m) | | ? | 667 (m) | | $B_1$ or $B_2$ |
| 266 (w) | | $B_1$ or $B_2$ | 717 (m) | 714 (s) | $E$ |
| | 314 (vs) | $A_2$ | 766 (m) | 760 (m) | $E$ |

The infrared frequencies 113, 137, 174, and 314 cm$^{-1}$ should be assigned to symmetry class $A_2$, as they are not found in the Raman spectrum. The frequency 334 cm$^{-1}$ cannot be assigned to this class with certainty, because the difference between this frequency and the 343 cm$^{-1}$ frequency in the Raman spectrum is very small (9 cm$^{-1}$) and is within the limits of experimental error. Therefore, if these frequencies coincide the frequency 334 cm$^{-1}$ should be assigned to class $E$. The infrared frequencies 119, 648, 714, and 760 cm$^{-1}$ should also be assigned to class E, as they are also present in the Raman spectrum. The fact that the last three frequencies are depolarized supports this. Owing to lack of data, nothing can be said about the degree of depolarization of the frequency 119 cm$^{-1}$. The remaining frequencies of vibration class E apparently lie in the uninvestigated long-wave regions (36-52 $\mu$ and above 110 $\mu$).

The frequencies 226, 343, 365, 409, 495, 590, and 667 cm$^{-1}$ should be assigned to vibration classes $B_1$ and $B_2$. It is difficult to distinguish between them, because the selection rules are the same. To separate the frequencies of these classes one requires a large crystal which must be investigated at different orientations with respect to the optical axis, which is impossible by the powder method. The assignment of the frequencies to the respective vibration classes is given in Table 22.

It must be pointed out that the frequency 204 cm$^{-1}$ cannot be assigned to any definite class owing to the lack of an infrared absorption spectrum in the far region. Assignment of the frequency 58 cm$^{-1}$, which is very low for a fundamental frequency, is also difficult; even if it is assigned to $A_1$ it is very close to the frequency 89 cm$^{-1}$ of the same type.

CHAPTER IV

# VIBRATIONAL SPECTRA OF SUBSTANCES
# WITH CHAIN LATTICE STRUCTURE

In Chapters II and III we presented the results of investigations of the vibrational spectra of a number of glass-forming oxides with molecular and coordination lattice structures.

An interesting effect was observed in these investigations: substances with a coordination structure showed a change of line breadth in the transition from the crystalline to the glassy state, while for substances with a molecular structure the line breadth remained almost unchanged in such transitions. It is interesting in this context to investigate substances having other structures, such as chains or layers. Unfortunately, there are no glass-forming oxides with such structures. We therefore used the glassy product formed by dehydration of orthophosphoric acid, with a chain lattice structure.

## §1.  Glassy Polyphosphate [60]*

It was shown by chemical investigations [61] that dehydration of orthophosphoric acid (and of ammonium dihydrogen phosphate) yields a mixture of various phosphoric acids and not phosphoric anhydride. The composition of the mixture depends greatly on the temperature at which the melt was kept. It was found [61] that above 350°C polymeric phosphoric acids with a chain structure are formed, and at temperatures above 450° a small amount of acids with a layer structure also appears. It was reported that the dehydration products obtained from melts at temperatures above 450°C pass fairly easily into the glassy state.

Numerous investigations of phosphorus compounds have shown that the main structural units are $PO_4$ tetrahedrons, which can give rise to various structures, from isolated tetrahedrons to spatial networks, dependent on the degree of bonding. It was noted in [62] that in many cases there is considerable similarity between oxygen compounds of phosphorus and of silicon (similar structural units—tetrahedrons—, similar X—O distances, equal type of polymerization, etc.), although there are some differences (different valences, leading to certain structural differences, especially in the case of substances having spatial network structure).

This fairly close similarity between silicon—oxygen and phosphorus—oxygen compounds is very important. The author of [62] correctly noted that "studies of the structure of crystalline phosphates and phosphate glasses by means of molecular spectroscopy are of definite interest, especially in view of the possible analogy with silicates."

Discussion of Results. Figure 37 shows a photograph and a microphotogram of the Raman spectrum of the substance investigated. It is seen that the line with a maximum at 1341 cm⁻¹ is asymmetric. On the shoulder of this line there is an unresolved line with a frequency of about 1300 cm⁻¹. It should be noted that the determination of the line at 1547 cm⁻¹ is very rough, because it is a fairly broad band.

---

*In this case the term "glassy polyphosphate" is applied to the product of thermal dehydration of orthophosphoric acid at about 650° and of thermal dehydration of ammonium dihydrogen phosphate at about 800°.

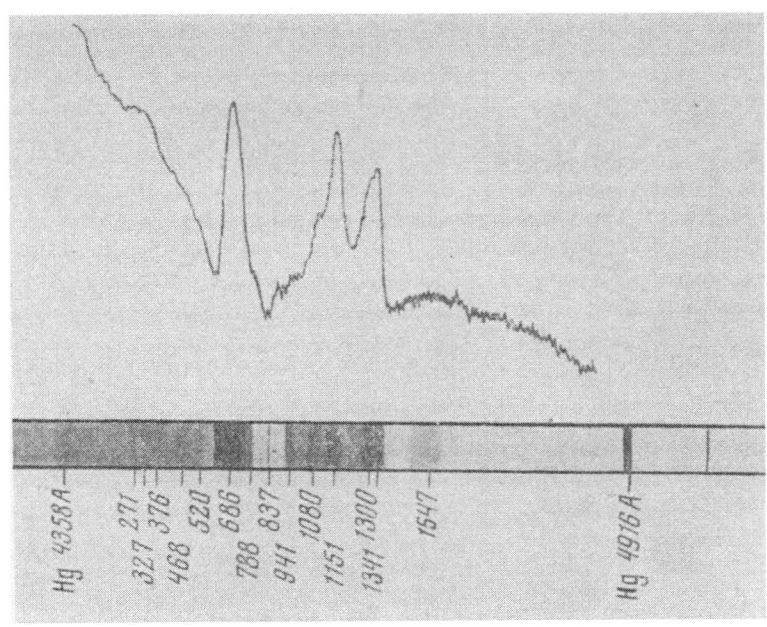

Fig. 37. Raman spectrum of glassy polyphosphate.

It is possible that this band contains several broad lines which merge into the band. The breadth of the band is clearly seen in the microphotogram.

The results show that nearly all the lines are broad and diffuse. Moreover, it is interesting to note that the Raman spectrum does not contain frequencies at 3600 and 2500 $cm^{-1}$, where vibrations characteristic of O−H and P−H bonds should appear.

All the results obtained on the Raman spectrum of the substance are presented in Table 23. The table does not give Raman spectrum data for the substance obtained from a melt of $(NH_4)H_2PO_4$, as it was found that the spectrum is identical with that of the substance from $H_3PO_4$ melts. Together with the frequencies the table contains assessments of intensities and, for the strongest lines, the results obtained in determinations of the degree of depolarization.

The fact that the Raman spectrum of this substance contains no frequencies in the 3600 and 2500 $cm^{-1}$ regions indicates that either the compound does not contain hydrogen atoms or the bonds between hydrogen atoms and the acidic residues are ionic in character and therefore not revealed in the Raman spectrum. The

TABLE 23. Raman Spectrum of Glassy Polyphosphate

| RS, $cm^{-1}$ | Depolar-ization | RS, $cm^{-1}$ | Depolar-ization |
|---|---|---|---|
| 271 (w) | — | 837 (vw) | — |
| 327 (w) | — | 941 (w) | — |
| 376 (w) | — | 1080 (w) | — |
| 468 (w) | — | 1151 (vs) | pol |
| 520 (w) | — | 1300 (m) | — |
| 686 (vs) | pol | 1341 (vs) | pol |
| 788 (w) | — | 1547 (w) | — |

first alternative can be rejected, because it contradicts the results of [61], where it was shown by chemical methods that the product contains phosphoric acids which contain hydrogen atoms. With the second altermative the Raman frequencies are due to vibrations of atoms in the anion, i.e., oxygen and phosphorus.

Thilo et al. [61] concluded from their investigations that the specimen obtained from orthophosphoric acid at 650° predominantly contains polyphosphoric acids which are composed of chain anions formed from tetrahedrons linked by common oxygen atoms.

The chain phosphoric acids may be schematically represented as follows:

$$
\begin{array}{ccc}
& O & O \\
& \| & \| \\
-O- & P & -O- P -O- \\
& | & | \\
& O & O \\
& | & | \\
& H & H
\end{array}
$$

The authors also note that, in addition to these acids, the product contains small amounts of other acids with different numbers of tetrahedrons in the chain.

In addition to chains, the authors report the presence of small amounts of layer acids of the type

$$
\begin{array}{cc}
| & | \\
O & O \\
| & | \\
O = P - O - P = O \\
| & | \\
O & O \\
| & | \\
O = P - O - P = O \\
| & | \\
O & O \\
| & |
\end{array}
$$

We were unable to find any data in the literature concerning the arrangement and parameters of the phosphoric acid chains. Therefore, by analogy with silicon–oxygen compounds, we will assume that the chain structure of the phosphorus–oxygen compounds is similar to that of silicon–oxygen compounds.

Calculation and Interpretation of the Frequencies of the Raman Spectrum of Polyphosphate. In accordance with the above reasoning, we can apply in the present instance the calculations performed by Prima and Stepanov in the case of silicon–oxygen compounds [63]. As our product consists mainly of chain phosphoric acids, we used the calculations performed for valence vibrations of the chains, modified appropriately for our case. Since valence vibrations usually appear in the frequency range roughly above 500 cm$^{-1}$, we confine ourselves to this region of the spectrum.

Using the designations $\varepsilon_x = 1/m_x$, $\varepsilon_y = 1/m_y$, where $m_x$ and $m_y$ are the masses of the atoms, and the subscripts x and y correspond to phosphorus and oxygen atoms, and denoting by $k_1$ and $k_2$ the elastic constants for vibrations of P–O$_{free}$ and P–O$_{bonded}$ bonds respectively, we obtain the following equations for the chain frequencies.

Symmetry Class A$_1$. The vibrations are active both in the Raman and in the infrared absorption spectrum. The lines are polarized.

$$
v_1^2 v_2^2 = \left( \frac{1}{3} \varepsilon_y^2 + \frac{8}{9} \varepsilon_x \varepsilon_y \right) k_1 k_2;
$$

$$
v_1^2 + v_2^2 = \left( \frac{2}{3} \varepsilon_x + \varepsilon_y \right) k_1 + \frac{2}{3} \left( \varepsilon_x + \frac{1}{3} \varepsilon_y \right) k_2;
$$

$$
v_3^2 = \left( \frac{4}{3} \varepsilon_x + \varepsilon_y \right) k_1.
$$

**Symmetry Class $B_2$.** The vibrations are both in the Raman and in the infrared absorption spectrum. The line is depolarized.

$$v_4^2 = \frac{4}{3}\left(\varepsilon_x + \frac{4}{3}\varepsilon_y\right)k_2.$$

**Symmetry Class $B_1$.** The vibrations are active both in the Raman and in the infrared absorption spectrum. The lines are depolarized.

$$v_5^2 v_6^2 = \left(\frac{16}{9}\varepsilon_y^2 + \frac{50}{27}\varepsilon_x\varepsilon_y\right)k_1 k_2;$$

$$v_5^2 + v_6^2 = \left(\frac{2}{3}\varepsilon_x + \varepsilon_y\right)k_1 + \frac{2}{3}\left(\varepsilon_x + \frac{8}{3}\varepsilon_y\right)k_2;$$

$$v_7^2 = \left(\frac{4}{3}\varepsilon_x + \varepsilon_y\right)k_1.$$

**Symmetry Class $A_2$.** The vibrations are active in the Raman spectrum only. The line is depolarized.

$$v_8^2 = \frac{2}{3}\left(2\varepsilon_x + \frac{1}{3}\varepsilon_y\right)k_2.$$

To simplify the calculations and to facilitate comparison of the theoretical and experimental results and tentative assignment of the frequencies, we first performed calculations with the assumption that all the quasi-elastic constants are equal, i.e., $\kappa_1 = \kappa_2 = \kappa$. As a result, the following expressions were obtained for the frequencies:

$$\left.\begin{array}{l} v_1 = 0.165\sqrt{k} \\ v_2 = 0.303\sqrt{k} \\ v_3 = 0.324\sqrt{k} \end{array}\right\} \text{Class } A_1 \qquad \left.\begin{array}{l} v_5 = 0.275\sqrt{k} \\ v_6 = 0.375\sqrt{k} \\ v_7 = 0.324\sqrt{k} \end{array}\right\} \text{Class } B_1$$

$$v_4 = 0.392\sqrt{k} \quad \text{Class } B_2 \qquad v_8 = 0.238\sqrt{k} \quad \text{Class } A_2$$

The force constant can be determined from one frequency, for example $v_2 = 1151$ cm$^{-1}$, which should be assigned to a totally symmetric vibration of class $A_1$, as it is polarized and of high intensity. It is quite justifiable to take other frequencies of class $A_1$ as the basis, as they are easiest to separate out of the spectrum. From the frequency 1151 cm$^{-1}$ we found the value of $\kappa = 14.42 \cdot 10^6$ cm$^{-2}$ for the force constant. Using this value, we calculated the other frequencies from the equations given above. The agreement with experiment is even better if we take an average elastic constant based on three totally symmetric vibrations of class $A_1$, which are the easiest to identify with respect to intensity and polarization. These frequencies $v_1 = 686$ cm$^{-1}$, $v_2 = 1151$ cm$^{-1}$, and $v_3 = 1341$ cm$^{-1}$. We then have for the average value of the elastic constant

$$k = \frac{k_1' + k_2' + k_3'}{3} = 16{,}27 \cdot 10^6 \text{ cm}^{-2},$$

where $\kappa_1'$, $\kappa_2'$, $\kappa_3'$ are the force constants found from the corresponding class $A_1$ frequencies.

From this average value for the force constant we find all the frequencies in the Raman spectrum of the chain. Now, with a tentative knowledge of the assignment of the frequencies to vibration classes, we can per-

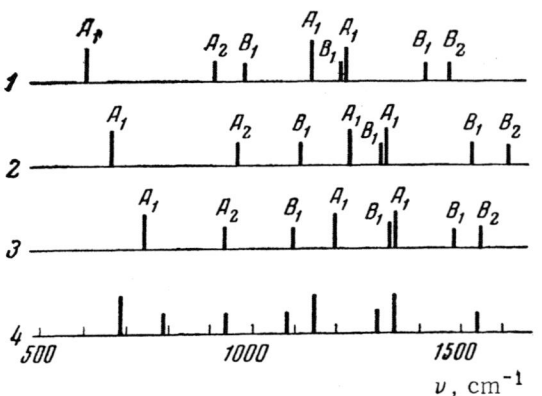

Fig. 38. Experimental and calculated Raman frequencies of glassy polyphosphate.

form the calculation with the two constants $\kappa_1$ and $\kappa_2$ taken into account from the equations given above. The elastic constants were calculated from the frequencies $\nu_3 = 1341$ cm$^{-1}$ and $\nu_8 = 940$ cm$^{-1}$.

The following numerical values were found for these constants: $\kappa_1 = 17.13 \cdot 10^6$ cm$^{-2}$ and $\kappa_2 = 15.59 \cdot 10^6$ cm$^{-2}$. These elastic constants were used for calculating the remaining Raman line frequencies.

Figure 38 shows schematic diagrams of the Raman spectra of the specimen, based on calculations and experimental data. The spectrum (1) in the first line was calculated from the above equations with the assumption that all the quasielastic constants are equal ($\kappa_1 = \kappa_2 = \kappa$). The constant was calculated from the frequency 1151 cm$^{-1}$. The theoretical spectrum (2) in the second line was obtained with the same assumption but the constant was obtained by averaging of three constants, obtained from the frequencies 685, 1151, and 1341 cm$^{-1}$, covering the entire range of valence vibration frequencies. Spectrum (3) in the third line was calculated with two constants, $\kappa_1$ and $\kappa_2$, taken into account. The fourth line gives the experimental spectrum (4) on the same scale. The lines relating to class $A_1$ vibrations are indicated by taller lines in Fig. 38, because lines corresponding to totally symmetric vibrations usually have higher intensities than lines for the other vibration classes.

The comparison of theory and experiment shows that, despite the crude assumptions, the calculations give fairly satisfactory agreement between the theory and the experimental results. As was to be expected, the best agreement between theory and experiment was obtained in the third case, when the difference between the force constants $\kappa_1$ and $\kappa_2$ was taken into account. This satisfactory agreement between theory and experiment in all three cases is apparently associated with the fact that we calculated the valence force constants from experimental results and thereby took into account, to some extent, the influence of deformational and rotational vibrations.

Figure 38 shows that the line for frequency 788 cm$^{-1}$, observed experimentally, remains unexplained by our calculations. This line should probably be assigned to a vibration associated with a layer structure, the appearance of which was reported in [61]. The frequency 837 cm$^{-1}$ is too weak to be assigned to fundamental vibrations, and therefore it is not shown in Fig. 38. It is most likely attributable to component frequencies or overtones which, with rare exceptions, are usually of low intensity.

Thus, we investigated the Raman spectra of glassy polyphosphates obtained by dehydration both of orthophosphoric acid and of ammonium dihydrogen phosphate.

The rough calculation of the valence vibration frequencies confirmed the findings of [61] concerning the chain structure of glassy polyphosphate. Unfortunately, the specimen was so strongly hygroscopic that it was impossible to obtain its infrared absorption spectrum, which might have given additional information on the structure of the substance. It must be pointed out that the lines of the polyphosphate spectrum are very broad and resemble in this respect the spectral lines of glassy oxides having a coordination lattice structure. From this it may be concluded that broadening of the lines in the transition from the spectrum of the crystalline to the spectrum of the glassy form should occur will all substances having lattice structures other than molecular.

# CONCLUSION

The following conclusions may be drawn from this investigation of the vibrational spectra of glass-forming oxides.

It was shown in Chapter II that the spectroscopic results obtained for the oxides $As_2O_3$, $Sb_2O_3$, and $As_2O_5$ are reliably consistent with a molecular lattice structure. It must be pointed out in this connection that in the case of $As_2O_3$ and $Sb_2O_3$ only one of the crystalline forms, the cubic, has a molecular lattice structure. The other crystalline form of these oxides has a coordination lattice structure.

Results of investigations of the oxides $Sb_2O_5$, $GeO_2$, $TeO_2$ and analysis of x-ray diffraction data showed that the vibrational spectra of these oxides can be satisfactorily interpreted if a coordination lattice structure is assumed. The unit cell characterizing the structure of these oxides was determined from our results and literature data.

Investigations of the vibrational spectra of a number of glass-forming oxides ($As_2O_3$, $GeO_2$, $TeO_2$) in the glassy state showed that the fundamental frequencies in the spectrum of a definite crystalline modification of a given oxide are also present in the spectrum of the glassy form. The fact that the frequencies in the crystal spectrum are retained in the spectrum of the glass apparently indicates that the complexes which characterize the lattice structure of the crystalline form and determine its vibrational spectrum (molecules in substances with molecular lattices, and unit cells in substances with nonmolecular lattice structure) also persist in the glassy state.

All this supports the view [4] that all glass-forming oxides both in the crystalline and in the glassy states, which by the Zachariasen theory should have a coordination lattice structure in both states, can be subdivided into two groups: with a molecular lattice structure and with a coordination lattice structure. On the basis of our results and literature data, of all the glass-forming oxides in the crystalline state the oxides, $P_2O_3$, $P_2O_5$, $As_2O_3$, $B_2O_3$, $B_2O_3$, $As_2O_5$, and $Sb_2O_3$, can quite definitely be assigned to the first group, and $SiO_2$, $GeO_2$, $TeO_2$, $V_2O_5$, $Nb_2O_5$, and $Ta_2O_5$ to the second.

In the glassy state, only $B_2O_3$ and $As_2O_3$ can with certainty be classed with substances having a molecular structure. Other oxides of this type have either been studied very little or have not been prepared at all. The glassy forms of $SiO_2$, $GeO_2$, and $TeO_2$ have a coordination structure. The remaining oxides of this type have not been obtained in the glassy state, although glasses have been obtained from $V_2O_5$ containing admixtures of other substances. It is interesting to note that glass-forming oxides with a molecular structure are characteristic mainly for elements in Group V of the periodic table. There is a gradual transition from oxides with a molecular structure to oxides with a coordination structure with increasing atomic weight of the elements in this group. This transition is associated with intensified intermolecular interaction, which results in higher melting points. The melting points of respective forms of glass-forming oxides of the types $X_2O_3$ or $X_2O_5$ rise in the series P—As—Sb. For oxides of the first type in the $P_2O_3$—$As_2O_3$—$Sb_2O_3$ series the melting point rises from 24°C for $P_2O_3$ to 656°C for $Sb_2O_3$.

Intensification of intermolecular interaction leads to peculiarities in the vibrational spectra with disturbances of the selection rules. However, whereas in this series of oxides the molecules retain their individuality despite the increased interaction, in the series $P_2O_5$—$As_2O_5$—$Sb_2O_5$ the last oxide, in contrast to the first two, has a coordination lattice structure. This is because the intermolecular forces have reached the magnitude of the intramolecular forces, so that molecular formations can no longer be distinguished in this compound.

It should be noted that oxides with a molecular lattice structure are formed by elements of Group III as well as Group V of the periodic table. Only one glass-forming oxide, $B_2O_3$, is known in this group; admittedly, in the glassy form only.

Investigation of the spectra of oxides in the crystalline and glassy states revealed some interesting peculiarities. The first of these is a weakening of the spectral lines in the transition from the crystalline to the glassy form. The lines become weaker not only in the transition from the crystalline to the glassy form, but also from the glassy form to the liquid. For example, a study of the Raman spectrum of glassy boric anhydride showed that the line intensities in the spectrum of glassy $B_2O_3$ decrease with rise of temperature, i.e., with decrease of viscosity. Thus, to judge by the variations of the spectral line intensities of glass-forming oxides, the glassy form is intermediate between crystals and the liquid. However, as glasses have many physicochemical properties both of crystals and of liquids [6], a glass cannot be characterized as a supercooled liquid [64] or merely as a substance with an ordered structure. A glass is most likely a system having features both of order and disorder. The second peculiarity is the difference in the behavior of line breadths in the transition from the spectra of crystals to spectra of glasses for oxides with different lattice structures.

It is quite clear that for substances with molecular lattice structure this transition should have very little effect on line breadth, because it affects mainly the intermolecular bonds, which are much weaker than the intramolecular forces. Consequently, changes of intermolecular bonding have very little influence on bonds within the molecules, and therefore large changes of line breadth cannot be expected. In fact, if we compare the spectra of the glassy and crystalline (cubic) $As_2O_3$, we find no sharp changes in line frequencies or in line breadths. Glassy boric anhydride, which also has a molecular structure, gives very narrow lines. In the case of substances with a coordination structure the lines of the vibrational spectrum of the glassy form are considerably broadened in comparison with the crystalline form ($GeO_2$, $TeO_2$, $SiO_2$). Transition from the crystalline to the glassy state appears to disturb the regularity of the lattice structure and this, of course, results in some breakdown in the structure of the unit cell; however, this is not large enough to alter the frequency spectrum substantially, but large enough to alter the line breadths. The spectrum of glassy polyphosphate with a chain lattice structure also has broad lines. Thus, broad lines are found in the spectra of substances which do not have purely molecular lattice structure.

It may be concluded from all the foregoing that it is possible to deduce tentatively from changes of line breadth in the transition from the spectrum of the crystalline form to that of the glassy form whether the given substance has a molecular or nonmolecular lattice structure. Therefore the subdivision of glass-forming oxides into substances with molecular and nonmolecular lattices should be based not on the breadth of the lines in the infrared absorption spectrum (as was done in [4]), but on changes of spectral line breadth in the transition from the crystalline to the glassy form of the given oxide. It must also be pointed out that the subdivision of oxide structures in accordance with melting points is also highly arbitrary. It should be remembered that when a definite melting point is ascribed to an oxide the fact that it may exist in several forms with very different lattice structures is often disregarded (for example, in [4]).

# SUMMARY

The principal results of the work are as follows:

1. A simplified technique was devised for obtaining Raman spectra of powders.

2. The Raman and infrared absorption spectra of the following oxides were investigated: $As_2O_3$, $Sb_2O_3$, $As_2O_5$, $Sb_2O_5$, $TeO_2$, $GeO_2$ (two forms) in the crystalline state, and $As_2O_3$, $GeO_2$, $TeO_2$ in the glassy state.

3. The vibrational spectra of the crystalline and glassy forms of the oxides $As_2O_3$, $GeO_2$, $TeO_2$ are similar in many main respects, indicating a similarity of their structural units which determine the vibrational spectra of these substances.

4. By a study of the vibrational spectrum it was shown that arsenious anhydride has a molecular structure and consists of $As_4O_6$ molecules having symmetry of the point group $T_d$. The force constants and vibration frequencies of the $As_4O_6$ molecule were calculated, and the vibrational spectrum of arsenious anhydride was reliably interpreted with the aid of these results.

5. A similar result was obtained for antimonous anhydride, consisting of $Sb_4O_6$ molecules with symmetry of the point group $T_d$. The force constants for this molecule were found and the vibrational spectrum was calculated.

6. Analysis of the frequencies of the vibrational spectrum of arsenic pentoxide showed satisfactory agreement with a molecular lattice structure consisting of $As_4O_{10}$ molecules having symmetry of the point group $T_d$.

7. The vibrational spectrum of $Sb_2O_5$ is at variance with a molecular structure and consistent with a coordination lattice structure, with a unit cell having symmetry of the space group $O_h$.

8. The vibrational spectra of two crystalline and one glassy form of $GeO_2$ were investigated. The results for the soluble and the glassy forms are in fairly good agreement with a unit cell consisting of three $GeO_2$ groups and belonging to the $D_3$ space group.

9. The vibrational spectrum of $TeO_2$ was studied. From the results and analysis of x-ray diffraction data a model of a unit cell was chosen, containing four $TeO_2$ groups and belonging to space group $D_4$. The glassy form probably has a similar unit cell structure.

10. Study of the Raman spectrum of glassy polyphosphate and calculations of the vibration frequencies showed fairly satisfactory agreement with the assumption that it has a chain lattice structure.

11. Changes in the vibrational spectra in the transition from crystalline to glassy forms lead to the conclusion that the change of line breadth in the transition is a rough indication of the structure of the substance.

In conclusion, I offer my sincere gratitude to my scientific supervisor, Doctor of Physicomathematical Sciences N. N. Sobolev, for suggesting the subject, for his constant interest, and for valuable advice in discussion of the results and in the writing of the dissertation.

I am also grateful to T. G. Baranova for help in the calculations.

# LITERATURE CITED

1. C. Schaefer and F. Matossi, Infrared Spectra [Russian translation] ONTI, 1935.
2. V. A. Kolesova, Zh. Eksperim. i Teor. Fiz., 26:124 (1954).
3. W. H. Zachariasen, J. Am. Chem. Soc., 54:3841 (1932).
4. T. A. Sidorov, Tr. Fiz. Inst. Akad. Nauk, 12:225 (1960).
5. A. A. Lebedev, Tr. Gos. Optich. Inst., 8 (10) (1921).
6. The Structure of Glass, Collection, edited by A. A. Lebedev, Izd. AN SSSR, 1955.
7. V. P. Zlomanov, I. I. Takanaeva, and A. V. Novoselova, Zh. Neorgan. Khim., 5:1632 (1960).
8. É. N. Lotkova, V. V. Obukhov-Denisov, N. N. Sobolev, and V. P. Cheremisinov, Opt. i Spektroskopiya, 1:773 (1956).
9. T. A. Sidorov and N. N. Sobolev, Opt. i Spektroskopiya, 1:393 (1956).
10. T. A. Sidorov and N. N. Sobolev, Opt. i Spektroskopiya, 3:560 (1957).
11. T. A. Sidorov and N. N. Sobolev, Opt. i Spektroskopiya, 2:711 (1957).
12. T. A. Sidorov and N. N. Sobolev, Opt. i Spektroskopiya, 2:713 (1957).
13. V. I. Malyshev, M. N. Markov, and A. A. Shubin, Dokl. Akad. Nauk, 86:273 (1952).
14. V. I. Malyshev, M. N. Markov, and A. A. Shubin, Izv. AN SSSR, seriya fizich., 17:654 (1953).
15. M. N. Markov, Zh. Tekhn. Fiz., 24:1864 (1954).
16. A. I. Demeshina and V. N. Murzin, Opt. i Spektroskopiya, 13:826 (1963).
17. M. L. Sosinskii, Izv. AN SSSR, seriya fizich., 17:621 (1953).
18. V. P. Cheremisinov, PTÉ, 2:122 (1956).
19. M. M. Sushchinskii, Tr. Fiz. Inst. Akad. Nauk, 12:54 (1960).
20. V. V. Obukhov-Denisov, N. N. Sobolev, and V. P. Cheremisinov, Opt. i Spektroskopiya, 8:505 (1960).
21. A. A. Fridrichs, Sprechsaal Keramik, Glas, Email, 61:261 (1928).
22. A. W. Laubengauer and D. S. Morton, J. Am. Chem. Soc., 54:2303 (1932).
23. B. F. Ormont, Structure of Inorganic Substances, Moscow—Leningrad, GTTL, 1950.
24. V. P. Cheremisinov, Opt. i Spektroskopiya, 7:454 (1959).
25. J. H. Schulman, J. Amer. Chem. Soc., 65:878 (1943).
26. L. R. Maxwell, S. B. Hendricks, and L. S. Deming, J. Chem. Phys., 5:626 (1937).
27. V. M. Borisov and M. G. Gorshtein, Zh. Obshchei. Khim., 22:1903 (1952).
28. O. Borgen and J. Kroph, Moc. Acta, Chem. Scand., 10:265 (1956).
29. F. A. Miller, G. L. Carlson, F. F. Bentley, and W. H. Jones, Spectrochim. Acta, 16:135 (1960).
30. D. H. Zijp, Vibrational Spectrum and Analysis of Molecules of General Formula $X_4Y_6Z_4$ Possessing $T_d$ Symmetry, Amsterdam, 1960.
31. G. Herzberg, Infrared and Raman Spectra of Polyatomic Molecules [Russian translation] IL, 1949.
32. L. Countre-Mathieu, J. P. Mathieu, J. Cramer, and H. Poulet, J. Chim. Phys., 48:1 (1951).
33. N. N. Sobolev and V. P. Cheremisinov, Opt. i Spektroskopiya, 9, 446:(1960).
34. C. R. Fresenius, Anleitung zur qualitativen chemischen Analyse, 17 Aufl. Braunschweig, 318, 1919.
35. J. T. Randall and H. P. Rooksby, J. Soc. Glass Techn., 17:290 (1930).
36. L. Debray, J. Prakt. Chem., 98:151 (1866).
37. R. M. Bororth, J. Am. Ceram. Soc., 45:1626 (1923).
38. K. E. Almin and A. Westgren, Ark. Kem. Min., 15B, (22):6 (1942).
39. A. Bystrom, Ark. Kem. Min., 25A, (13):16 (1948).

40. F. A. Miller and C. H. Wilkins, Analyt. Chem., 24:1253 (1952).

41. M. Parodi, Compt. Rend., Acad. Sci., 205:906 (1937).

42. V. V. Obukhov-Denisov, N. N. Sobolev, and V. P. Chermisinov, Izv. AN SSSR, seriya fizich., 1083 (1958).

43. U. Dehlinger, Z. Kristallogr., 66:117 (1928).

44. B. V. Nekrasov, Course of General Chemistry, Moscow, Goslitizdat, 1960.

45. V. M. Goldschmidt, Naturwissenschaften, 14:295 (1926).

46. V. H. Zachariasen, Z. Kristallogr., 67:226 (1928).

47. J. H. Müller, Proc. Amer. Philos. Soc., 65:183 (1926).

48. J. H. Müller, and H. R. Blank, J. Amer. Chem. Soc., 46:2338 (1924).

49. W. R. Davey, J. Opt. Soc. Amer., 5:479 (1921).

50. E. R. Lippincott, J. Res. Nat. Bur. Standards, 61:61 (1958).

51. B. D. Saksena, Proc. Indian. Acad. Sci., A12:93 (1940).

52. D. Krishnamurti, Proc. Indian. Acad. Sci., A47:276 (1958).

53. R. S. Krishnan, Nature, 155:457 (1945).

54. V. P. Chermisinov and V. P. Zlomanov, Opt. i Spektroskopiya, 12:208 (1962).

55. T. Ito and H. Sowada, Z. Kristallogr., 102A:13 (1939).

56. V. M. Goldschmidt, Skrifter Norske Violenshapi-Akad. Oslo, I: Mat. Naturv. Kl. (1):9 (1926).

57. B. Stelik and A. Balok, Collect. Czechosl. Chem. Communs, 14:595 (1949).

58. V. P. Zlomanov, A. V. Novoselova, A. S. Pashinkin, Yu. P. Simakov, and K. I. Semenenko, Zh. Neorgan. Khim. 3:1473 (1958).

59. J. H. Brady, J. Chem. Phys., 27:300 (1957).

60. V. V. Obukhov-Denisov and V. P. Cheremisinov, Opt. i Spektroskopiya, 12:723 (1962).

61. E. Thilo and R. Sauer, J. Prakt.Chem., 4, 5-6:324 (1957).

62. V. A. Kolesova, Opt. i Spektroskopiya, 2:165 (1957).

63. A. M. Prima and B. I. Stepanov, Opt. i Spektroskopiya, 4:734 (1958).

64. G. Tammann, The Glassy State [Russian translation] ONTI, 1935.

# CALCULATION OF CROSS SECTIONS FOR EXCITATION OF ATOMS AND IONS BY ELECTRON IMPACT

L. A. Vainshtein

## §1. Introduction

A whole series of approximate methods has been proposed for calculation of the cross sections for collisions of electrons with atoms and ions. So far, however, there is no method which gives sufficiently reliable results and which can be recommended for wide application. To obtain a clearer idea of the capabilities of various methods and some common properties of the cross sections we undertook calculations of the cross sections for the excitation of a number of transitions in different approximations. This paper gives an account of some details of the method of calculations and the results obtained.

All the calculations were performed numerically on an electronic computer, and we confined ourselves to methods which required not more than one hour of machine time for one ten-point excitation function (at a machine speed of 20,000 operations per second). This ruled out straightaway the consideration of energetically unattainable virtual states (at least by accurate solution of the corresponding system of equations).

Throughout the following account we use atomic units with the rydberg for energy. Cross sections are expressed in units of $\pi a_0^2$. In addition, the results can conveniently be expressed in the so-called threshold units

$$u = \frac{k_0^2 - \Delta\varepsilon}{\Delta\varepsilon} = \frac{k^2}{\Delta\varepsilon}; \quad x = \sqrt{u} = \frac{k^2}{\sqrt{\Delta\varepsilon}}, \tag{1}$$

where $k_0$ and $k$ are the wave numbers of the incident and scattered electron, $\Delta\varepsilon$ is the excitation energy, $u$ and $x$ are the energy and momentum of the scattered electron in the threshold units. The functions $\sigma(u)$ or $\sigma(x)$ have the same scale along the x-axis for transitions of the most diverse type. In the following account the zero subscript denotes the initial state and the absence of a subscript indicates the final state.

## §2. Some General Formulas

The calculation of cross sections entails expansion in partial waves in most cases. One of the few exceptions is the Born approximation, in which the cross section (in units of $\pi a_0^2$) can be put in the form:

$$\sigma = \frac{3}{g_0 k_0^2} \sum_{M_0 M} \int_{k_0-k}^{k_0+k} |<a_0| \sum_i e^{iqr} |a>|^2 \frac{dq}{q^3}, \tag{2}$$

where $M_0$ and $M$ are the magnetic quantum numbers of the atom. Expanding $e^{iqr}$ in multipoles we obtain

$$\sigma = \frac{8}{k_0^2} \sum_{\varkappa=|l_0-l|}^{l_0+l} \bar{c}_\varkappa \int_{k_0-k}^{k_0+k} \rho_\varkappa^2(q) \frac{dq}{q^3}, \tag{3}$$

where

$$\rho_\varkappa(q) = \int_0^\infty P_{a_0}(r) P_a(r) j_\varkappa(qr) dr, \tag{4}$$

$$c_\varkappa = (2\varkappa + 1)(2L + 1) \begin{Bmatrix} l_0 & L_0 & L_p \\ L & l & \varkappa \end{Bmatrix}^2 (2l_0 + 1)(2l + 1) \begin{pmatrix} l_0 & l & \varkappa \\ 0 & 0 & 0 \end{pmatrix}^2 \delta_{S_0 S}. \tag{5}$$

Here $a = \alpha L_p S_p\, l\, \tfrac{1}{2} LS$ is the set of quantum numbers of the atom, $P_a(r)$ is the radial function of the optical electron of the atom; $g_0 = (2L_0 + 1)(2S_0 + 1)$ is the statistical weight of the initial state; $j_\varkappa$ is a spherical Bessel function. We note that when $L_0 = 0$, $C_\varkappa = (2\varkappa + 1)/(2L_p + 1)$ and $\varkappa = L$.

In the case of expansion in partial waves $\sigma = \Sigma \sigma_{\tilde{l}_0 \tilde{l}}$, where $\tilde{l}$ denotes the orbital quantum number of the partial wave of the outer electron. This infinite sum converges very slowly when $k^2 \geq \Delta\varepsilon$. Hence, in fact, the total cross section has to be calculated from the formula

$$\sigma = \sigma^B + \sum_{\tilde{l}_0 \tilde{l}}^{\tilde{l}_m} (\sigma_{\tilde{l}_0 \tilde{l}} - \sigma^B_{\tilde{l}_0 \tilde{l}}), \tag{6}$$

where the superscript "B" denotes the Born cross section and $\sigma^B$ is calculated from formulas (3)-(5). It is usually sufficient to take $\tilde{l}_m = 4 - 6$.

In the first approximation of the distorted wave method the partial cross section is

$$\sigma_{\tilde{l}_0 \tilde{l}} = \frac{4}{k_0^3 k} \sum_{S_T L_T} \frac{(2S_T + 1)(2L_T + 1)}{2(2S_0 + 1)(2L_0 + 1)} \, |<\Gamma_0 | U | \Gamma>|^2, \tag{7}$$

where $\Gamma = ak\tilde{l}\,\tfrac{1}{2} L_T S_T$ is the set of quantum numbers of the system of atom and outer electron; $L_T S_T$ are the total moments of the system. The matrix element in expression (7) includes direct and exchange transitions:

$$<\Gamma_0 | U | \Gamma> = \sum_\varkappa 2\alpha^\varkappa_{\Gamma_0 \Gamma} \iint \frac{r^\varkappa_<}{r^{\varkappa+1}_>} F_{\Gamma_0}(r) P_{a_0}(r_1) F_\Gamma(r) P_a(r_1)\, dr dr_1$$

$$- \sum_\varkappa 2\beta^\varkappa_{\Gamma_0 \Gamma} \iint \frac{r^\varkappa_<}{r^{\varkappa+1}_>} \left(1 - \delta_{\varkappa 0} \frac{k^2 - \varepsilon_0}{2} r_>\right) F_{\Gamma_0}(r) P_{a_0}(r_1) F_\Gamma(r_1) P_a(r)\, dr\, dr_1. \tag{8}$$

The coefficients $\alpha$ and $\beta$ were determined in [1] (see [2] also).

The summation is conducted over the values of $\varkappa$ which satisfy the triangle rules in the j-symbols contained in $\alpha$ and $\beta$; $\varepsilon$ is the energy parameter corresponding to the radial function $P_a(r)$. Henceforth we use semiempirical functions for which the condition $k^2 - \varepsilon_0 = k_0^2 - \varepsilon$ is fulfilled.

The radial function $F_\Gamma$ describes the elastic scattering of the outer electron in the field of the unperturbed atom and satisfies the equation (see [1]):

$$\left[\frac{d^2}{dr^2} - \frac{\tilde{l}(\tilde{l} + 1)}{r^2} - U_\Gamma(r) + k^2\right] F_\Gamma = 0 \tag{9}$$

with the boundary conditions:

$$F_\Gamma(0) = 0, \qquad F_\Gamma \underset{r\to\infty}{\sim} \sin\left(kr - \frac{\tilde{l}\pi}{2} + \frac{z}{k}\ln 2kr + \eta_\Gamma\right), \tag{10}$$

where $z = Z - N \neq 0$, if we are dealing with ion excitation ($Z$ is the charge of the nucleus, $N$ is the number of electrons). For simplicity we will henceforth speak of atom excitation irrespective of the value of $z$.

The potential $U_\Gamma$ contained in (9) can be divided into two parts: $U_\Gamma = U^C_\Gamma + U_{\Gamma\Gamma}$, where $U^C_\Gamma$ is the interaction with the atomic core [for hydrogenlike ions $U^C_\Gamma = 2(z + 1)/r$], and $U_{\Gamma\Gamma}$ is the interaction with the optical electron. When exchange is taken into account $U_{\Gamma\Gamma}$ includes the integral operator:

$$U_{\Gamma\Gamma} F_\Gamma = \sum_\varkappa 2\alpha^\varkappa_{\Gamma\Gamma} F_\Gamma \int_0^\infty \frac{r^\varkappa_<}{r^{\varkappa+1}_>} P_a^2\, dr_1 - \sum_\varkappa 2\beta^\varkappa_{\Gamma\Gamma} P_a \int_0^\infty \frac{r^\varkappa_<}{r^{\varkappa+1}_>} \left(1 - \delta_{\varkappa 0} \frac{k^2 - \varepsilon}{2} r_>\right) P_a F_\Gamma\, dr_1. \tag{11}$$

Generally speaking, the same applies to $U^C_\Gamma$. In the present calculations, however, the exchange part $U^C_\Gamma$ is always omitted.

The numerical calculation of integrals (8) entails an additional difficulty: the first integral, which can be written in the form

$$I' = \int_0^\infty F_{\Gamma_0}(r) \, F_\Gamma(r) \, U_{\Gamma_0\Gamma}(r) \, dr,$$

converges slowly and must be calculated to too high values of r. It is more convenient to conduct the integration up to a lower (but still sufficiently large) value $r_m$, and to replace the remainder by the asymptotic expression:

$$I = \int_0^{r_m} F_{\Gamma_0} F_\Gamma U_{\Gamma_0\Gamma} \, dr + \frac{1}{2} U_{\Gamma_0\Gamma}(r_m) \left( \frac{\sin \varphi_+}{k_0 + k} - \frac{\sin \varphi_-}{k_0 - k} \right), \tag{12}$$

where

$$\varphi_\pm = (k_0 \pm k) \, r_m - (\tilde{l}_0 \pm \tilde{l}) \, \frac{\pi}{2} + (\eta_0 \pm \eta).$$

## §3. Atomic Wave Functions

The atomic radial functions $P_a$ were obtained by a semiempirical method with exchange included [3] by numerical solution of the integrodifferential equation

$$\left[ \frac{d^2}{dr^2} - \frac{l(l+1)}{r^2} - \frac{1}{\omega} U_a \left( \frac{r}{\omega} \right) + \varepsilon \right] P_a = 0, \tag{13}$$

where $U_a(x)$ is a potential of the type (11), including exchange with one or two outer shells of the atom. The parameter $\omega$ describes the effective compression of the atomic core. In [3] it was chosen so that $\varepsilon = \varepsilon_{exp}$—the experimental value of the energy level. In this case, however, $\omega$ is slightly different for the initial and final states. Consequently, the atomic functions with the same $l$ are nonorthogonal and superfluous terms due to interaction with the nucleus appear in the expression for the excitation cross section. To eliminate this fault we chose the eigenvalue of $\omega$ only for the initial state, and for the final state we chose $\varepsilon$. Although $\varepsilon \neq \varepsilon_{exp}$ in this case the difference does not exceed 10% and in most casses is 2-3%. It should be noted that the difference is much greater when exchange is neglected.

The integrodifferential equation (13) was solved by the method of successive approximations. The approximation neglecting exchange was taken as the zero one. Four to eight iterations were sufficient to ensure good convergence.

In all the calculations conducted in this work the potential and radial functions of the atomic core were calculated by means of simple analytical functions of the Slater type.

## 4. Solution of Integrodifferential Equation for the Outer Electron

In the case of unbound states the method of successive approximations often converges very badly. In addition, it requires a great amount of machine time. Hence, the integrodifferential equation for the wave function of the outer electron was solved by means of the noniteration method suggested by Percival and Marriott [4]. After slight modifications to suit the particular case this method is as follows.

Equations (9) with (11) taken into account can be written in the form (the subscripts $\Gamma$ and a are omitted for brevity):

$$LF \equiv \left[ \frac{d^2}{dr^2} - \frac{l(l+1)}{r^2} - U^0 + k^2 \right] F + \sum_\varkappa 2\beta^\varkappa P \left[ r^{-\varkappa-1} \int_0^r r_1^\varkappa PF \, dr_1 + r^\varkappa \left( I_\varkappa - \int_0^r r_1^{-\varkappa-1} PF \, dr_1 \right) \right], \tag{14}$$

where $U^0$ includes the direct interaction [terms with $\alpha^\varkappa$ in (11)]; $I_\varkappa$ are as yet unknown constants, which depend on the required functions F in the following way:

$$I_{\varkappa}[F] = \int\limits_{0}^{\infty} r_1^{-\varkappa-1}\left(1 - \delta_{\varkappa 0}\frac{k^2 - \varepsilon}{2}r_1\right) P\,(r_1)\,F\,(r_1)\,dr_1. \tag{15}$$

If those constants are known, equation (14) can be solved by the usual numerical methods, since L does not include integrals over $r_1 > r$.

Let $\varphi$ be a solution of an equation of type (14), but with $I_{\varkappa} = 0$, and $\varphi_{\nu}$ solutions of the equation with $I_{\varkappa} = \delta_{\varkappa\nu}\mu_{\nu}$, where $\mu_{\nu}$ are certain specified constants. Their selection is restricted only by the condition that $\varphi$ and $\varphi_{\nu}$ must differ sufficiently (from the viewpoint of errors of the numerical computation) from one another (in this case the results do not depend on $\mu_{\nu}$). If $\varphi$ and $\varphi_{\nu}$ are found, the solution of equation (8) can be sought in the form

$$F = \varphi + \sum_{\nu} c_{\nu}\varphi_{\nu}, \tag{16}$$

and coefficients $c_{\nu}$ are given by a system of algebraic equations

$$\sum_{\nu} \{\mu_{\nu}\delta_{\nu\varkappa} - I_{\varkappa}[\varphi_{\nu}]\}\,c_{\nu} = I_{\varkappa}[\varphi]. \tag{17}$$

In particular, when the sum over $\varkappa$ in equation (14) reduces to one term,

$$F = \varphi + \frac{I_{\varkappa}[\varphi]}{\mu_{\varkappa} - I_{\varkappa}[\varphi_{\varkappa}]}. \tag{18}$$

The described method is much more convenient than the method for successive approximations. However, we should mention the case where it does not come off. It may happen that function P is also a solution of equation (14). In particular, this occurs in the case of antisymmetric scattering on hydrogen-like atoms with $\tilde{l} = l$ (for instance, antisymmetric 0−1(−) and 2−1(−) electron transitions on excitation of the s-p transition in H). It can be shown that in this case the solution of system (17) is not determinate, but the function $\varphi$ itself is a solution of equation (14). Hence, we can put $c_{\nu} = 0$ and $F = \varphi$. In this case, however, other solutions of equation (14) satisfying the boundary conditions (10) will be functions $F = \varphi + c'P$ with any c'. This indeterminacy, generally speaking, does not affect the physical results, since the term c'P makes no contribution to the matrix element of the transition. Nevertheless, when numerical methods are used, this sometimes leads to very large computation errors. Fortunately, the partial cross section in this case is generally fairly small.

§5.  Results

Tables 1-3 give the cross sections for excitation of some transitions in the hydrogen atom, the hydrogen-like $C^{5+}$ ion, and the Na atom. The following symbols are used: B—the Born approximation; Coul.—the Coulomb approximation, i.e., consideration of the distortion only by the Coulomb field of the ion ($U_{\Gamma} = -2z/r$, exchange neglected); DW—distorted wave approximation with exchange neglected; EDW—distorted wave approximation with full consideration of exchange. BSC—Born approximation including strong coupling.* The machine times required for the complete calculation of one excitation function (including the calculation of atomic functions and potentials) were: 2-3 min (B), 7-8 min (Coul.), 20 min (DW), and 60-80 min (EDW). The tables give the values of $(z+1)^4\,\sigma(\varkappa)$, where $\varkappa$ is the momentum of the scattered electron in threshold units. In this case the Born cross sections for H and $C^{5+}$ are the same.

Tables 4-5 give the partial cross sections $\sigma_{\tilde{l},\tilde{l}}$, calculated in various approximations for the H 1s−2p, 1s−2s transitions. The symbol (+ −) corresponds to transitions of the outer electron with symmetrical ($S_T = 0$) and antisymmetrical ($S_T = 1$) scattering.

---

*See section 7 below.

182

TABLE 1. Cross Sections for Excitation of H Atom (in units of $\pi a_0^2$)*

| $x$ | 1s–2s | | | 1s–2p | | | | 1s–3s | | | 1s–3p | | | 2s–3s | | | 2s–3p | | |
|---|---|---|---|---|---|---|---|---|---|---|---|---|---|---|---|---|---|---|---|
| | B | DW | EDW | B | DW | EDW | BSC | B | DW | EDW | B | DW | EDW | B | DW | EDW | B | DW | EDW |
| 0.04 | $278^{-1}$ | $329^{0}$ | $612^{-1}$ | $842^{-1}$ | $117^{-2}$ | $219^{0}$ | $705^{-1}$ | $515^{-2}$ | | $207^{-1}$ | $155^{-1}$ | | $138^{0}$ | $120^{1}$ | $180^{1}$ | $921^{0}$ | $417^{-2}$ | $248^{-1}$ | $326^{1}$ |
| 0.08 | $552^{-1}$ | $492^{0}$ | $131^{0}$ | $189^{0}$ | $106^{-1}$ | $330^{0}$ | $140^{0}$ | $102^{0}$ | | $304^{-1}$ | $310^{-1}$ | | $188^{0}$ | $239^{1}$ | $395^{1}$ | $180^{1}$ | $305^{-1}$ | $483^{-1}$ | $822^{1}$ |
| 0.12 | $816^{-1}$ | $543^{0}$ | $175^{0}$ | $251^{0}$ | $384^{-1}$ | $403^{0}$ | $212^{0}$ | $151^{-1}$ | | $444^{-1}$ | $462^{-1}$ | | $157^{0}$ | $355^{1}$ | $690^{1}$ | $310^{1}$ | $102^{0}$ | $830^{-1}$ | $918^{1}$ |
| 0.16 | $107^{0}$ | $591^{0}$ | $195^{0}$ | $333^{0}$ | $993^{-1}$ | $480^{0}$ | $283^{0}$ | $197^{-1}$ | | $457^{-1}$ | $612^{-1}$ | | $141^{0}$ | $467^{1}$ | $111^{2}$ | $499^{1}$ | $232^{0}$ | $130^{0}$ | $931^{1}$ |
| 0.20 | $130^{0}$ | $701^{0}$ | $206^{0}$ | $413^{0}$ | $213^{0}$ | $507^{0}$ | $356^{0}$ | $240^{-1}$ | $129^{0}$ | $519^{-1}$ | $760^{-1}$ | $320^{0}$ | $141^{0}$ | $574^{1}$ | $167^{2}$ | $948^{1}$ | $439^{0}$ | $204^{0}$ | $962^{1}$ |
| 0.40 | $216^{0}$ | $714^{0}$ | $298^{0}$ | $783^{0}$ | $137^{1}$ | $655^{0}$ | $710^{0}$ | $393^{-1}$ | $104^{0}$ | $860^{-1}$ | $141^{0}$ | $344^{0}$ | $331^{0}$ | $998^{1}$ | $458^{2}$ | $210^{2}$ | $275^{1}$ | $451^{1}$ | $985^{1}$ |
| 0.60 | $248^{0}$ | $525^{0}$ | $337^{0}$ | $106^{1}$ | $148^{1}$ | $940^{0}$ | $100^{1}$ | $445^{-1}$ | $864^{-1}$ | $719^{-1}$ | $187^{0}$ | $319^{0}$ | $327^{0}$ | $122^{2}$ | $435^{2}$ | $326^{2}$ | $668^{1}$ | $237^{2}$ | $136^{2}$ |
| 0.80 | $244^{0}$ | $421^{0}$ | $308^{0}$ | $123^{1}$ | $141^{1}$ | $122^{1}$ | $118^{1}$ | $431^{-1}$ | $702^{-1}$ | $575^{-1}$ | $212^{0}$ | $292^{0}$ | $271^{0}$ | $127^{2}$ | $307^{2}$ | $270^{2}$ | $109^{2}$ | $192^{2}$ | $275^{2}$ |
| 1.20 | $194^{0}$ | $274^{0}$ | $221^{0}$ | $132^{1}$ | $130^{1}$ | $116^{1}$ | $129^{1}$ | $337^{-1}$ | $454^{-1}$ | $388^{-1}$ | $213^{0}$ | $228^{0}$ | $209^{0}$ | $110^{2}$ | $193^{2}$ | $172^{2}$ | $170^{2}$ | $197^{2}$ | $191^{2}$ |
| 1.60 | $144^{0}$ | $183^{0}$ | $156^{0}$ | $122^{1}$ | $115^{1}$ | $105^{1}$ | $122^{1}$ | $248^{-1}$ | $297^{-1}$ | $266^{-1}$ | $196^{0}$ | $176^{0}$ | $169^{0}$ | $863^{1}$ | $125^{2}$ | $119^{2}$ | $191^{2}$ | $192^{2}$ | $186^{2}$ |
| 2.00 | $107^{0}$ | $127^{0}$ | $110^{0}$ | $107^{1}$ | $997^{0}$ | $941^{0}$ | $106^{1}$ | $183^{-1}$ | $217^{-1}$ | $188^{-1}$ | $169^{0}$ | $149^{0}$ | $141^{0}$ | $659^{1}$ | $841^{1}$ | $776^{1}$ | $186^{2}$ | $177^{2}$ | $176^{2}$ |

*All the tables give the mantissa and the decimal power of the number. For instance, $278^{-1} \equiv 0.278 \cdot 10^{-1}$.

TABLE 2. Cross Sections for Excitation of $C^{5+}$ Ion [in units of $\pi a_0^2/(z+1)^4$]

| $x$ | 1s–2p | | | 1s–2s | | 1s–3p | | 2s–3p |
|---|---|---|---|---|---|---|---|---|
| | Coul. | DW | EDW | Coul. | EDW | Coul. | EDW | Coul. |
| 0.04 | $208^{1}$ | $212^{1}$ | $209^{1}$ | $543^{0}$ | $586^{0}$ | $350^{0}$ | | $277^{2}$ |
| 0.08 | $208^{1}$ | $212^{1}$ | $208^{1}$ | $541^{0}$ | $585^{0}$ | $349^{0}$ | | $277^{2}$ |
| 0.12 | $207^{1}$ | $212^{1}$ | $208^{1}$ | $537^{0}$ | $583^{0}$ | $348^{0}$ | | $277^{2}$ |
| 0.16 | $207^{1}$ | $211^{1}$ | $207^{1}$ | $532^{0}$ | $581^{0}$ | $347^{0}$ | $383^{0}$ | $276^{2}$ |
| 0.20 | $206^{1}$ | $210^{1}$ | $205^{1}$ | $526^{0}$ | $578^{0}$ | $346^{0}$ | $369^{0}$ | $274^{2}$ |
| 0.40 | $200^{1}$ | $205^{1}$ | $182^{1}$ | $478^{0}$ | $558^{0}$ | $335^{0}$ | $321^{0}$ | $262^{2}$ |
| 0.60 | $190^{1}$ | $191^{1}$ | $162^{1}$ | $415^{0}$ | $516^{0}$ | $318^{0}$ | $289^{0}$ | $251^{2}$ |
| 0.80 | $178^{1}$ | $177^{1}$ | $147^{1}$ | $350^{0}$ | $428^{0}$ | $297^{0}$ | $261^{0}$ | $246^{2}$ |
| 1.20 | $149^{1}$ | $144^{1}$ | $123^{1}$ | $240^{0}$ | $268^{0}$ | $244^{0}$ | $213^{0}$ | $201^{2}$ |
| 1.60 | $124^{1}$ | $120^{1}$ | $108^{1}$ | $166^{0}$ | $174^{0}$ | $199^{0}$ | $178^{0}$ | $182^{2}$ |
| 2.00 | $104^{1}$ | $101^{1}$ | $942^{0}$ | $119^{0}$ | $118^{0}$ | $165^{0}$ | $152^{0}$ | $173^{2}$ |

TABLE 3. Cross Sections for Excitation of $3_s$–$3_p$ in Transition in Na (in units of $\pi a_0^2$)

| $x$ | B | DW | EDW | BSC, two equations | BSC, three equations |
|---|---|---|---|---|---|
| 0.04 | 897¹ | 102³ | 207¹ | 684⁰ | 225¹ |
| 0.08 | 179² | 108³ | 456¹ | 165¹ | 362¹ |
| 0.12 | 265² | 128³ | 842¹ | 283¹ | 552¹ |
| 0.16 | 350² | 164³ | 151² | 408¹ | 730¹ |
| 0.20 | 431² | 557³ | 264² | 550¹ | 920¹ |
| 0.4 | 765² | 520² | 679² | 174² | 224² |
| 0.6 | 967² | 181³ | 729² | 336² | 379² |
| 0.8 | 105³ | 469² | 953³ | 495² | 527² |
| 1.2 | 102³ | 794² | 634² | 694² | 704² |
| 1.6 | 893² | 840² | 668² | 725² | 724² |
| 2.0 | 754² | 740² | 636² | 674² | 670² |

As is known, the Born approximation usually leads to overestimated cross sections in the region $\varkappa \sim 1$. It was noted in [5] that a consideration of the distortion of the incident and scattered wave leads to even larger cross sections, i.e., the disagreement with experiment becomes worse. The increase in the cross section is due to the anomalously large partial cross section with $\tilde{l} = 1$.

The results of the present work show that although consideration of exchange leads to a reduction in the total cross section, it is still of the same order as the Born cross section. Thus, the very laborious EDW method does not lead to any significant improvement in the results, at least for the H 1s–2p and 1s – 2s transitions, for which experimental data are available (Figs. 1-2).

As was to be expected, consideration of the exchange had a greater effect on the cross sections for neutral atoms. For ions, irrespective of the consideration of exchange, the excitation cross section was a maximum at the threshold and decreased monotonically with increase in energy.

Both the direct and exchange distortion affected the partial cross sections much more than the total cross section. This indicates that the partial wave method is not the best way of solving the problem. Unfortunately, no other methods are known at present.

The reduction of the cross section in the EDW method in comparison with the DW method is due mainly to the marked reduction in the partial cross sections with $\tilde{l} = 1$, $S_T = 0$, where exchange corresponds to repulsion.

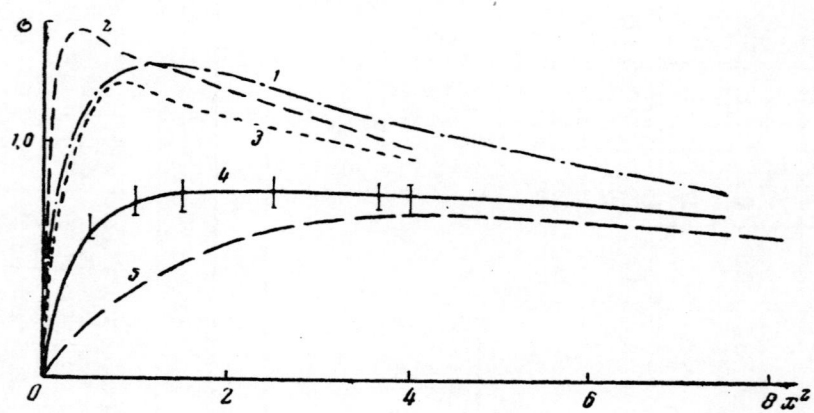

Fig. 1. Cross section for excitation of 1s–2p transition in hydrogen. 1) B; 2) DW; 3) EDW; 4) experiment [6]; 5) from formulas (21)-(22).

184

## TABLE 4. Partial Cross Sections for H 1s–2p Transition (in units of $\pi a_0^2$)

| x | $l_0=0;\, l=1$ | | | | $l_0=1;\, l=0$ | | | | $l_0=1;\, l=2$ | | | | $l_0=2;\, l=1$ | | | | $l_0=2;\, l=3$ | | | | $l_0=3;\, l=2$ | | | | $l_0=4;\, l=3$ | | | |
|---|---|---|---|---|---|---|---|---|---|---|---|---|---|---|---|---|---|---|---|---|---|---|---|---|---|---|---|---|
| | B | DW | EDW+ | EDW− | B | DW | EDW+ | EDW− | B | DW | EDW+ | EDW− | B | DW | EDW+ | EDW− | B | DW | EDW+ | EDW− | B | DW | EDW+ | EDW− | B | DW | EDW+ | EDW− |
| 0.04 | $333^{-3}$ | $245^{-5}$ | $109^{-1}$ | $810^{-5}$ | $843^{-1}$ | $242^{-3}$ | $181^{-2}$ | $216^{6}$ | $170^{-8}$ | $110^{-7}$ | | | $356^{-3}$ | $125^{-5}$ | $198^{-5}$ | $348^{-8}$ | $737^{-4}$ | $846^{-4}$ | $531^{-5}$ | $285^{-4}$ | $451^{-5}$ | $448^{-5}$ | $129^{-6}$ | $409^{-6}$ |
| 0.08 | $293^{-1}$ | $246^{-1}$ | $181^{-3}$ | $448^{-5}$ | $166^{0}$ | $440^{-3}$ | $381^{-2}$ | $308^{0}$ | $452^{-3}$ | $662^{-7}$ | | | $288^{-2}$ | $106^{-3}$ | $198^{-1}$ | $113^{-1}$ | $270^{-1}$ | $310^{-1}$ | $442^{-7}$ | $532^{-4}$ | $140^{-1}$ | $134^{-1}$ | $668^{-7}$ | $363^{-5}$ |
| 0.12 | $935^{-1}$ | $959^{-1}$ | $699^{-1}$ | $252^{-1}$ | $243^{0}$ | $582^{-3}$ | $623^{-2}$ | $315^{0}$ | $200^{-1}$ | $113^{-6}$ | | | $938^{-3}$ | $333^{-3}$ | $814^{-1}$ | $592^{-1}$ | $183^{-1}$ | $214^{-3}$ | $908^{-3}$ | $694^{-3}$ | $327^{-5}$ | $306^{-1}$ | $149^{-5}$ | $166^{-5}$ |
| 0.16 | $220^{-3}$ | $273^{-3}$ | $238^{-3}$ | $320^{-1}$ | $313^{0}$ | $635^{-3}$ | $928^{-2}$ | $233^{0}$ | $190^{-1}$ | $267^{-3}$ | | | $213^{-1}$ | $987^{-3}$ | $172$ | $764^{-1}$ | $776^{-1}$ | $924^{-3}$ | $217^{-1}$ | $269^{-1}$ | $206^{-1}$ | $195^{-1}$ | $459^{-1}$ | $286^{-1}$ |
| 0.20 | $415^{-3}$ | $647^{-3}$ | $553^{-3}$ | $784^{-1}$ | $374^{0}$ | $665^{-3}$ | $133^{-1}$ | $269^{0}$ | $165^{-1}$ | $122^{-1}$ | | | $394^{-1}$ | $210^{-3}$ | $211^{0}$ | $200^{-3}$ | $213^{-1}$ | $262^{-2}$ | $735^{-1}$ | $856^{-1}$ | $929^{-1}$ | $838^{-1}$ | $117^{-1}$ | $876^{-1}$ |
| 0.4 | $243^{-3}$ | $162^{-1}$ | $157^{-1}$ | $980^{-3}$ | $529^{0}$ | $368^{-3}$ | $447^{-1}$ | $170^{0}$ | $162^{-1}$ | $298^{-3}$ | | | $205^{0}$ | $129^{1}$ | $260^{0}$ | $133^{-1}$ | $414^{-1}$ | $636^{-1}$ | $310^{-1}$ | $144^{0}$ | $463^{-1}$ | $674^{-2}$ | $109^{-3}$ | $750^{-1}$ |
| 0.6 | $530^{-1}$ | $579^{-1}$ | $264^{-1}$ | $144^{0}$ | $479^{0}$ | $111^{-3}$ | $180^{-1}$ | $974^{-1}$ | $226^{-1}$ | $150^{-3}$ | | | $371^{0}$ | $108^{1}$ | $473^{0}$ | $722^{-1}$ | $149^{-1}$ | $269^{0}$ | $317^{-1}$ | $181^{0}$ | $121^{0}$ | $549^{-1}$ | $102^{-1}$ | $532^{-1}$ |
| 0.8 | $707^{-1}$ | $456^{-1}$ | $364^{-1}$ | $668^{-2}$ | $353^{0}$ | $235^{-1}$ | $210^{-2}$ | $459^{-1}$ | $729^{-5}$ | $290^{-3}$ | | | $416^{0}$ | $665^{0}$ | $566^{0}$ | $583^{-1}$ | $257^{0}$ | $448^{0}$ | $126^{0}$ | $175^{0}$ | $218^{0}$ | $157^{0}$ | $365^{-1}$ | $117^{0}$ |
| 1.2 | $666^{-1}$ | $213^{-1}$ | $118^{-1}$ | $657^{-1}$ | $150^{0}$ | $534^{-5}$ | $238^{-3}$ | $688^{-3}$ | $396^{-1}$ | $230^{-3}$ | | | $283^{0}$ | $2.22$ | $182^{0}$ | $131^{-1}$ | $286^{0}$ | $366^{0}$ | $167^{0}$ | $123^{0}$ | $186^{0}$ | $278^{0}$ | $879^{-1}$ | $155^{0}$ |
| 1.6 | $440^{-1}$ | $987^{-2}$ | $296^{-1}$ | $513^{-2}$ | $605^{-1}$ | $110^{-1}$ | $296^{-1}$ | $813^{-1}$ | $173^{-1}$ | $845^{-3}$ | | | $143^{0}$ | $757^{-1}$ | $473^{-1}$ | $114^{-1}$ | $187^{0}$ | $181^{0}$ | $172^{-1}$ | $675^{-1}$ | $121^{-1}$ | $206^{0}$ | $696^{-3}$ | $109^{0}$ |
| 2.0 | $259^{-1}$ | $475^{-2}$ | $111^{-2}$ | $300^{-2}$ | $250^{-1}$ | $154^{-1}$ | $301^{-1}$ | $730^{-1}$ | $241^{-3}$ | $240^{-3}$ | | | $685^{-1}$ | $283^{-1}$ | $144^{-1}$ | | $105^{0}$ | $816^{-1}$ | $321^{-1}$ | $346^{-1}$ | | $117^{0}$ | $391^{-1}$ | $629^{-1}$ |

## TABLE 5. Partial Cross Sections for H 1s–2s Transition (in units of $\pi a_0^2$)

| x | $l_0=l=0$ | | | | $l_0=l=1$ | | | | $l_0=l=2$ | | | | $l_0=l=3$ | | | | $l_0=l=4$ | | | |
|---|---|---|---|---|---|---|---|---|---|---|---|---|---|---|---|---|---|---|---|---|
| | B | DW | EDW+ | EDW− | B | DW | EDW+ | EDW− | B | DW | EDW+ | EDW− | B | DW | EDW+ | EDW− | B | DW | EDW+ | EDW− |
| 0.04 | $279^{-1}$ | $329^{0}$ | $369^{-1}$ | $119^{-1}$ | $335^{-1}$ | $105^{-3}$ | $847^{-6}$ | $245^{-1}$ | $141^{-7}$ | $361^{-7}$ | $300^{-5}$ | $316^{-6}$ | $387^{-11}$ | $569^{-11}$ | $135^{-10}$ | $192^{-8}$ | $204^{-12}$ | $207^{-12}$ | $268^{-12}$ | $101^{-11}$ |
| 0.08 | $552^{-1}$ | $482^{0}$ | $398^{-1}$ | $237^{-1}$ | $265^{-3}$ | $945^{-2}$ | $711^{-4}$ | $917^{-1}$ | $453^{-4}$ | $116^{-3}$ | $786^{-5}$ | $102^{-4}$ | $518^{-8}$ | $759^{-8}$ | $296^{-10}$ | $554^{-8}$ | $235^{-11}$ | $254^{-11}$ | $415^{-11}$ | $354^{-10}$ |
| 0.12 | $811^{-1}$ | $504^{0}$ | $361^{-1}$ | $353^{-1}$ | $876^{-3}$ | $386^{-1}$ | $265^{-1}$ | $140^{0}$ | $336^{-5}$ | $883^{-5}$ | $537^{-7}$ | $782^{-6}$ | $865^{-6}$ | $127^{-7}$ | $867^{-8}$ | $843^{-7}$ | $127^{-10}$ | $156^{-10}$ | $253^{-10}$ | $356^{-10}$ |
| 0.16 | $105^{0}$ | $478^{0}$ | $335^{-1}$ | $455^{-1}$ | $202^{-2}$ | $113^{0}$ | $765^{-1}$ | $162^{0}$ | $138^{-4}$ | $373^{-4}$ | $177^{-5}$ | $336^{-3}$ | $634^{-7}$ | $942^{-7}$ | $511^{-6}$ | $626^{-5}$ | $244^{-8}$ | $290^{-4}$ | $417^{-11}$ | $238^{-8}$ |
| 0.20 | $127^{0}$ | $437^{0}$ | $328^{-1}$ | $522^{-1}$ | $381^{-1}$ | $264^{0}$ | $201^{-3}$ | $174^{0}$ | $407^{-4}$ | $114^{-3}$ | $406^{-4}$ | $105^{-2}$ | $291^{-4}$ | $438^{-4}$ | $220^{-7}$ | $265^{-5}$ | $167^{-7}$ | $201^{-4}$ | $360^{-9}$ | $112^{-7}$ |
| 0.40 | $193^{0}$ | $289^{0}$ | $487^{-1}$ | $280^{-1}$ | $230^{-1}$ | $422^{0}$ | $413^{-1}$ | $185^{0}$ | $938^{-3}$ | $363^{-3}$ | $135^{-6}$ | $221^{-1}$ | $285^{-4}$ | $479^{-4}$ | $814^{-4}$ | $270^{-3}$ | $659^{-6}$ | $828^{-4}$ | $475^{-6}$ | $447^{-6}$ |
| 0.60 | $193^{0}$ | $204^{0}$ | $791^{-1}$ | $171^{-1}$ | $509^{-1}$ | $302^{0}$ | $557^{-1}$ | $157^{0}$ | $480^{-2}$ | $188^{-1}$ | $263^{-3}$ | $463^{-1}$ | $318^{-3}$ | $613^{-3}$ | $194^{-3}$ | $244^{-1}$ | $166^{-4}$ | $227^{-4}$ | $277^{-4}$ | $972^{-4}$ |
| 0.80 | $159^{0}$ | $143^{0}$ | $803^{-1}$ | $409^{-3}$ | $718^{-1}$ | $238^{0}$ | $538^{-1}$ | $105^{0}$ | $121^{-1}$ | $371^{-1}$ | $242^{-2}$ | $578^{-1}$ | $138^{-2}$ | $279^{-2}$ | $302^{-2}$ | $721^{-1}$ | $127^{-3}$ | $188^{-2}$ | $524^{-2}$ | $592^{-3}$ |
| 1.20 | $851^{-1}$ | $682^{-1}$ | $411^{-1}$ | $627^{-2}$ | $758^{-1}$ | $141^{0}$ | $499^{-1}$ | $487^{-1}$ | $263^{-1}$ | $520^{-1}$ | $901^{-2}$ | $503^{-1}$ | $632^{-2}$ | $113^{-1}$ | $104^{-1}$ | $160^{-1}$ | $123^{-1}$ | $183^{-1}$ | $971^{-3}$ | $333^{-3}$ |
| 1.60 | $425^{-1}$ | $337^{-1}$ | $187^{-1}$ | $651^{-1}$ | $559^{-1}$ | $792^{-1}$ | $323^{-1}$ | $281^{-1}$ | $301^{-1}$ | $461^{-1}$ | $113^{-1}$ | $338^{-1}$ | $115^{-1}$ | $172^{-1}$ | $294^{-1}$ | $174^{-1}$ | $358^{-2}$ | $505^{-1}$ | $640^{-3}$ | $619^{-3}$ |
| 2.00 | $220^{-1}$ | $178^{-1}$ | $883^{-1}$ | $510^{-1}$ | $367^{-1}$ | $451^{-1}$ | $183^{-1}$ | $178^{-1}$ | $263^{-1}$ | $348^{-1}$ | $988^{-2}$ | $224^{-1}$ | $136^{-1}$ | $150^{-1}$ | $398^{-1}$ | $150^{-1}$ | $579^{-1}$ | $753^{-1}$ | $137^{-1}$ | $734^{-2}$ |

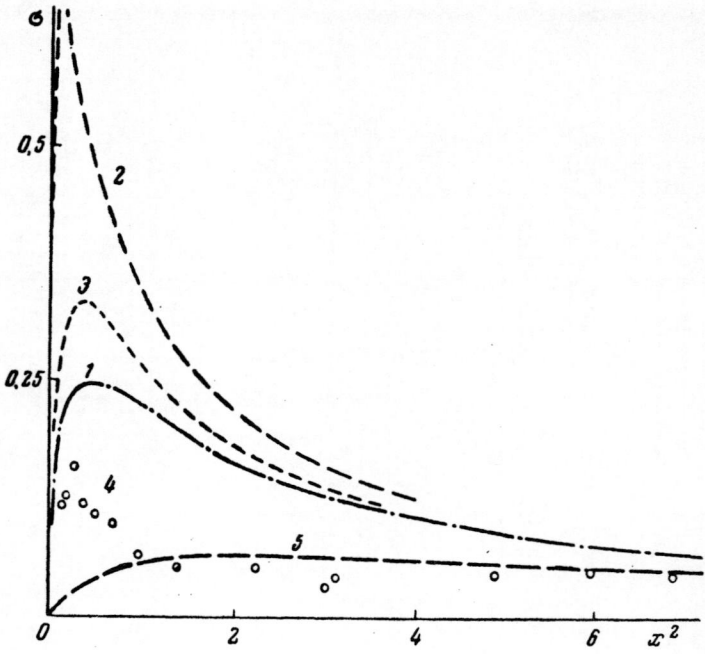

Fig. 2. Cross sections for excitation of 1s–2s transition in hydrogen. 1) B; 2) DW; 3) EDW; 4) experiment [7]; 5) from formulas (21)-(22).

## §6. Conclusions

Thus, the distorted wave method does not lead to any appreciable improvement in the accuracy of cross sections in comparison with the Born method, and in many cases even leads to a significant deterioration of the results. This can be explained in the following way. In the distorted wave approximation the cross section is increased by the attraction of the incident electron to the center of the atom. However, excitation of the atom depends not so much on the distance of the outer electron from the center of the atom, as on the distance between it and the atomic electron. The latter distance is reduced much less by polarization of the atom, and may even be increased.

Since the premises of the distorted wave method are inadequate for the problem of inelastic collision cross sections, it is advisable to turn to the Born method. However, this applies only to collisions with a neutral atom. In the case of collisions with positive ions the Coulomb attraction exceeds the repulsion of the electrons and the Coulomb approximation should probably be used. Unfortunately, the latter conclusion cannot be verified experimentally at present.

The distorted wave method includes, in addition to the attraction of the outer electron to the atom, the increase in the velocity of the electron close to the atom. If the attraction is largely compensated by the repulsion of the electrons, the effect of acceleration is not compensated in any way and may be significant. The two effects can be separated only on the basis of a quasi-classical approach. We will not dwell on this here.

## §7. Use of other Methods

The following method (BSC method), in which strong coupling without diagonal potentials is taken into account, is a very simple way of improving the accuracy of the Born approximation. The radial functions $F_\Gamma$ in this case are solutions of the system of equations

$$\left[ \frac{d^2}{dr^2} - \frac{\tilde{l}(\tilde{l}+1)}{r^2} + k^2 \right] F_\Gamma = \sum_{\tilde{l}'} U_{\Gamma\Gamma'} F_{\Gamma'}. \tag{19}$$

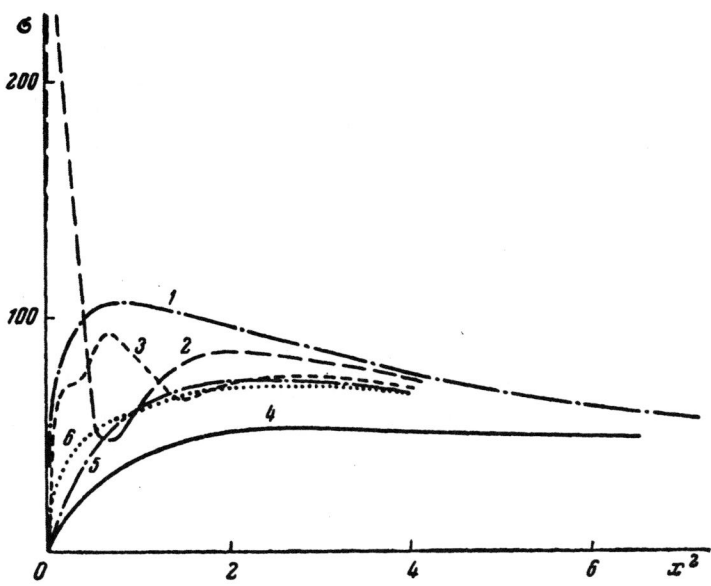

Fig. 3. Cross sections for excitation of Na 3s–3p transition. 1) B;
2) DW; 3) EDW; 4) experiment [8]; 5) BSG; 6) with allowance for strong coupling by Seaton's method [9].

In the case of s–s transitions $\tilde{l} = \tilde{l}_0$ and, hence, the system consists of two equations. In the case of s–p transitions, generally speaking, there are two possible values: $\tilde{l} = \tilde{l}_0 - 1$ and $\tilde{l}_2 = \tilde{l}_0 + 1$, i.e., the system comprises three equations:

$$\left[ \frac{d^2}{dr^2} - \frac{\tilde{l}_0(\tilde{l}_0 + 1)}{r^2} + k_0^2 \right] F_0 = U_{01}F_1 + U_{02}F_2,$$

$$\left[ \frac{d^2}{dr^2} - \frac{\tilde{l}_1(\tilde{l}_1 + 1)}{r^2} + k_1^2 \right] F_1 = U_{10}F_0 + U_{12}F_2,$$

$$\left[ \frac{d}{dr^2} - \frac{\tilde{l}_2(\tilde{l}_2 + 1)}{r^2} + k_1^2 \right] F_2 = U_{20}F_0 + U_{21}F_1.$$

$U_{12}$ and $U_{21}$ correspond to quadrupole transitions between two final states with equal energy $k_1^2 = k_2^2$. The results obtained in the BSC approximations are included in Tables 1 and 3 (see Fig. 3). The data obtained in the two- and three-equation approximations for Na are given.

As was to be expected, the effect of strong coupling is light for hydrogen and plays an important role close to the threshold for Na. The cross section is reduced, which considerably improves the agreement with experiment. A reduction of the cross section when strong coupling is taken into account can be expected also in the general case. In terms of the adiabatic approximation this can be attributed to separation of the interacting levels. From this viewpoint the increase in the cross section can be understood when allowance is made for strong coupling of channels with $\tilde{l}_{1,2} = \tilde{l}_0 \pm 1$, despite the fact that the true cross section of channel $\tilde{l}_2$ is generally very small. Repulsion of the components $(\tilde{l}_1)$, $(\tilde{l}_2)$ can lead to convergence of the levels $(\tilde{l}_1)$ and $(\tilde{l}_0)$.

The effect of strong coupling is significantly different for different partial waves and can be significant even at low values of the Born cross section. Figure 3 also shows the function $\sigma(\varkappa)$, calculated from the Born cross sections by conversion from the R-matrix to the S-matrix [9]. This conversion is in the nature of a renormalization and, as is clear, describes the strong coupling effect well in the region of the maximum. Close to the threshold, however, it is inadequate.

In conclusion we will discuss one model, recently suggested by L. Presnyakov, L. Sobel'man, and the author [10], for calculation of the cross sections for inelastic collisions of electrons with atoms. We will not dwell on its basis here but will confine ourselves to an account of the results. The matter reduces to replacement of the plane-wave system $\Psi_0 = \varphi_{a_0}(\mathbf{r}_1) e^{i\mathbf{k}_0\mathbf{r}_2}$ in the Born function by a function $g(\mathbf{r}_1, \mathbf{r}_2)$ describing the scattering of the outer and atomic electrons by one another and their common center of gravity by the nucleus:

$$g = \frac{\pi/k_0}{\sinh \pi/k_0} e^{ik_0(\mathbf{R}+\rho)} F\left(\frac{i}{k_0}, 1, ik_0R - i\mathbf{k}_0\mathbf{R}\right) F\left(-\frac{i}{k_0}, 1, ik_0\rho - i\mathbf{k}_0\rho\right),$$

where

$$\rho = \frac{1}{2}(\mathbf{r}_2 - \mathbf{r}_1), \quad \mathbf{R} = \frac{1}{2}(\mathbf{r}_2 + \mathbf{r}_1) \tag{20}$$

(F is a degenerate hypergeometric function). Substituting this function in the matrix element

$$\left\langle \varphi_a e^{i\mathbf{k}\mathbf{r}_2} \left| \frac{1}{r_{12}} \right| \Psi_0(\mathbf{r}_1, \mathbf{r}_2) \right\rangle,$$

we obtain in place of (3) the formula

$$\sigma = \frac{8}{k_0^2} \sum_{\varkappa} c_{\varkappa} \int\limits_{k_0-k}^{k_0+k} |P_{\varkappa}(q) f(q)|^2 \frac{dq}{q^3}, \tag{21}$$

and

$$f(q) = \frac{\pi/k_0}{\sinh \pi/k_0} F\left(-\frac{i}{k_0}, \frac{i}{k_0}, 1; \left[\frac{qk_0}{q^2+qk_0}\right]^2\right), \tag{22}$$

where F is a hypergeometric function. We note that expression (21) is obtained only with an approximate calculation of one of the integrals. The results of the numerical calculations from formulas (21)-(22) show that the cross sections are usually underestimated.[*] In this connection is must be noted that almost all the approximate methods used so far lead to overestimated results. Figures 1 and 2 show the cross sections obtained from formulas (21)-(22).

## Literature Cited

1. L. A. Vainshtein and I. I. Sobel'man, Zh. Eksperim. i Teor. Fiz., 39:767 (1960).
2. I. Percival and M. Seaton, Proc. Cambridge, Philos. Soc., 53:654 (1957).
3. L. A. Vainshtein, Opt. i Spektroskopiya, 3:313 (1957); Izv. AN SSSR, seriya fizich., 22:671 (1958).
4. R. Marriott, Proc. Phys. Soc., 72:121 (1958).
5. L. A. Vainshtein, Opt. i Spektroskopiya, 11:301 (1961).
6. W. Fite and R. Brackmann, Phys. Rev., 112:1151 (1955); W. Fite, R. Stebbings, and R. Brackmann, Phys. Rev., 116:356 (1959).
7. R. Stebbings, W. Fite, D. Hummer, and R. Brackmann, Phys. Rev., 124:2051 (1961).
8. G. Haft, Z. Phys., 82:73 (1933); W. Christoph, Ann. Physik, 23:51 (1935).
9. M. Seaton, Proc. Phys. Soc., 77:184 (1961).
10. I. I. Sobel'man, Introduction to Atomic Spectroscopy, §46, Fizmatgiz, 1963.
11. L. A. Vainshtein, L. P. Presnyakov, and I. I. Sobel'man, Zh. Eksperim. i Teor. Fiz., 45, (6) (1963).

[*]Note added in proof [of the Russian edition]. Further investigation showed that introduction of an effective charge into $g(\mathbf{r}_1, \mathbf{r}_2)$ can lead to a very good agreement of the cross sections for H 1s–2p and 1s–2s transitions with the experimental data [11].